DYNAMIC MODELS IN PHYSICS

A WORKBOOK OF COMPUTER SIMULATIONS USING ELECTRONIC SPREADSHEETS

VOLUME ONE:
MECHANICS

DYNAMIC MODELS IN PHYSICS

A Workbook of Computer Simulations Using
Electronic Spreadsheets

Volume One:
Mechanics

Frank Potter
University of California, Irvine

Charles W. Peck
California Institute of Technology

David S. Barkley, Editor

N. Simonson & Company
Marina del Rey, California

Sponsoring Editor: David S. Barkley
Copy Editor: Betty Duncan-Todd
Cover Designer: Tarabochia & Hunt

Library of Congress Cataloging-in-Publication Data

Potter, Frank, 1944–
Dynamic models in physics: a workbook of computer simulations using electronic spreadsheets / Frank Potter; Charles W. Peck; David S. Barkley, editor.
p. cm.
Bibliography: v. 1. p.
Contents: v. 1. Mechanics.
ISBN 0-9622556-1-0 (v. 1)
1. Physics—Computer simulation. 2. Electronic spreadsheets.
I. Peck, Charles, W., 1934– . II. Barkley, David S. III. Title.
QC52.P67 1989
530' .01' 13—dc20 89-11602
CIP

N. Simonson & Company
13450 Maxella Avenue
Suite G225
Marina del Rey, California 90292-5631
Phone: (213) 301-2847

We dedicate this book to our wives
Patricia and Kathleen,
and to our children
David and Steven,
Cynthia, David, Brian, and Steven

N. Simonson & Company Titles of Related Interest

D. E. Atkinson, S. G. Clarke, and D. C. Rees
Dynamic Models in Biochemistry (1987)

D. E. Atkinson, D. C. Brower, and R. McClard
Dynamic Models in Chemistry (1989)

PREFACE

This special property of digital computers, that they can mimic any discrete machine, is described by saying that they are universal *machines. The existence of machines with this property has the important consequence that ... it is unnecessary to design various new machines.*
— A. M. Turing

Students of the sciences must often integrate large amounts of information into simple, coherent frameworks of insight. They are then expected to use this insight to solve new problems in which the driving principles may be obscure. Such insight is acquired by actively and repetitively engaging the mind in numerous problem-solving explorations — explorations in the laboratory, in assigned problem sets, and in independently constructed thought experiments. Students must dissect many different kinds of problems and explore many alternative solutions until the barriers to insight are overcome and the student says, "Ah Hah!"

The pocket calculator changed science instruction by permitting the design of problem sets that were not contrived around the limitations of pencil, paper, and slide rule.

The personal computer may again change science instruction by permitting the design of problem sets that enable students to build and interrogate simulations of realistic physical situations and to explore these simulations with a full complement of graphical tools.

We believe that the emphasis should be as much on the building of models as on their interrogation, but traditional software tools are not designed for rapid, one-time prototyping. We have therefore chosen to design this book around a powerful, ubiquitous, nonprocedural programming language that permits rapid and intuitive construction of sophisticated simulations — the electronic spreadsheet. Other software tools may eventually make their way to the marketplace in forms useful to undergraduate students, but the intuitive man-machine interface that constitutes the electronic spreadsheet will have a place in personal computing for decades to come.

Electronic spreadsheet programs can serve as construction kits with which to build and explore alternative representations of ideas. Students who build spreadsheet models that correspond to their understanding of physical events can correct errors in their understanding as they refine their models to make them work correctly. They can ask questions of their simulations that may not have been anticipated by their textbook authors and instructors and can verify their understanding of a textbook table or graph by reproducing it from basic principles.

As we designed and wrote this book we found that many unexpected questions arose — questions that could be immediately answered by interrogating the model itself. We believe that this ability to enter into extended dialogs with computer simulations will be as valuable to students as it has been for us.

We hope and expect that early introduction to model building and interrogation will provide students with a tool than can be used throughout a career — a simple device with which to prototype and clarify ideas.

Scope of the Book

This book has been designed for students taking a first course in calculus-level physics. Much of the material drills students on elementary topics, but numerous opportunities exist for the motivated student to explore issues deeply. Indeed, personal computers and computer simulations naturally promote deep study and exploration.

The models in this book exploit numeric methods not usually covered in introductory physics. We believe that personal computers have created a place for these methods in introductory physics, but we are also aware that unnecessary side excursions must be avoided when the primary goal of the student is to learn physics. The material has been kept as simple and straightforward as possible. Models in this book exploit methods for differentiating and integrating functions and for solving ordinary differential equations. We hope you will find, as we have, that the introduction of these methods will help you to clarify lessons in physics that are otherwise quite difficult. For example, the central importance of Newton's second law of motion in mechanics becomes clear when students can solve any second-order differential equation and effortlessly graph the results. Students can track the position and velocity of an object at the same time that they track its kinetic and potential energy — even when Newton's second law has no analytical solution. The effect of a dissipative force on the total mechanical energy of a system becomes instantly clear!

The text also includes a series of clearly marked, brief discussions, called "Spreadsheet Techniques," designed to increase spreadsheet ease of use and efficiency. These techniques will often reduce the time needed to set up, modify, and explore models but they are not necessary to successfully complete any of the exercises.

Using the Book

Dynamic Models in Physics is meant to supplement the material found in introductory physics books.

It is organized as a series of exercises to be performed on personal computers running electronic spreadsheet software. There are no special requirements with respect to the brand of personal computer or electronic spreadsheet program provided that the software has the capability of generating simple *xy*-plots. The models in this book are described using a notation similar to that found in Lotus 1-2-3, Supercalc, Excel, etc. (i.e., columns are identified by letters and rows are identified by numbers). Chapters 1 and 2 are designed to help students with no prior computer experience acquire the facility needed to make effective use of this book. Chapter 2 outlines all basic elements of intelligent spreadsheet use, but exercises in subsequent chapters do introduce additional software features where appropriate. The "Table of Contents" identifies not only the physics introduced in each exercise but also the spreadsheet tools introduced.

Students build each computer simulation by following the book's detailed directions while sitting at a personal computer. The exercises follow the same progression found in most major physics textbooks but are sufficiently modular that they may be sampled out of order where appropriate.

Each exercise begins with a brief introduction to the conceptual and quantitative aspects of the topic to be explored and then provides detailed directions for building one or more spreadsheet models on virtually any personal computer and spreadsheet program. Exercises are divided into a "Lessons" section and a "Problems" section.

Lessons are tightly controlled, and students are held firmly by the hand as they interrogate the model, complete tables, and graph results. Problems require varying degrees of student initiative and often permit the interested student to enter domains of genuine, independent exploration.

This book is a beginning and will grow and prosper only with your help, advice, and comments. Please write us with your experiences as you explore the book.

Acknowledgments

We have benefited from the experience and advice of many people as we have designed and written this book.

The idea for such books grew out of discussions with Neil Patterson and Peter Renz at W. H. Freeman & Company. An earlier book, *Dynamic Models in Biochemistry*, pioneered our understanding of the use of spreadsheets to learn and to think and we are deeply indebted to its authors — Daniel Atkinson, Steven Clarke, and Douglas Rees. We are also indebted to Diane Bowen, Merry Finley and Jane Gillen of the Benjamin/Cummings Publishing Company for their help and support as we learned the intricacies of the publishing business.

It was initiative and resourcefulness of Professor Geoffrey Fox that flooded the Caltech campus with personal computers — offering one of us the opportunity to abandon the friendly slide rule and to perform "incidental" physics calculations using the far more powerful and just as friendly electronic spreadsheet program.

David S. Barkley, Editor
Frank Potter
Charles W. Peck

CONTENTS

1
GETTING STARTED 1

2
MOTION 9

Introduction

2.1 Uniform Acceleration 12

Physics: Kinematics of Uniform Acceleration

Tools: Building Spreadsheet Models; Spreadsheet Graphics

2.2 Rates of Change 26

Physics: Velocity and Acceleration from Derivatives

Tools: Numerical Differentiation; Data Tables; Keyboard Macros

2.3 Vectors 35

Physics: Vector Operations

Tools: ATAN2 Function; Formula Composition

2.4 Nonuniform Acceleration 46

Physics: Kinematics of Nonuniform Acceleration

Tools: Simple Numerical Integration; Speeding Up Model Building

2.5 Velocity by Simpson's Rule 56

Physics: Kinematics of Nonuniform Acceleration

Tools: Accurate Numerical Integration; Naming Cells; Conditionals

3
FORCES **61**

Introduction

3.1 The Second Law of Motion: Euler's Method 64

Physics: Laws of Motion

Tools: Solving Differential Equations Numerically

3.2 The Second Law of Motion: Runge-Kutta Method 74

Physics: Laws of Motion

Tools: Compact Runge-Kutta Models;

Solving Differential Equations Numerically; Function Macros

3.3 Trajectories 97

Physics: Differential Equations in Two Dimensions

Tools: Correct Geometric Pictures;

Multi-dimensional Runge-Kutta Solutions

3.4 Orbits 107

Physics: Motion in Response to a Gravitational Field

Tools: Self-Adusting Time Intervals

4
WORK, ENERGY AND MOMENTUM **121**

Introduction

4.1 The Conservation of Energy 127

Physics: Conservation of Total Mechanical Energy

4.2 The Potential Energy Function:
Energy Conservation in Two Dimensions 137

Physics: Conservation of Total Mechanical Energy

Tools: Visualizing Three-Dimensional Surfaces

4.3 The Conservation of Energy:
Nonconservative Systems 153

Physics: Friction and Other Dissipative Forces

Tools: Discontinuous Functions

4.4 The Energy Integral 168

Physics: Determining Motion from Energy Considerations

Tools: Finding Oscillation Periods: Lookup Tables

4.5 The Conservation of Momentum: 180
Collisions in One Dimension

Physics: Determining Motion from Conservation Laws

4.6 The Conservation of Momentum: Collisions in Multiple Dimensions 200

Physics: Determining Motion from Conservation Laws

5
ROTATIONAL MOTION **209**

Introduction

5.1 An Introduction to Uniform Rotary Motion 218

Physics: Building Intuition About Rotary Motion

Tools: Building Models Using Existing Components

5.2 The Center of Mass And The Inertia Tensor 241

Physics: Calculating the Center of Mass and the Inertia Tensor

5.3 The Inertia Tensor of a Uniformly Rotating Body 252

Physics: The Role of the Inertia Tensor in Rotational Dynamics

5.4 Diagonalizing the Inertia Tensor 263

Physics: Using the Inertia Tensor to Find Principal Axes

Tools: Random Number Generators

5.5 Fixed-Axis Rotation with Variable Angular Velocity 290

Physics: Pulleys and Related Objects

5.6 General Rotational Motion 309

Physics: Tennis Rackets, Gyroscopes, and Tops

6
PERIODIC MOTIONS **333**

Introduction

6.1 The Driven Harmonic Oscillator 337

Physics: Simple Harmonic Motion and Related Motion

6.2 Coupled Oscillators 350

Physics: Coupled Oscillators and Normal Mode Behavior

6.3 Discrete Systems with a Periodic Restoring Force 363

Physics: Oscillators with Stable and Unstable Solutions

APPENDIX, ANSWER KEY, AND ENDNOTES

Making *xy*-plots with spreadsheets 382

Selected Answers for Lessons 385

Endnotes 397

DYNAMIC MODELS IN PHYSICS

A WORKBOOK OF COMPUTER SIMULATIONS USING ELECTRONIC SPREADSHEETS

VOLUME ONE: MECHANICS

CHAPTER 1

GETTING STARTED

If you wish to understand Nature, you must pay attention to the details. — L. da Vinci

This book about introductory physics uses electronic spreadsheets to create computer simulations that can help you to better understand the behavior of Nature. Until recently the ability to simulate physical events with digital computers was limited to those people with special computer facilities. The introduction of affordable computing hardware and especially the introduction of powerful yet easy-to-learn nonprocedural programming languages such as the electronic spreadsheet now make remarkable computing powers accessible to all students. Your goal is to use this book to apply these capabilities to the subjects found in your introductory physics course.

You will find that your personal computer and your electronic spreadsheet program are capable of representing all numeric operations associated with introductory physics (e.g., vectors, Newton's second law, the inertia tensor). These representations can be built and modified far more quickly than with traditional programming languages such as Pascal, C, or FORTRAN. These representations can be shaped, experienced, and explored in forms that are closer to the way you actually think and work with pencil and paper. The experience of designing and interrogating an electronic spreadsheet model has been compared with the experience of sitting at a potter's wheel. You can shape ideas as you might shape lumps of clay.

In addition, personal computers and electronic spreadsheets encourage you to explore nontraditional problems that have been too difficult to handle analytically or to tackle without sophisticated computer programs — a falling object in a retarding medium such as atmosphere, for example. The air density is changing during the motion, thus the drag is changing. This physical fact generates a mathematical problem that is very difficult to handle analytically but easy to handle numerically using a spreadsheet program.

A new model world is accessible, one much closer to the real world.

The Electronic Spreadsheet

The first electronic spreadsheet program *Visicalc* appeared in 1978 and quickly became the principal calculating tool of personal computer users in virtually every walk of life. With spreadsheets it is possible to

- Build models of dynamic events in a small fraction of the time that it would take with a procedural language such as C, Pascal, or FORTRAN.

- Change and refine models quickly and intuitively in order to explore alternative scenarios that might otherwise go unnoticed.

- Graph the results of your explorations on-screen and automatically change the graph by changing model assumptions.

You may already know how to use an electronic spreadsheet program. If you've never worked with such a program, prepare yourself for an exciting experience. You will be building sophisticated models in a few hours.

The concept is simple. A large, two-dimensional array of rectangular cells is displayed on a computer terminal with column and row headings that provide each cell with a unique name. Although a few exceptions remain, the vast majority of spreadsheet programs have adopted the convention of identifying columns with letters and rows with numbers. Each cell in the spreadsheet is named according to its column and row location (Figure 1.1).

Spreadsheet cells may be blank or may contain either *labels*, *values*, or *formulas*. Collections of labels, values, and formulas work together to generate *models*. Figure 1.2 is a model that calculates the height above the ground at time t of a ball thrown vertically into the air from a height y_0 and with an initial, vertical velocity of v_0 (Equation 1.1). The acceleration of gravity at Earth's surface is represented by g:

$$y = y_0 + v_0 t + \frac{gt^2}{2} \qquad (1.1)$$

- *Labels* are text that can describe a spreadsheet; identify the entries in a column, row, or cell; or name regions within a spreadsheet. They are not used in numeric calculations. Figure 1.2 contains a label in cell A7 that points to the right and says "cell B7 contains a value for time."

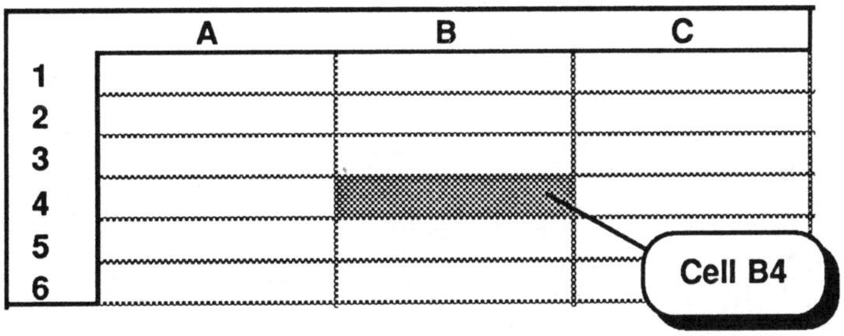

Figure 1.1

Electronic spreadsheet programs appear on your computer screen as a rectangular array of cells. Each cell is uniquely named according to the column and row in which it is located. Cell B4, for example, is the only cell that resides in both column B and row 4. A few spreadsheet programs exist that do not label columns with letters and rows with numbers. Even if your program has its own convention, you will still be able to adapt the models of this book for your program.

- *Values* are data to be used by your model. Cell B7 in Figure 1.2 contains a value of 5 seconds (s).

  Most spreadsheets can perform calculations on numbers and strings of letters. In this book, your calculations will be numeric.

- *Formulas* perform calculations. Consider, for example, an object with the equation of motion $y = y_0 + v_0 t + (gt^2/2)$. Figure 1.2 represents this expression as a spreadsheet formula (cell B9) that reads =B3+(B4*B7)+(B5*B7*B7/2). The expression B4*B7 within this formula indicates that the contents of cell B4 should be multiplied by the contents of cell B7. (The symbol * is often used by computers to signify multiplication. The other basic operations — addition, subtraction, division, and exponentiation — are represented by +, –, / and ^, respectively.) Although your spreadsheet program may represent this formula slightly differently, all formulas are easy to read and understand.

  Cells that contain formulas usually display the result of the calculation rather than the formula itself. Of course, if you wish to look behind the calculated result to the formula, your spreadsheet will always have a method for doing so.

The principal elegance of spreadsheet models resides in their response to change. The equation of motion in Figure 1.2 represents a ball thrown vertically upward from the surface of the Earth. The model in Figure 1.2 can be modified to represent conditions on the moon by changing the value of acceleration from –9.8 m s^{-2} to –1.6 m s^{-2}. The calculated position of the ball at $t = 2$ s is immediately updated to its new value (Figure 1.3).[1]

Figure 1.2

Each cell in a spreadsheet array may be blank or contain either a *label*, *value*, or *formula*. Labels are usually passive signposts that tell you where you are (e.g., cell A7) or provide other identifying information (e.g., cell A1). Values are usually numbers required by the model for its calculations (e.g., cell B5). Formulas are algebraic expressions that use the addresses of other cells as variables (e.g., cell B9).

Labels, values, and formulas are entered into cells by first pointing to the cell with your computer's pointing device (typically either an arrow key on your keyboard or a mouse) and then typing in the required string of characters.

	A	B	C
1	Uniform Acceleration		
2			
3	y0 =	2	m
4	v0 =	10	m/s
5	g =	-9.8	m/s^2
6			
7	t =	5	s
8			
9	y =	=B3+(B4*B7)+(B5*B7*B7/2)	m

	A	B	C
1	Uniform Acceleration		
2			
3	y0 =	2	m
4	v0 =	10	m/s
5	g =	-9.8	m/s^2
6			
7	t =	5	ε
8			
9	y =	-70.5	m

→

	A	B	C
1	Uniform Acceleration		
2			
3	y0 =	2	m
4	v0 =	10	m/s
5	g =	-1.6	m/s^2
6			
7	t =	5	s
8			
9	y =	32	m

Change the value of cell B3 from –9.8 to –1.6

Figure 1.3

Electronic spreadsheet models are dynamic. When the value of a cell is changed, all formulas that use that value are recalculated. When the contents of cell B5 is changed from –9.8 to –1.6, the calculated result displayed in cell B9 also changes. A model that calculates the position of a ball thrown on the Earth becomes a model of a ball thrown on the Moon.

Most modern spreadsheet programs also include the capability to represent model behavior graphically. Suppose, for example, that you wish to plot the trajectory of an object (i.e., a plot of y vs t). You might modify the model in Figure 1.2 to calculate y at a number of different time points (Figure 1.4a) and then plot y vs t (Figure 1.4b) using special commands built into your spreadsheet program (see the Appendix). Most spreadsheet programs with built-in graphics capability can plot ordered pairs of numbers such as (t,y) and this capability is used extensively in this book. We refer to all such plots as xy-*plots*. Many spreadsheet programs use some other designation (e.g., scatter plot). For aesthetic reasons, this book always represents *xy*-plots with smooth, connected lines (Figure 1.5). You may wish to mimic this format using the formatting commands available in your spreadsheet program.

[NOTE: The *line chart* option available in many spreadsheet programs may or may not generate a true *xy*-plot. Some spreadsheet programs distribute values uniformally across the horizontal axis even when the values themselves are not uniformly distributed. See the Appendix for more details.]

Using Electronic Spreadsheets

Now is the time to get out the instruction manual that came with your electronic spreadsheet program, turn on your computer, and start to work.

Many spreadsheets come with tutorials that will get you up to speed quickly. You may find that the power of your program is so tantalizing that you will spend hours seeking out exotic features and hidden opportunities. For all of the exercises in this book, however, you need only learn how to:

- Point to a cell and write into it.

Figure 1.4

Electronic spreadsheet models are often designed to expedite graphical exploration.

(a) This model calculates position *y* for a number of different values of time *t*.

(b) This *xy*-plot was generated automatically by a spreadsheet program.

(a)

	A	B	C
1	Uniform Acceleration		
2			
3	y0 =	2	m
4	v0 =	10	m/s
5	g =	-9.8	m/s^2
6			
7	t (s)	y (m)	
8	0	2	
9	0.2	3.804	
10	0.4	5.216	
11	0.6	6.236	
12	0.8	6.864	
13	1	7.1	
14	1.2	6.944	
15	1.4	6.396	
16	1.6	5.456	
17	1.8	4.124	
18	2	2.4	

(b)

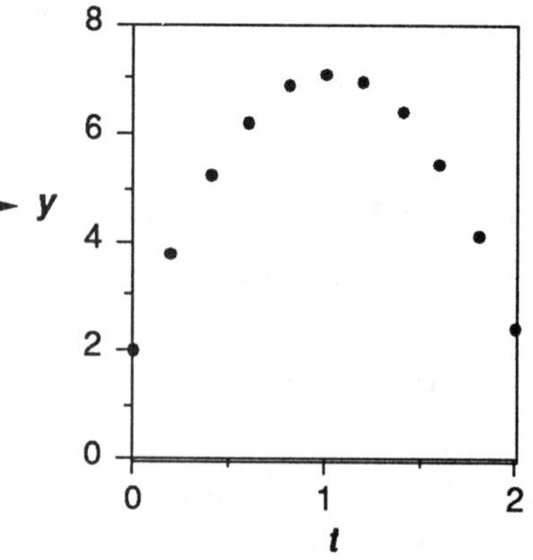

Figure 1.5

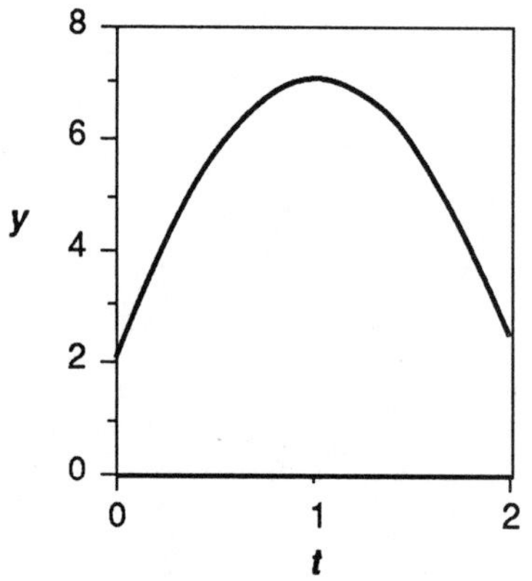

xy-plots can be conveniently represented by smooth curves, and this format will be the standard format of this book. Your spreadsheet program may be able to format *xy*-plots in a number of different ways.

You point to a cell by using either the arrow keys on your keyboard or an alternative pointing device such as a mouse. You write to a cell by typing the appropriate string of characters from your keyboard.

There are special functions (called built-in functions) on spreadsheets that permit you to calculate logarithms, exponentials, sinusoids, sums of lists of numbers, etc. Take a quick look at these and get a sense of how to enter them. Formulas that contain such built-in functions are easy to spot — e.g., =A3+SUM(B1:B10). The precise form used by your program may be slightly different.

When preparing formulas, it is not necessary to manually type in addresses such as B3. Instead, you can simply point to cell B3 with your cursor (using the arrow keys, mouse, or other pointing device) when you come to the appropriate place in the construction of your formula. Take the time to explore and practice thoroughly this technique using the exercises in your spreadsheet manual or tutorial. This elegant feature makes flawless formula building both easy and intuitive.

- Edit a cell.

Spreadsheets provide simple editing capabilities that allow you to change a cell's contents without retyping it entirely.

- Format a cell.

  You will discover that spreadsheets permit you to define the manner in which cell contents are displayed. Do you want your label left- or right-justified? Do you want your values displayed in floating point or scientific notation and to how many decimal places?

- Copy a cell or range of cells to another location using absolute and relative variables.

  Once you've written a complex formula, you need not rewrite it into the next 100 cells. Copy it instead.

- Edit a spreadsheet.

  A spreadsheet is a highly malleable medium, and you will sometimes want to rearrange a spreadsheet without completely rewriting it. Spreadsheets provide simple yet powerful editing capabilities that allow you to insert or delete rows and columns, move groups of cells to new locations, etc.

- Save and recall a model from disk.

  When you turn off your computer, it forgets. Disks are used to save work between sessions.

- Graph a model.

  You should learn how to make simple *xy*-plots using your spreadsheet program.

You do not need to know about such optional features as keyboard macros or data-list sorting, finding, or extracting. Where, however, such advanced features might be useful for you, we will point them out in specially marked spreadsheet techniques. Although unnecessary, you may find that your models can be built more concisely and with greater ease with the help of such advanced features.

If you have never used an electronic spreadsheet program, turn now to the training and reference manuals that were distributed with your program and spend an hour or so learning the basics. Once you are comfortable with the basic functions of your program, proceed to the next chapter. Exercise 2.1, "Uniform Acceleration," emphasizes the features of electronic spreadsheets helpful in building simulations of physical situations. When you finish this exercise, you should have an excellent grasp on the power and uses of these kinds of programs. Subsequent exercises will place more emphasis on physics.

The "Table of Contents" identifies both the physics and the spreadsheet tools that are introduced in each exercise.

CHAPTER 2

MOTION

When Galileo rolled balls down an inclined plane a new light burst upon all investigators of Nature — Immanuel Kant

Aristotle (384 - 322 B.C.) started it. He proposed that each object possessed a natural place. If a stone were lifted from the ground and released, motion would ensue because the stone strove to return to its natural place. No external agent was needed to cause the motion, and heavy objects returned to their natural place faster than light objects.

In 1590 Galileo (1564 - 1642) published *De Moto*, a record of experiments designed in part to test Aristotle's predictions. Galileo's most famous comments on his experimental results appeared in his book *Dialogues Concerning Two New Sciences* (1638):

Salviati: I greatly doubt that Aristotle ever tested by experiment whether it be true that two stones, one weighing ten times as much as the other, if allowed to fall at the same instant from a height of one hundred cubits, would so differ in speed that when the heavier had reached the ground the other would not have fallen more than ten cubits. ...

Sagredo: ... I, who have made the test, can assure you that a cannon ball weighing one hundred pounds or more will not reach the ground by as much as a span ahead of a musket ball weighing only half a pound. ...

Simplicio: Your discussion is really admirable; yet I do not find it easy to believe that a birdshot falls as swiftly as a cannon ball.

Salviati: Why not say a grain of sand as rapidly as a grindstone? ...

There are no more famous characters in scientific literature than the three above. Salviati, of course, is Galileo speaking.

Once Galileo had determined that different objects fell the same distance in the same time, he investigated the details of the motion. Obviously, an object speeds up during the fall. But just how does it speed up? Is there a simple pattern? Is the speed proportional to the distance of fall? Is the speed proportional to the elapsed time?

There were no accurate clocks in the late 1500s, and so Galileo could only use his pulse, choir chanters, or dripping water to mark equal intervals of time.[1] Because a falling object moves too swiftly for reasonable measurements, he "diluted gravity" by rolling a ball down a long, straight incline. He controlled the degree of "dilution" by adjusting the angle of the incline with respect to the horizontal. A remarkable pattern emerged, and modern science began.

Galileo discovered that objects rolling down an inclined plane increase their speed in proportion to the elapsed time. This pattern is true always and at all angles — including the case of a freely falling object where the angle of incline is 90° and there is no rolling (Table 2.1).

Elapsed Time (s)	Total Distance (ft)	Average Speed (ft s^{-1})	Average Acceleration (ft s^{-2})
0	0		
		1 x 16	
1	1 x 16		2 x 16
		3 x 16	
2	4 x 16		2 x 16
		5 x 16	
3	9 x 16		2 x 16
		7 x 16	
4	16 x 16		2 x 16
		9 x 16	
5	25 x 16		2 x 16
		11 x 16	
6	36 x 16		

Table 2.1

The speed of a freely falling body is directly proportional to elapsed time. Notice that the values in each column depend on the values in the column to the left in a beautiful way. Can you predict the values in a table extended to an elapsed time of 8 s?

Galileo had begun the study of *kinematics* — the study of the motion of an object in terms of its displacement, speed (or velocity), and acceleration. In this chapter, you will continue his explorations. You will use your spreadsheet program to study the kinematics of accelerating objects and plot their movements.[2] Because kinematics is concerned with the directions as well as the magnitudes of displacements, velocities, accelerations, you will also learn about vectors. Vectors are indispensable throughout physics.

EXERCISE **2.1**

UNIFORM ACCELERATION

TO GET THE MOST OUT OF THIS EXERCISE, YOU SHOULD ALREADY KNOW

The definitions of the terms:
electronic spreadsheet
cell
label
value
formula
relative variable
absolute variable

How to build a simple model with your electronic spreadsheet program by writing labels, values, and formulas directly into cells and by copying the contents of one cell or a range of cells into other cells.

How to write formulas with both absolute and relative variables and know how such formulas behave when copied using your spreadsheet program's COPY command.

This exercise tests your ability to build a simple electronic spreadsheet model. It also introduces you to the format of this book and its style of model presentation.

This first application to physics introduces the kinematics of an object with uniform (constant) acceleration. The mathematics used to describe the motion of such an object is uncomplicated. Acceleration is the first derivative of velocity with respect to time (i.e., the rate of change of velocity with time) (Equation 2.1):

$$\frac{dv}{dt} = a \quad \text{or} \quad \int_{v_0}^{v} dv = \int_{0}^{t} a\, dt \tag{2.1}$$

where v_0 is initial velocity, v is velocity at time t, and a is acceleration.

When a is constant,

$$v - v_0 = at - a \cdot 0 \quad \text{or} \quad v = v_0 + at \tag{2.2}$$

Similarly, velocity is the first derivative of position with respect to time (i.e., the rate of change of position with time) and when a is a constant you get Equation 2.3:

$$\frac{ds}{dt} = v_0 + at \quad \text{or} \quad \int_{s_0}^{s} ds = \int_{0}^{t} (v_0 + at)dt \tag{2.3}$$

and

$$s - s_0 = v_0 t + \frac{at^2}{2} \quad \text{or} \quad s = s_0 + v_0 t + \frac{at^2}{2} \tag{2.4}$$

where s_0 is initial position.

Motion with uniform accelertion can be used to approximate many real-life situations—the movement of an automobile stopping with constant breaking force or the motion of an brick falling freely near the surface of the Earth, for example.

If Equation 2.4 represents an object falling freely near the surface of the Earth, then s is height (meters, m), t is time (seconds, s), s_0 is the initial height above the Earth's surface (m); v_0 is the velocity at $t = 0$ (m s^{-1}), and a is the constant acceleration of the Earth's gravitational field (–9.8 m s^{-2}). The very small changes in the value of a with s may be ignored when the value of s is very much less than the radius of the Earth. For the present, the effects of air resistance are also ignored.

This example is used extensively in the exercises that follow. You must remember to pay proper attention to the signs of these initial values. In the exercises that follow, upward directions are positive, downward directions are negative. Thus, a ball thrown upward from a height of 2 m will have a positive s_0 and positive v_0.

For the remainder of this exercise, you should be sitting next to your personal computer with your spreadsheet program loaded and running.

Throughout this book, each exercise is prefaced by introductory text and followed by *Lessons*, *Problems*, and *Models*.

Lessons are intended to guide you step by step through the construction and interrogation of computer simulations. In many ways, *Lessons* are similar to instructions and questions that you might find in a laboratory manual. The exercise, however, is performed inside your computer rather than at your laboratory bench.

Problems are designed to give your imagination and problem-solving skills a freer reign. You will build, modify, and interrogate models with considerably less guidance than in *Lessons*. Some *Problems* are meant to be difficult. Others simply offer a new twist on an old theme.

Models contain model descriptions and directions for their construction.

LESSONS

1. Turn to the end of this exercise and briefly examine Model 2.1.

   Most models consist of two parts:

   - A figure that defines the contents of the labels, values, and formulas contained in the model

   - A legend containing verbal instructions for building and using the model

   The figure requires a bit of explanation. All figures that describe the structure of a spreadsheet model have two parts: A formula block and a value block. Many figures also include one or more formula set blocks.

   - *Formula* blocks show the actual contents of each cell. We built our models using the electronic spreadsheet program Excel; if your computer is running that program, they should appear precisely as shown. If you are using a different program, small changes may be required. The differences are very small and you will soon find yourself making the translations automatically and without conscious thought.[1]

   - *Value* blocks show you how your screen will appear when the model is complete and correct. Use value blocks to *verify* your model after it has been built. Compare the calculated results on your screen with the calculated results in the value block. Complex formulas can contain errors when first prepared or copied.

   - *Formula Set* blocks are used whenever the formula contains too many characters to fit comfortably into its cell in the formula block.

     Formula Set I of Model 2.1 says, "prepare this formula in cell B12 (the prototype cell). COPY the formula into cells B12 - B52."[2] Examine Figure 2.1 carefully to see how Formula Sets are organized. Bold-faced variables of the formula set are *absolute* variables and do not change from cell to cell when a

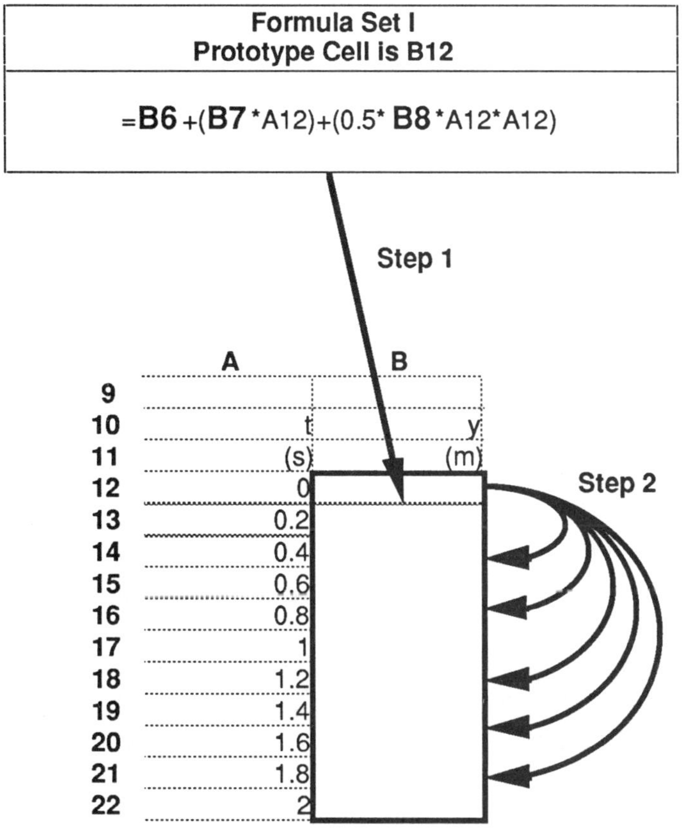

Figure 2.1

Formula sets are used to describe formulas that will not conveniently fit into the formula block of a model description.

Step 1: The formula in a formula set is constructed in the prototype cell. (Absolute variables are displayed in bold face, relative variables in plain face.)

Step 2: The formula is copied from the prototype cell to all member cells of the formula set. The distinction between *relative* and *absolute* is relevant only during a COPY (and not a MOVE/CUT and PASTE) operation.

formula is copied.[3] Plain-faced variables of the formula set are *relative* variables and change their address depending on their location after a copy operation. The first three instances of Formula Set I in Model 2.1 translate to

Cell	Formula
B12	=B6+(B7*A12)+(0.5*B8*A12*A12)
B13	=B6+(B7*A13)+(0.5*B8*A13*A13)
B14	=B6+(B7*A14)+(0.5*B8*A14*A14)

Notice that only the cell address A12 (in the formula in cell B12) changes when copied to cells B13 and B14. Refer to your program reference manual if you don't understand the difference between relative and absolute addresses.

[NOTE: Variables can be relative in one direction and absolute in the other. The variable **B**3, for example, copies "absolutely" when copying across columns but "relatively" when copying across rows. You'll get some experience with this a bit later if you don't understand now.]

2. Different spreadsheet users have different model-building styles. Until you develop your own, use the sequence outlined below:

   a. Enter all labels from Model 2.1 (FORMULAS) into your computer. When complete your screen should look like Figure 2.2a.

      Labels alone can give you a general sense of your screen layouts. You then have an opportunity to edit this layout before proceeding.

   b. Enter all values from Model 2.1 (FORMULAS) into your computer (don't forget the initial time value of zero in cell A12). When complete your screen should look like Figure 2.2b.

   c. Prepare the formula contained in cell A13 (Model 2.1, FORMULAS). You may find that your spreadsheet program uses a different syntax than we have chosen for this book. Your spreadsheet program may or may not use = to identify formulas. If =A12+B5 doesn't work, try +A12+B5. If neither character string works, look in your spreadsheet instruction manual for help.[4]

      Although you can literally type the appropriate formula into cell A12, you should try to get into the habit of preparing formulas by *pointing* — a feature found in most spreadsheet programs. Once you have started a formula by typing the first character (=, +, etc.), all cell addresses can be entered by simply moving your cursor to the cell you wish to reference. Only the operators need be typed from the keyboard. Conversion of cell addresses from

Figure 2.2

(a) Labels are signposts that organize your model into regions. Many spreadsheet model builders prefer to enter all labels first.

(b) Values are the data used by your model to calculate results. If you enter values before formulas, you will find it easier to correct formula mistakes "on the fly."

(a)

	A	B	C
1	Uniform Acceleration		
2			
3	y = y0 + v0*t + 0.5*a*t^2		
4			
5	t (incr) =		s
6	y0 =		m
7	v0 =		m/s
8	a =		m/s^2
9			
10	t	y	
11	(s)	(m)	
12			
13			
14			
15			
16			
17			

(b)

	A	B	C
1	Uniform Acceleration		
2			
3	y = y0 + v0*t + 0.5*a*t^2		
4			
5	t (incr) =	0.2	s
6	y0 =	10	m
7	v0 =	30	m/s
8	a =	-9.8	m/s^2
9			
10	t	y	
11	(s)	(m)	
12	0		
13			
14			
15			
16			
17			

relative to absolute is accomplished using a function or other special key sequence. After a little practice, you will find that this method of preparing formulas is not only intuitive but also almost foolproof.

d. Cells A13 - A52 contain formulas to calculate a range of values for time *t*. The range actually starts with cell A12 and its initializing value of zero. All subsequent cells in the range add a constant increment *t* (incr) (cell B5) to the time in the previous cell of the range. Study Model 2.1 (FORMULAS) until you understand these formulas and how they function. This trick is used throughout this book.

Notice that the formula in cell A13 is similar to the formulas that exist in cells A14 - A52. You should use your program's COPY command to avoid excessive typing. You must take into account, however, the fact that formula in cell A13, =A12+B5, contains both a *relative* variable (A12) and an *absolute* variable (B5). You know that B5 is an absolute variable because it remains unchanged throughout the range A12 - A52. If this variable appeared in a formula set it would be bold-faced.

If your spreadsheet program declares absolute variables by placing the special character $ in front of both the letter part and the number part of the cell address (=A12+B5), then simply enter the formula including these special characters and execute your COPY command. If your program uses a different conven-

	A	B	C
1	Uniform Acceleration		
2			
3	y = y0 + v0*t + 0.5*a*t^2		
4			
5	t (incr) =	0.2	s
6	y0 =	10	m
7	v0 =	30	m/s
8	a =	-9.8	m/s^2
9			
10	t	y	
11	(s)	(m)	
12	0		
13	0.2		
14	0.4		
15	0.6		
16	0.8		
17	1		

The formulas in cells A13 - A52 must generate the values displayed. The syntax of the formulas will be very similar to those displayed at the right.

=A12+B5
=A13+B5
=A14+B5
=A15+B5
=A16+B5

Figure 2.3

The syntax of formulas varies among the different manufacturers of spreadsheet programs. The formulas in the figure were written for the spreadsheet program Excel. If you have studied your program's instruction manual, you should know what changes to make (if any) in the figure formulas so that they will run on your computer.

tion to declare absolute and relative variables, check your manual. Your program may simply ask you which variables are relative and which are absolute when you execute the COPY command.

Your screen should now look like Figure 2.3.

e. Examine Model 2.1 (FORMULAS) and confirm that the formulas in cells B12 - B52 (Formula Set I) represent the equation of motion for a particle under uniform acceleration.

   Enter the formula in Formula Set I into cell B12 and copy down to cell B52.

   Your screen should now look like Model 2.1 (VALUES).

f. Verify the correctness of your model by comparing the on-screen values with the values of Model 2.1 (VALUES).

3. Whenever instructions call for you to build and verify a model, build the model by using the formula block and formula sets to enter labels, values, and formulas into the appropriate cells. Verify the model by comparing your on-screen values with the values in the values block.

4. Your model is now complete and ready to be explored. (You may wish to save it on a diskette before proceeding.)

   You will usually use your spreadsheet program's on-screen graphics capability to generate *xy*-plots of your data (e.g., Figures 1.4 and 1.5 in Chapter 1). Although all full-feature spreadsheet programs include graphics capability, the commands for generating such graphics vary considerably from program to program. The Appendix contains information for generating *xy*-plots using several common programs, but your best source of information is the user manual or tutorial that came with your program.

5. You should at least quickly review the contents of the Appendix before proceeding. Although this book displays *xy*-plots as smooth curves, you may find it safer to plot your data using discrete points. The line chart option on some graphics packages cannot be used to represent ordered pairs of numbers.[5]

Figure 2.4

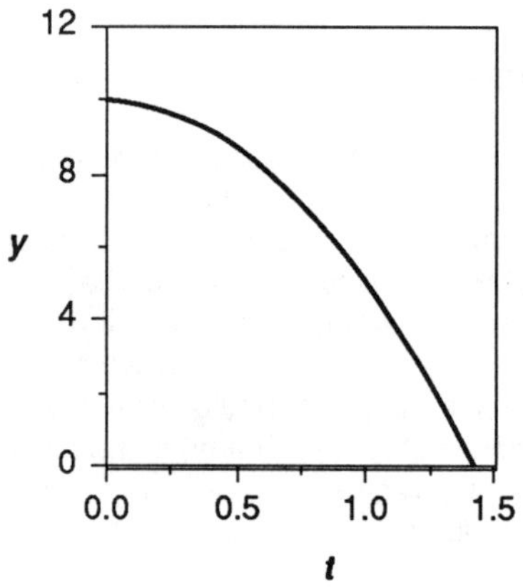

XY-PLOT

Data
Horizontal (*x*) axis: cells A12 - A19 (time elapsed)
Vertical (*y*) axis: cells B12 - B19 (*y*-position)

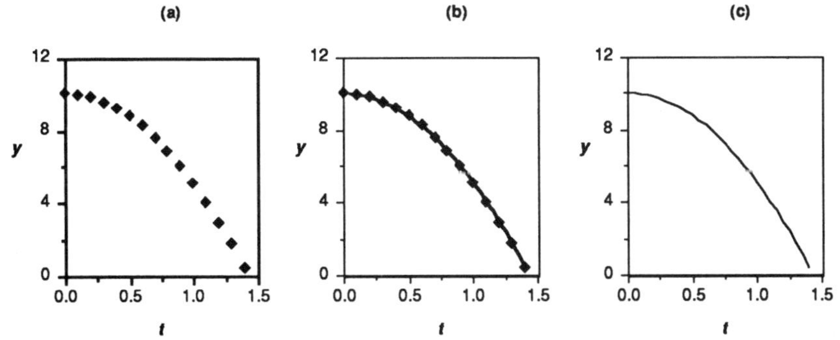

Figure 2.5

Your spreadsheet program may be able to format your *xy*-plots in several ways.

(a) The default value for an *xy*-plot is often data points alone.

(b) Some programs permit you to connect the data points with lines.

(c) Some programs permit you to draw lines alone.

a. Suppose that a ball is dropped from rest with $v_0 = 0$ from a platform 10 m above the ground. Plot the height of the ball above the ground as a function of time by building and verifying a plot of height y vs elapsed time t (Figure 2.4). Use t (incr) = 0.2.

Figure 2.4 asks you to build an *xy*-plot of the data in your model. The elapsed time t in cells A12 - A19 is represented on the x-axis, and the height y in cells B12 - B19 is represented on the y-axis. *Notice that the data in cells A20 - A52, B20 - B52 are not plotted. Data beyond the time when the ball hits the ground are meaningless in this model.*

b. Figure 2.4 shows how *xy*-plots are described in this book. Figure 2.5 shows several of the equivalent representations that may appear on your screen. Determine which of these formats are supported by your spreadsheet program.

6. Now for some physics! When your model is initialized for free fall from rest ($v_0 = 0$), only downward motion is possible (Figure 2.4). Other initial conditions, however, are possible:

a. Change the initial velocity (v_0) to represent an upward throw of 30 m s^{-1}. As the values of your model recalculate, your on-screen graphic should also change (Figure 2.6).

b. Because the data plotted for Lesson 6a include only rows 12 - 19 of your model, your plot doesn't really give you a feel for the total behavior of the ball under the new conditions. Change the range of values plotted on the x-axis by including only those values that represent the behavior of the ball before it hits the ground (Figure 2.7).

Figure 2.6

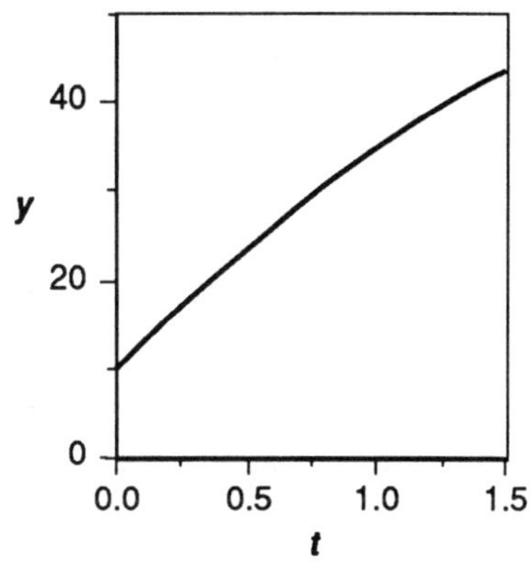

XY-PLOT

Data
Horizontal (x) axis: cells A12 - A19 (time elapsed)
Vertical (y) axis: cells B12 - B19 (y-position)

Figure 2.7

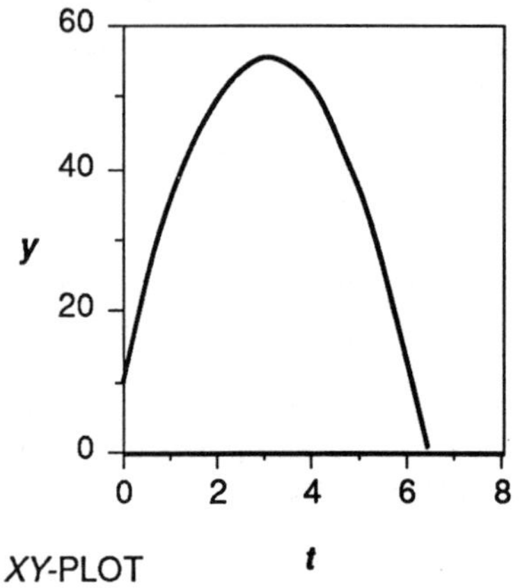

XY-PLOT

Data
Horizontal (*x*) axis: cells A12 - A46 (time elapsed)
Vertical (*y*) axis: cells B12 - B46 (*y*-position)

Figure 2.8

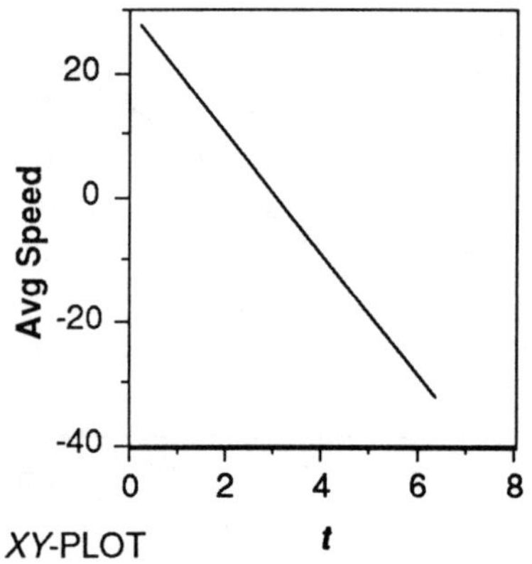

XY-PLOT

Data
Horizontal (*x*) axis: cells A13 - A46 (time elapsed)
Vertical (*y*) axis: cells C13 - C46 (average speed)

c. By direct observation of the appropriate graph, determine the maximum height of a ball thrown vertically upward from the ground ($y_0 = 0$) at an initial velocity of 30 m s^{-1}.

d. Suppose that a ball is thrown upward from the ground at an initial velocity of 25 m s^{-1}. Will the maximum height of the ball be greater or less than that in Lesson 6c? Confirm your intuition by direct observation of the appropriate graph.

7. The *average speed* and *average acceleration* can be calculated for the times and positions present in columns A and B. Build and verify Model 2.2 by adding to Model 2.1.

a. Examine the new formulas in column C and D (Formula Sets II and III).

At each time *t*, there will be a change in position in column B. This change can be calculated by subtracting the previous position in column B from the present position. Thus, the average speed in cell C13 is calculated by the formula =(B13-B12)/B5 (or a close equivalent, depending on your spreadsheet program). Remember, cell B5 is an *absolute* variable.

Similarly, at each time *t* there will be a change in average speed in column C calculated by subtracting the previous average speed from the present average speed. Thus, the average acceleration in cell D14 is calculated by the formula =(C14-C13)/B5 (or its equivalent for your spreadsheet program).

b. Plot average speed vs t using initial values of *t* (incr) = 0.16 s, $y_0 = 10$ m, $v_0 = 25$ m s^{-1}, $a = -9.8$ m s^{-2} (Figure 2.8). Explain in simple English why average speed vs *t* is a linear plot but *y*-position vs *t* is parabolic.

c. Plot average acceleration vs *t* (Figure 2.9). Explain in words how uniform acceleration leads to a linear increase in average velocity and to a nonuniform change in position with time.

d. You may be able to plot position, average speed, and average acceleration vs time elapsed on the same graph (keeping in mind that the numbers on the *y*-axis have different dimensions for each property). If your spreadsheet program has this capability, plot these three properties together and examine the triple plot carefully (Figure 2.10). Interpret and explain each graph and its relation to the other two graphs.

e. If you examine your plot (Figure 2.10) carefully, you may notice that the the maximum height of the thrown ball doesn't quite align with the point associated with zero average velocity. Why?

You will shortly see how to generate spreadsheet models that permit you to calculate *instantaneous speed* (*dy/dt*) to any desired degree of accuracy.

PROBLEMS

1. Model 2.3, an extension of Model 2.1, simultaneously calculates the trajectories of objects thrown on Earth, Venus, Mars, and Jupiter.

   a. Examine Formula Set I and notice that variable A12 is represented with only A in boldface. **A**12 means that, when copied, the variable will be treated as an *absolute* variable when copied across columns and as a *relative* variable when copied across rows. In cell C13, for example, the variable points to cell A13 — the column letter remains constant but the row number changes.

   Similarly, variable B8 is represented as B**8**. The variable points to cell D8 in cell C13 (two columns over and zero rows down).

   b. Build and verify this model. Design an on-screen graphic that will simultaneously plot the trajectories of an object thrown with the same height and speed on Venus, Earth, Mars, and Jupiter.

2. Imagine that you are in communication with research teams on Earth and on the Moon (the uniform acceleration of gravity on the Moon is about one-sixth the acceleration value here on Earth).

   a. Instruct both of your teams to drop an object from 100 m above the surface. Use Model 2.1 to simulate the behavior of the object on both bodies and plot the height above the surface, *y* vs elapsed time *t*, in both instances. By inspecting your plots, determine the length of time it will take each object to fall to the surface.

   b. Now throw the object upward at 25 m s^{-1}. By examining the appropriate graphs, determine the maximum heights and the total times of travel of both bodies.

Figure 2.9

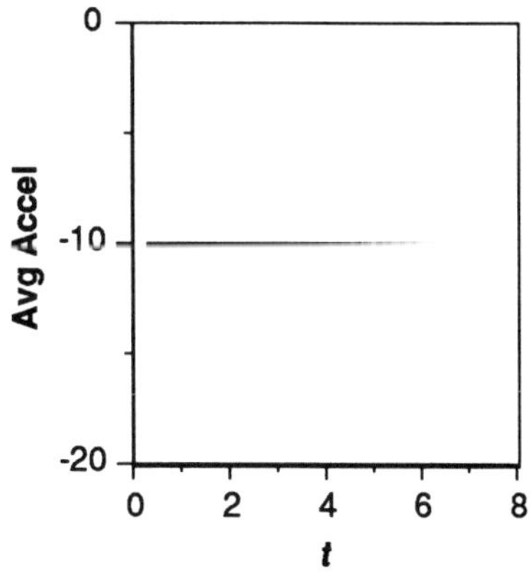

XY-PLOT

Data
Horizontal (*x*) axis: cells A14 - A46 (time elapsed)
Vertical (*y*) axis: cells D14 - D46 (average acceleration)

Figure 2.10

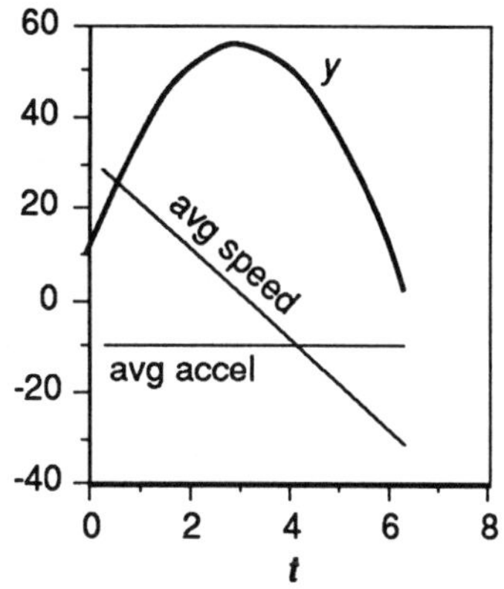

XY-PLOT

Data
Horizontal (*x*) axis: cells A12 - A46 (time elapsed)
Vertical (*y*) axis: cells B12 - B46 (*y*-position), C13 - C46 (average speed), D14 - D46 (average acceleration)

c. Ask your teams to determine by trial and error the value of the initial speed that will result in a maximum height of 200 m. Simulate this experiment on your computer.[6] [HINT: The spreadsheet function MAX can make this task easier.]

3. A glass of juice sliding across a long, horizontal table where the friction force is constant experiences uniform acceleration. Suppose that you are in a laboratory with a flat surface with a retarding frictional acceleration of -1.27 m s^{-2}. Use Model 2.1 to simulate an experiment in which a glass is sent from one end of the table with an initial velocity of 10 m s^{-1}.

   a. Plot position vs elapsed time and by inspection determine whether the glass will fall off a table 13.6 m long.

   b. By trial and error, determine the value of v_0 that brings the glass to rest at just 1 cm ± 1 mm from the far end of the table.

   c. What range of values for v_0 causes the glass to come to rest *within* 1 cm of the far end of the table?

SPREADSHEET TECHNIQUE

Formulas of Convenience

It is often useful to add to a model at some convenient place formulas that highlight features of data. You might insert the formulas given below into either Model 2.1 or Model 2.2, for example, to directly read the maximum height of a thrown object (cell E7). You can also implement the trial-and error technique by changing *t* (incr) — cell B5 of either model — until cell E6 is near zero. Cell E5 now directly displays the total travel time of the thrown object.

	D	E
5	Last t =	=A52
6	Last y =	=B52
7	Max y =	=MAX(B12:B52)

MODELS

FORMULAS

	A	B	C
1	Uniform Acceleration		
2			
3	y = y0 + v0*t + 0.5*a*t^2		
4			
5	t (incr) =	0.2	s
6	y0 =	10	m
7	v0 =	30	m/s
8	a =	-9.8	m/s^2
9			
10	t	y	
11	(s)	(m)	
12	0	Formula Set I	
13	=A12+B5		
14	=A13+B5		
51	=A50+B5		
52	=A51+B5		

VALUES

	A	B	C
1	Uniform Acceleration		
2			
3	y = y0 + v0*t + 0.5*a*t^2		
4			
5	t (incr) =	0.2	s
6	y0 =	10	m
7	v0 =	30	m/s
8	a =	-9.8	m/s^2
9			
10	t	y	
11	(s)	(m)	
12	0	10	
13	0.2	15.804	
14	0.4	21.216	
51	7.8	-54.116	
52	8	-63.6	

Formula Set I Prototype Cell is B12
=**B6**+(**B7***A12)+(0.5* **B8***A12*A12)

Model 2.1

IDENTIFY the spreadsheet with labels placed in rows 1 - 3.

ENTER data into cells B5 - B8. Cell B5 contains the value of the time increment *t* (incr) that will be used to generate the range of elapsed time values in column A.[7] Cells B6 and B7 are the initial values of position y_0 and velocity v_0, respectively. Cell B8 contains the constant acceleration value.

READ data from cells A12 - A52 (time *t*) and B12 - B52 (position *y*).

NOTICE that the third term of the equation of motion described by Formula Set I (0.5*B8*A12*A12) multiples cell A12 by itself rather than taking the square of A12 (as in 0.5*B8*A12^2). Some electronic spreadsheets use the same algorithm to calculate fractional powers (e.g., $10^{2.3}$) that they use to calculate integer powers (e.g., 10^2). This algorithm can lead to small errors in the last decimal position (e.g., 1.999999999 vs 2) and requires more calculation time. For this reason, many spreadsheet modelers prefer to multiply out expressions containing small integer exponents.

Model 2.2

IDENTIFY the spreadsheet with labels placed in rows 1 - 3.

ENTER data into cells B5 - B8. Cell B5 contains the value of the time increment *t* (incr) that will be used to generate the range of elapsed time values in column A. Cells B6 and B7 are the initial values of position y_0 and velocity v_0, respectively. Cell B8 contains the constant acceleration value.

READ data from cells A12 - A52 (time *t*), B12 - B52 (position *y*), C13 - C52 (average speed), and D14 - D52 (average acceleration).

NOTICE that the formulas in cells D14 - D52 can be easily generated by copying from cells C14 - C52.

FORMULAS

	A	B	C	D
1	Uniform Acceleration			
2				
3	y = y0 + v0*t + 0.5*a*t^2			
4				
5	t (incr) =	0.2	s	
6	y0 =	10	m	
7	v0 =	25	m/s	
8	a =	-9.8	m/s^2	
9				
10	t	y	Avg Speed	Avg Accel
11	(s)	(m)	(m/s)	(m/s^2)
12	0	Formula Set I		
13	=A12+B5		Formula Set II	
14	=A13+B5			Formula Set III

	A	B	C	D
51	=A50+B5			
52	=A51+B5			

VALUES

	A	B	C	D
1	Uniform Acceleration			
2				
3	y = y0 + v0*t + 0.5*a*t^2			
4				
5	t (incr) =	0.2	s	
6	y0 =	10	m	
7	v0 =	25	m/s	
8	a =	-9.8	m/s^2	
9				
10	t	y	Avg Speed	Avg Accel
11	(s)	(m)	(m/s)	(m/s^2)
12	0	10		
13	0.2	14.804	24.02	
14	0.4	19.216	22.06	-9.8

	A	B	C	D
51	7.8	-93.116	-50.46	-9.8
52	8	-103.6	-52.42	-9.8

Formula Set I **Prototype Cell is B12**
=**B6** +(**B7** *A12)+(0.5* **B8** *A12*A12)

Formula Set II **Prototype Cell is C13**
=(B13-B12)/ **B5**

Formula Set III **Prototype Cell is D14**
=(C14-C13)/ **B5**

FORMULAS

	A	B	C	D	E	F
1	Uniform Acceleration					
2						
3	y = y0 + v0*t + 0.5*a*t^2					
4						
5	t (incr) =	0.2				s
6	y0 =	10				m
7	v0 =	30				m/s
8	a =	-9.8	-8.82	-3.92	-26.46	m/s^2
9						
10	t	y (Earth)	y (Venus)	y (Mars)	y (Jupiter)	
11	(s)	(m)	(m)	(m)	(m)	
12	0	Formula Set I				
13	=A12+B5					
14	=A13+B5					
15	=A14+B5					

	A	B	C	D	E	F
51						
52						

VALUES

	A	B	C	D	E	F
1	Uniform Acceleration					
2						
3	y = y0 + v0*t + 0.5*a*t^2					
4						
5	t (incr) =	0.2				s
6	y0 =	10				m
7	v0 =	30				m/s
8	a =	-9.8	-8.82	-3.92	-26.46	m/s^2
9						
10	t	y (Earth)	y (Venus)	y (Mars)	y (Jupiter)	
11	(s)	(m)	(m)	(m)	(m)	
12	0	10	10	10	10	
13	0.2	15.804	15.8236	15.9216	15.4708	
14	0.4	21.216	21.2944	21.6864	19.8832	
15	0.6	26.236	26.4124	27.2944	23.2372	

	A	B	C	D	E	F
51	7.8	-54.116	-24.3044	124.7536	-560.913	
52	8	-63.6	-32.24	124.56	-596.72	

Formula Set I Prototype Cell is A1
=**B6**+(**B7***A12)+(0.5*B**8***A12*A12)

Model 2.3

IDENTIFY the spreadsheet with labels placed in rows 1 - 3.

ENTER data into cells B5 - B7 and B8 - E8. Cell B5 contains the value of the time increment t (incr) that will be used to generate the range of elapsed time values in column A. Cells B6 and B7 are the initial values of position y_0 and velocity v_0, respectively. Cells B8 - E8 contain the constant acceleration values for each of the four planets under study.

READ data from cells A12 - A52 (time t), B12 - B52 (position y [Earth]), C12 - C52 (position y [Venus]), D12 - D52 (position y [Mars]), E12 - E52 (position y [Jupiter]).

NOTICE that Formula Set I contains variables that are absolute when copied in one direction but relative when copied in the other direction.

EXERCISE **2.2**

RATES OF CHANGE

TO GET THE MOST OUT OF THIS EXERCISE, YOU SHOULD ALREADY KNOW

The definitions of the terms:
derivative
velocity
acceleration

How to obtain the derivative of a function. You should also understand that the first derivative of a function is the slope of the line tangent to a graph of the function at a point.

Rates of change play a major role in physics. The rate of change of position with time, velocity, and the rate of change of velocity with time, acceleration, are two familiar examples. In physics it is usually the *instantaneous* rates of change that are at issue.

The mathematical description of instantaneous rates of change is contained within the differential calculus — independently discovered in the 1600s by Sir Isaac Newton in England and Gottfried von Leibnitz in Germany.

Consider a function of time that describes an object that can move only along a line (Equation 2.5):

$$s = f(t) \tag{2.5}$$

where s is a position along the line and $f(t)$ is a functional expression of that position in terms of time t.

The instantaneous rate of change of position with time v can be obtained by using differential calculus: differentiate s with respect to t (Equation 2.6):

$$v = \frac{ds}{dt} \tag{2.6}$$

Calculus provides a series of rules for differentiating algebraic and other expressions. For example, suppose that the position of an object is given as

$$s = bt^2 \tag{2.7}$$

where b is a constant. Then, its velocity is

$$v = \frac{ds}{dt} = 2bt \tag{2.8}$$

The rules for differentiating algebraic expressions are found in any elementary calculus or physics book.

The symbolic manipulation of algebraic expressions to obtain their first or higher-order derivatives constitutes an *analytic* method of solution.

The *numeric* method of approximating derivatives does not require the symbolic manipulation of expressions. It is, however, based on the analytic definition of a derivative. If $s = f(t)$, then

$$\frac{ds}{dt} = \lim_{\Delta t \to 0} \frac{\Delta s}{\Delta t} = \lim_{\Delta t \to 0} \frac{f(t + \Delta t) - f(t)}{\Delta t} \tag{2.9}$$

where Δt is the elapsed time and Δs is the change in position during the elapsed time Δt.

If you ignore the analytic limiting process, you can use Equation 2.9 to approximate the value of ds/dt for specific values of t by just choosing a small value of Δt. Then,

$$v(t) = \frac{ds}{dt} \cong \frac{f(t + \Delta t) - f(t)}{\Delta t} \tag{2.10}$$

How small? That's the hard question. In numeric work, you often choose several progressively smaller values of Δt until the approximation for ds/dt doesn't change "too much" for the problem at hand. Your good judgment is required.

Numeric methods of finding derivatives are especially suitable in computer environments and very handy when the problem is difficult to handle analytically. In fact, they are very handy for checking that you have successfully performed the correct analytic differentiation of a complicated function.

In this exercise, you use Equation 2.9 to build a model capable of finding an approximation for the rate of change ds/dt of any smooth function $s = f(t)$ for any value, or range of values, of t.[1]

LESSONS

1. Build and verify Model 2.4. Study the model and verify that it represents Equation 2.9.

   a. Suppose that an object moves under constant acceleration according to the equation $s = 4.9t^2$, where s is position in meters and t is time in seconds. Complete the table below to find the velocity of the object when $t = 2$ s. Give your answer to two decimal places.

Δt	$\Delta s/\Delta t$
2	________
1	________
0.5	________
0.1	________
0.01	________
0.001	________
0.0001	________
0.00001	________

SPREADSHEET TECHNIQUE

Data Tables

Data tables (or "what if" tables) are a feature of many spreadsheet programs. Data tables are like robots that query a model many times and write down the result of each interrogation. The attached data table, for example, interrogates Model 2.4 eight separate times by placing the values in cells A13 - A20 into cell B3 one at a time and writing down the contents of cell B10 in cells B13 - B20. Most "single-valued" data tables have the layout shown here. Different spreadsheet programs have different specific commands for identifying a spreadsheet region as a data table.

Model 2.4 could, of course, be laid out so that data tables are unnecessary (e.g., Model 2.5), but sometimes the required calculations make this difficult or impossible. Attach this data table to Model 2.4 (if your spreadsheet program permits it) and plot $\Delta s/\Delta t$ vs Δt.

2 Return this result in the column below

	A	B
12		=B10
13	2	29.4
14	1	24.5
15	0.5	22.05
16	0.1	20.09
17	0.01	19.649
18	0.001	19.6049
19	0.0001	19.60049
20	0.00001	19.600049

1 Place these values into cell B3

Enter your table into a convenient place in your spreadsheet. then plot $\Delta s/\Delta t$ as a function of Δt. What is the shape of the plot? What happens to $\Delta s/\Delta t$ as Δt decreases? Why?

b. To two decimal places, what is the velocity of the object at 4 s? 10 s?

2. Modify Model 2.4 such that $s = 10 + 20t - 4.9t^2$. (You need only enter the formula once into cell B7 and then copy it to cell B8). How fast is the object traveling at $t = 0$? $t = 2$? $t = 12$?

3. Build and verify Model 2.5. Model 2.5 extends Model 2.4 such that s and ds/dt can be calculated for a range of t values. Column F of the model contains an analytic expression for ds/dt and column G shows the difference between the numeric (column E) and the analytic (column F) solution.

A special feature of this model (and many other models in this book) is the contents of cells B3 and C3. Cell C3 contains the formula (Equation 2.9) used throughout the model to calculate $s(t)$, and cell B3 contains a dummy value of t. If the equation of motion for $s(t)$ changes, you need only change the spreadsheet formula in cell C3 and copy it into all appropriate spreadsheet locations.[2]

SPREADSHEET TECHNIQUE

Keyboard Macros

Keyboard macros are supported by many spreadsheet programs. Macros usually consist of one or more cells (on the same or a different spreadsheet file) that contain instructions for executing an entire sequence of keystrokes. The entire sequence can be invoked easily by executing a single command. Some programs even have a *learn* option that automatically writes the commands for you as you execute the sequence manually. The keystroke macro shown here was generated automatically by the program Excel by recording the keystrokes associated with copying the contents of cell C3 into cells B11 - B37, C11 - C37. Automatically recording keystrokes is often the most convenient way of writing macros because, as you can see, the instructions themselves can be rather complex. If you enjoy programming and wish to learn how to write macros, practice first with some simple macros in the instruction manual of your spreadsheet program. You may then wish to incorporate macros into the models in this book.

	A
1	Copy
2	=ACTIVATE("Model 2.5")
3	=SELECT("R3C3")
4	=COPY()
5	=SELECT("R11C2")
6	=VPAGE(1)
7	=SELECT("R11C2:R37C2")
8	=PASTE()
9	=VLINE(-1)
10	=SELECT("R11C4")
11	=VPAGE(1)
12	=SELECT("R11C4:R37C4")
13	=PASTE()
14	=RETURN()

Figure 2.11

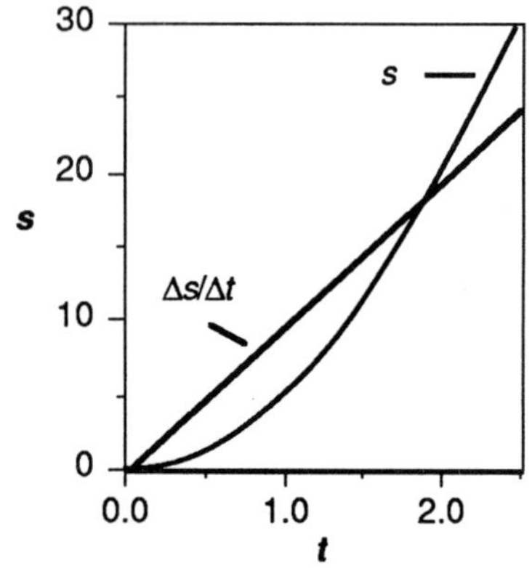

XY-PLOT

Data
Horizontal (*x*) axis: cells A11 - A37 (*t*)
Vertical (*y*) axis: cells B11 - B37 (*s*), E11 - E37 (Δ*s*/Δ*t*)

a. Explore Model 2.5 and verify that it accurately represents Equation 2.9 for a range of values of *t*. What is the purpose of cells B5 and B6?

b. Suppose that an object moves under constant acceleration according to the equation $s = 4.9t^2$ where *s* is position in meters and *t* is elapsed time in seconds. Use an on-screen graphic similar to Figure 2.11 to plot position and velocity vs elapsed time. $\Delta s/\Delta t$ is very close to ds/dt when $\Delta t = 0.0001$.

c. Explore the influence of Δt on the accuracy of $\Delta s/\Delta t$ as an approximation of ds/dt by comparing the numeric solution (column E) with the analytic solution (column F). Column G calculates the difference between these two values. For this function, the error is independent of *t*, but this is not generally the case.

PROBLEMS

1. A chandelier in a cathedral begins to swing after a mild earthquake. The horizontal displacement *s* from its equilibrium position obeys Equation 2.11:

$$s(t) = \sin t \tag{2.11}$$

where *t* is elapsed time in seconds.

a. Use a plot similar to Figure 2.11 to examine the displacement and velocity of the chandelier at increments of 0.15 s. At what point in the swing of the chandelier is it traveling fastest? At what two points in the swing of the chandelier does the chandelier have a zero velocity?

b. Analytically, if $s = \sin t$, then $ds/dt = \cos t$. How large a value of Δt still supports a difference between $\Delta s/\Delta t$ and ds/dt of less than 0.001? [HINT: You can quickly search the error column for the largest error with the formula =MAX(MAX(G11:G37),-MIN(G11:G37)).]

c. Plot the error vs $s(t)$. Note that in this case the error is proportional to s itself. How is the proportionality constant related to Δt?

2. An explosion is a chemical reaction in which the number of molecules that have reacted by time t grows very rapidly. An exponential function is often used to model this phenomenon. Suppose that

$$N(t) = e^t \tag{2.12}$$

where $N(t)$ is the number of molecules that have reacted by time t (in microseconds).

   a. Plot the number of interactions vs time (for a range of 10 microseconds) using a plot similar to Figure 2.11. Can you see why exponential functions offer reasonable descriptions of explosions?

   b. Use the numeric method (column E) to obtain the rate of change of the number of interactions vs time. Are you surprised?

   c. Use calculus to show that $dN/dt = e^t$ when $N(t)$ satisfies Equation 2.12. Does this make sense given what you saw in Problem 2b?

   d. Plot the error vs $N(t)$. Again notice that the error is proportional to the value of the function. What is the value of the proportionality constant?

   This property of the error (i.e., the error is proportional to the value of the function) is unique to sinusoidal and exponential functions.

3. Suppose that you are asked to follow the behavior of an object, located on a vertical track 10 m above the ground and traveling up at 20 m s^{-1} at $t = 0$. The acceleration of gravity is –9.8 m s^{-2}.The motion for such an object is given given by Equation 2.13:

$$s = s_0 + v_0 t + \frac{at^2}{2} \tag{2.13}$$

where s_0 is the position of the object (m) at $t = 0$, v_0 is the velocity (m s^{-1}) at $t = 0$, and a (m s^{-2}) is the value of the constant acceleration of gravity (–9.8 m s^{-2}).

   a. Plot the position of the object as a function of time. When does the object reach its maximum height? When does it hit the ground?

b. Plot the velocity of the object as a function of time. When does the object have zero velocity? When does the object attain its maximum velocity?

4. Consider the object described in Problem 3. Recalling that acceleration is dv/dt (or d^2s/dt^2) add a column to your model that will permit you to plot acceleration vs time. Does your plot confirm that the acceleration is constant with time?

5. Consider the function

$$s = \sqrt{t} \tag{2.14}$$

Explore this function using t (incr) $= 0.01$ and $\Delta t = 0.001$. Notice that this function does not have a derivative at $t = 0$ although the numeric method happily produces one. Plot error vs s. What happens to the error as s decreases?

6. Sometimes it can be tedious to calculate a derivative analytically. Numeric methods, however, work as quickly and simply on complex functions as they do on simple functions.

Find the value of x (to two significant figures) at the steepest slope on the curve

$$f(x) = e^{-\left(\frac{1}{1-x^2}\right)^3} \tag{2.15}$$

for $0 < x < 1$.

Would you like to try this analytically?

7. Search your physics textbook for problems that can be explored using Model 2.4. Use calculus in each case to find an analytic expression for ds/dt and, in each case, find the maximum value of Δt that keeps $\Delta s/\Delta t$ within 0.1% of ds/dt.

MODELS

FORMULAS

	A	B	C
1	Rate of Change		
2			
3	Δt =	0.01	s
4	t =	0	s
5	t + Δt =	=B4+B3	s
6			
7	s(t) =	=4.9*B4*B4	m
8	s(t + Δt) =	=4.9*B5*B5	m
9			
10	Δs/Δt =	=(B8-B7)/B3	m/s

VALUES

	A	B	C
1	Rate of Change		
2			
3	Δt =	0.01	s
4	t =	0	s
5	t + Δt =	0.01	s
6			
7	s(t) =	0	m
8	s(t + Δt) =	0.00049	m
9			
10	Δs/Δt =	0.049	m/s

Model 2.4

BUILD and verify Model 2.4.

ENTER data using cells B3 (Δt) and B4 (t). The formula for any smooth function can entered into cell B7 and then copied to B8.

READ output from the model at cell B10 ($\Delta s/\Delta t$). $\Delta s/\Delta t$ approaches ds/dt as Δt approaches a very small number.

NOTICE that our spreadsheet can incorporate Greek letters such as Δ into labels. Your spreadsheet may or may not be able to do this.

Model 2.5

BUILD and verify Model 2.5.

ENTER data using cells B5 - B7. Cells B5 and B6 determine the range of values of t to be calculated. Cell B7 specifies the value of Δt used to approximate ds/dt. The formula for any smooth function can be entered into cell C6 and copied into cells B11 - B37, and D11 - D37.

READ output from the model at column E ($\Delta s/\Delta t$) and column F (ds/dt). $\Delta s/\Delta t$ approaches ds/dt as Δt approaches a very small number.

NOTICE that cells B3 and C3 are not necessary to run the model but simply store a copy of formulas used in columns B and D (Equation 2.9). The relative variable of the formula in cell C3 points to the contents of the cell to its left (cell B3). Because the formulas in columns B and D, $s(t)$ and $s(t + \Delta t)$, are both driven by the contents of the columns to their left, t and $t + \Delta t$, the contents of cell C3 may be used as a "master copy" of the expression to be evaluated. Thus, any time you wish to study a new function, simply prepare an expression for the new function in cell C3, test that you have made no mistakes by changing values in cell B3 (watching C3), and then copy it to the appropriate cells in columns B and D. If your software supports macros, you can write a macro that performs this COPY automatically. This technique is a convenient way to build models that solve many different problems.

FORMULAS

	A	B	C	D	E	F	G
1	Rate of Change						
2							
3	s = 4.9t^2	0.1	=4.9*B6*B6				
4							
5	t (start) =	0	s				
6	t (incr) =	0.1	s				
7	Δt =	0.0001	s				
8							
9	t	s(t)	t + Δt	s(t + Δt)	Δs/Δt	ds/dt	Error
10	(s)	(m)	(s)	(m)	(m/s)	(m/s)	(m/s)
11	=B5	Formula Set II	Formula Set III	Formula IV	Formula Set V	Formula Set VI	Formula Set VII
12–37	Formula Set I						

VALUES

	A	B	C	D	E	F	G
1	Rate of Change						
2							
3	s = 4.9t^2	0.1	0.049	<== Pattern for formula			
4							
5	t (start) =	0	s				
6	t (incr) =	0.1	s				
7	Δt =	0.0001	s				
8							
9	t	s(t)	t + Δt	s(t + Δt)	Δs/Δt	ds/dt	Error
10	(s)	(m)	(s)	(m)	(m/s)	(m/s)	(m/s)
11	0	0	0.0001	4.9E-08	0.00049	0	0.00049
12	0.1	0.049	0.1001	0.0491	0.98049	0.98	0.00049
13	0.2	0.196	0.2001	0.1962	1.96049	1.96	0.00049

	A	B	C	D	E	F	G
36	2.5	30.625	2.5001	30.6275	24.5005	24.5	0.00049
37	2.6	33.124	2.6001	33.1265	25.4805	25.48	0.00049

Formula Set I Prototype Cell is A12
=A11+**B6**

Formula Set II Prototype Cell is B11
=4.9*A11*A11

Formula Set III Prototype Cell is C11
=A11+**B7**

Formula Set IV Prototype Cell is D11
=4.9*C11*C11

Formula Set V Prototype Cell is E11
=(D11-B11)/**B7**

Formula Set VI Prototype Cell is F11
=9.8*A11

Formula Set VII Prototype Cell is G11
=E11-F11

EXERCISE **2.3**

VECTORS

In the mid-1800s Sir William Rowan Hamilton laid the groundwork for the mathematics of vectors with his study of quaternions, but the development of the subject for the specific needs of physics was not done until the late-1800s by J. W. Gibbs of Yale University. Today, physicists describe and explore a wide range of physical properties using vector notation — forces, displacements, velocities, accelerations, momenta, and many other physical quantities. You will find that many vector applications involve combinations of just two or three vectors in the two-dimensional plane; however, because the world of classical physics is acted out in three dimensions, you also will explore three-dimensional vectors in this exercise.

Consider a vector $\boldsymbol{A}$ in the x-, y-plane of magnitude A and angle ϕ with respect to the x-axis. This vector also can be represented by the vector sum of a vector $A_x\boldsymbol{i}$ along the x-axis and another vector $A_y\boldsymbol{j}$ along the y-axis. The two numbers A_x and A_y are called the x- and y-components of the vector $\boldsymbol{A}$ (Figure 2.12a). Similarly, a three-dimensional vector can be represented by the three-component vector: $A_x\boldsymbol{i} + A_y\boldsymbol{j} + A_z\boldsymbol{k}$ (Figure 2.12b).[1]

There are only a few vector operations required for the exercises in this book, and if necessary you should review them in your primary physics textbook.

TO GET THE MOST OUT OF THIS EXERCISE, YOU SHOULD ALREADY KNOW

The definitions of the terms:
scalar quantity
vector quantity
vector magnitude
vector sum
vector dot product
vector cross product

Figure 2.12

Vectors represent quantities that have both magnitude and direction.

(a) The two-dimensional vector ***A*** can be described either by a magnitude A and direction ϕ or as the sum of the two vectors $A_x\boldsymbol{i}$ and $A_y\boldsymbol{j}$, directed along the x- and y-axes. The numbers A_x and A_y are called the components of vector ***A***.

(b) A three-dimensional vector ***A*** may similarly be expressed as the sum of the three vectors $A_x\boldsymbol{i}$, $A_y\boldsymbol{j}$, and $A_z\boldsymbol{k}$, directed along the x-, y-, and z-axes. The numbers A_x, A_y, and A_z are the components of ***A***.

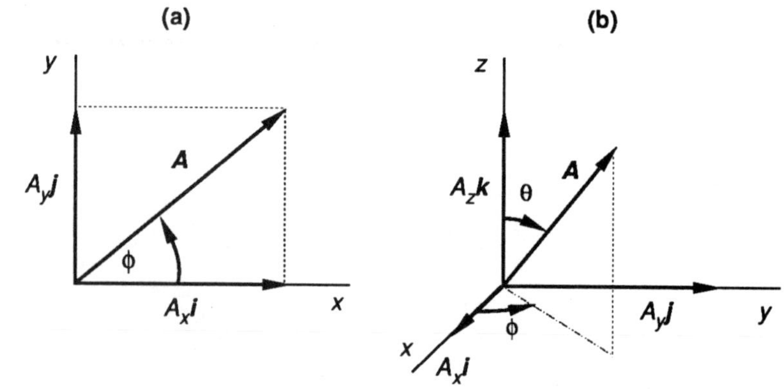

Given the angle ϕ and magnitude A of a vector in two dimensions ***A***, the components A_x and A_y may be resolved using Equations 2.16 and 2.17:

$$A_x = A \cos \phi \qquad (2.16)$$

$$A_y = A \sin \phi \qquad (2.17)$$

If the components A_x and A_y are known, ϕ and A can be found using Equations 2.18 and 2.19:

$$\phi = \arctan \frac{A_y}{A_x} \qquad \text{with} \qquad -\pi < \phi \leq \pi \qquad (2.18)$$

$$A = +\sqrt{A_x^2 + A_y^2} \qquad (2.19)$$

The magnitude of a vector in three dimensions is a simple extension of Equation 2.19 (Equation 2.20):

$$A = +\sqrt{A_x^2 + A_y^2 + A_z^2} \qquad (2.20)$$

The two angles ϕ and θ in Figure 2.12b can be calculated from the formulas

$$\phi = \arctan\left(\frac{A_y}{A_x}\right) \quad \text{with} \quad -\pi < \phi \leq \pi \qquad (2.21)$$

$$\theta = \left(\frac{A_z}{|A|}\right) \quad \text{with} \quad 0 < \theta \leq \pi \qquad (2.22)$$

Figure 2.13

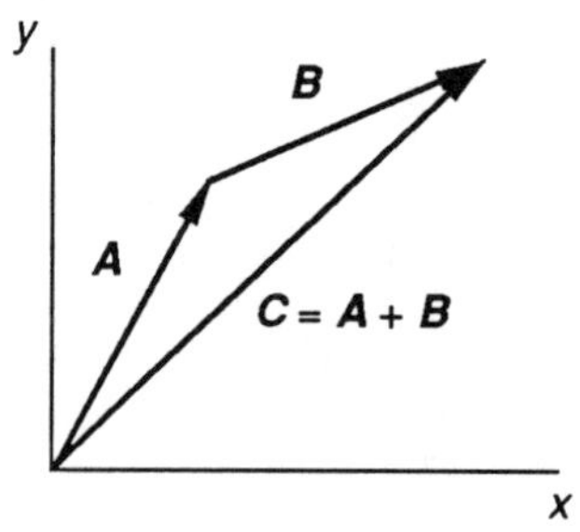

The sum of two vectors ***C*** = ***A*** + ***B*** can be represented graphically by placing the tail of ***B*** at the head of ***A*** and then drawing a new vector ***C*** from the tail of ***A*** to the head of ***B***.

The sum of two vectors ***A*** and ***B*** in three dimensions can be found using Equation 2.23 (Figure 2.13). Two-dimensional vectors are treated by setting $A_z = B_z = 0$:

$$\boldsymbol{A} + \boldsymbol{B} = (A_x + B_x)\boldsymbol{i} + (A_y + B_y)\boldsymbol{j} + (A_z + B_z)\boldsymbol{k} \qquad (2.23)$$

The angle ψ between vectors $\boldsymbol{A}$ and $\boldsymbol{B}$ can be determined using Equation 2.24:

$$\cos\psi = \frac{A_xB_x + A_yB_y + A_zB_z}{AB} \qquad (2.24)$$

with A and B of Equation 2.24 found by using Equation 2.20.

The dot product of two vectors in three dimensions $\boldsymbol{A}$ and $\boldsymbol{B}$ is a simple numeric quantity (also called a scalar) and is found using either Equation 2.25 or 2.26, depending on the information available (Figure 2.14).

$$\boldsymbol{A}\cdot\boldsymbol{B} = AB\cos\psi \qquad (2.25)$$

$$\boldsymbol{A}\cdot\boldsymbol{B} = A_xB_x + A_yB_y + A_zB_z \qquad (2.26)$$

The cross product of two vectors in three dimensions, $\boldsymbol{A}$ and $\boldsymbol{B}$, is a vector quantity (Figure 2.15). The magnitude of the cross product may be found using Equation 2.27. The components of $\boldsymbol{A} \times \boldsymbol{B}$ may be found using Equation 2.28:

$$|\boldsymbol{A} \times \boldsymbol{B}| = AB\sin\psi \qquad (2.27)$$

$$\boldsymbol{A} \times \boldsymbol{B} = (A_yB_z - A_zB_y)\boldsymbol{i} + (A_zB_x - A_xB_z)\boldsymbol{j} + (A_xB_y - A_yB_x)\boldsymbol{k} \qquad (2.28)$$

As you will see, spreadsheet technique "formula composition" enables you to use Equations 2.26 and 2.28 almost exclusively in your model building.

In this exercise you build simple spreadsheet models that perform each of the calculations identified above. Their use in physics becomes apparent in subsequent lessons.

LESSONS

In the lessons that follow, you build slowly toward Model 2.6. However, you can use these lessons to refresh your memory of vector operations by not referring to the description of Model 2.6 unless you get stuck.

1. Figure 2.16 displays a portion of an electronic spreadsheet model for resolving the components of a two-dimensional vector $\boldsymbol{A}$, given an angle and a magnitude. It can also perform the reverse calculation

Figure 2.14

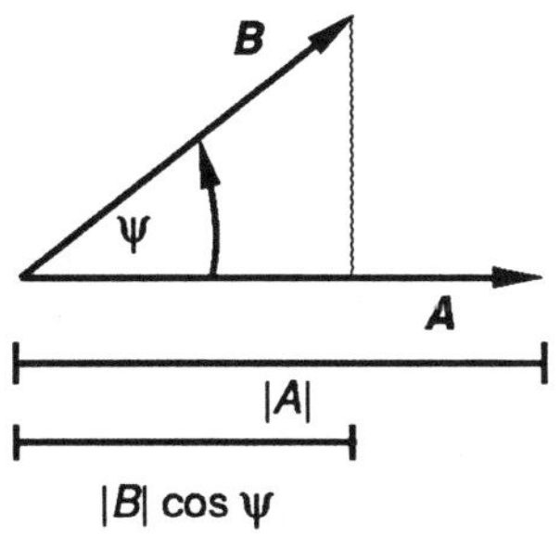

The dot product of two vectors $\boldsymbol{A}\cdot\boldsymbol{B}$ is a scalar quantity and may be represented as the component of $\boldsymbol{B}$ that lies in the direction of $\boldsymbol{A}$, $B\cos\psi$, multiplied by the magnitude of $\boldsymbol{A}$, A.

Figure 2.15

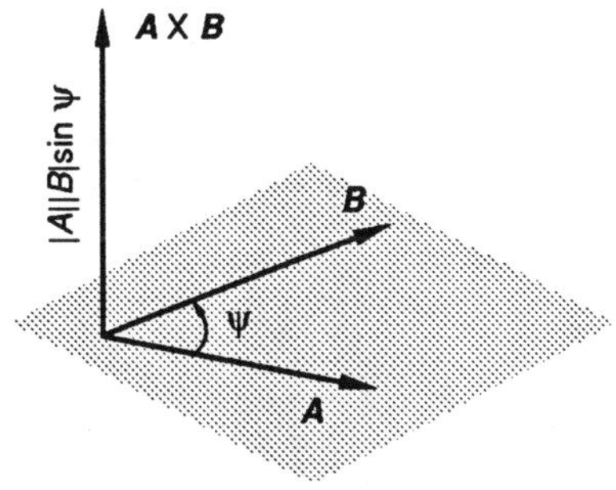

The cross product of two vectors, $\boldsymbol{A} \times \boldsymbol{B}$, is a vector that is perpendicular to the plane containing $\boldsymbol{A}$ and $\boldsymbol{B}$.

and find the magnitude and angle of a vector, given the two vector components A_x and A_y.

Because most spreadsheet programs calculate trigonometric functions in radians, cell C6 contains the formula for translating from degrees to radians — =(C5*PI())/180.[2]

a. Cells B9 - C9 contain formulas for the vector components A_x and A_y. What are these formulas? ($A_z = 0$ always in this lesson.)

SPREADSHEET TECHNIQUE

ATAN2 Function

Using the functions arcsine, arccosine, and arctangent requires care. You must be very sensitive into which quadrant your spreadsheet program places the resulting angle.

Usually, the built-in function ACOS produces an angle in the range $[0,\pi]$, and the function ASIN produces an angle in the range $[-\pi/2,\pi/2]$. The function ATAN, however, is not so standard, and your spreadsheet program may give results in the range $[0,\pi]$ or $[-\pi/2,\pi/2]$. This can be a real problem because these angular ranges cover only half the plane. As an example of the problem, notice that the two vectors

$$\mathbf{A} = 10\mathbf{i} + 10\,\mathbf{j}$$
$$\mathbf{B} = -10\mathbf{i} - 10\mathbf{j}$$

both have the ratio $A_y/A_x = +1$, but $\mathbf{A}$ has an angle of $\phi = 45°$ (with respect to the x-axis) and $\mathbf{B}$ has an angle of $\phi = -135°$ (or $+225°$). Using the ATAN function to calculate ϕ would produce the same angle for both vectors.

Many spreadsheet programs have a second arctangent function named ATAN2 that nicely handles this problem. Its range of values is usually $[-\pi, \pi]$. Thus, ATAN2(10,10) produces 45° and ATAN2(–10,–10) gives –135°. You should experiment with your spreadsheet program to learn how it treats arc functions and pay special attention to those exceptional cases for vectors on or very near the coordinate axes.

The argument for ATAN is the value of *y/x*

ATAN(arg)

ATAN2 requires two arguments: one for the *x*-direction and one for the *y*-direction. Is the argument order the same in your spreadsheet program?

ATAN2(*x*-direction, *y*-direction)

	A	B	C	D	E	F	G
1	Vector Operations						
2							
3	VECTOR COMPONENTS X-Y				VECTOR ANGLE AND MAGNITUDE		
4	Magnitude =		150			i	j
5	Angle(deg) =		53		Vector =	90	120
6	Angle(rad) =		0.925				
7					Magnitude =		150
8		i	j	k	Angle(rad) =		0.9273
9	Vector =	90.272	119.8	0	Angle(deg) =		53.13

Figure 2.16

Spreadsheet calculations for a single, two-dimensional vector.

b. Cells G7 - G9 contain formulas for vector magnitude and angle. What are these formulas? Cell G8 uses the ATAN2 function instead of ATAN (see spreadsheet technique "ATAN2 Function"). If your spreadsheet does not support ATAN2, use ATAN instead (but remember that the angles produced will be in the wrong half of the plane for some vectors).

c. Build the model described by Figure 2.16 and verify that it functions correctly. (You can use Model 2.6 (VALUES) if you wish.)

d. Complete the table below:

A	ϕ (deg)	A_x	A_y
5	0	______	______
5	30	______	______
5	45	______	______
5	90	______	______
10	30	______	______
10	90	______	______
______	______	10	10
______	______	20	20
______	______	10	0
______	______	0	10
______	______	0	0

(Notice that for the last case the angle ϕ is undefined. We've included it so that you can learn how your spreadsheet program handles this kind of error.)

2. You can extend your model by including cells that add the components of two vectors in three dimensions and find the angle between

	A	B	C	D	E	F	G	H
11	ADDITION				ANGLE BETWEEN A AND B			
12		i	j	k		i	j	k
13	Vector A =	200	0	0	Vector A =	200	10	0
14	Vector B =	90	120	0	Vector B =	90	120	0
15	Sum =	290	120	0	Angle(rad) =		0.8773	
16	Magnitude =		313.85		Angle(deg) =		50.268	

Figure 2.17

Addition of two, three-dimensional vectors.

Figure 2.18

More three-dimensional vector operations with two vectors.

	A	B	C	D	E	F	G	H
18	DOT PRODUCT				CROSS PRODUCT			
19		i	j	k		i	j	k
20	Vector A =	200	10	0	Vector A =	200	10	0
21	Vector B =	90	120	0	Vector B =	90	120	0
22	Dot Product =		19200		Cross Product =	0	0	23100
23					Magnitude =		23100	

them (Figure 2.17). Notice that this portion of your model begins at row 11.

a. Cells B15 - D15 contain formulas for the resultant vector components when vector ***A*** and vector ***B*** are added together. What are these formulas?

b. Cell C16 contains a formula for the magnitude of the vector sum. What is this formula?

c. Cells G15 - 16 contain the angle between vectors ***A*** and ***B*** in radians and degrees, respectively. What formulas belong in these cells?

d. Build the model described by Figure 2.17 and verify that it functions correctly.

e. Suppose that $\boldsymbol{C} = \boldsymbol{A} + \boldsymbol{B}$ and that ψ is the angle between ***A*** and ***B***.

Complete the table below:

Vector *A* A_x	A_y	A_z	Vector *B* A_x	A_y	A_z	Vector *C* C	ψ
0	1	0	0	1	0	_____	_____
0	0	1	1	0	0	_____	_____
5	0	3	1	1	1	_____	_____
3	5	2	–3	–5	2	_____	_____
3	5	2	–3	–5	–2	_____	_____
1	1	1	0	0	0	_____	_____

Figure 2.19

Two-dimensional vector components on a spreadsheet.

	A	B	C	D
1	Vector Operations			
2				
3	VECTOR COMPONENTS X-Y			
4	Magnitude =		150	
5	Angle(deg) =		53	
6	Angle(rad) =		0.925	
7				
8		i	j	k
9	Vector =	90.272	119.8	0

3. Extend your model further by including cells for the dot product and cross product of two vectors (Figure 2.18). Notice that this portion of your model begins at row 18.

   a. Cell C22 contains a formula to calculate the dot product of two vectors. What is this formula?

   b. Cells F22 - H22 contain formulas for the resultant vector components of the cross product of two vectors. What are these formulas?

   c. Cell G23 contains a formula to calculate the magnitude of $\boldsymbol{A} \times \boldsymbol{B}$. What is this formula? (Notice that you can prepare this formula by simply copying from cell C16.)

   d. Build the model described by Figure 2.18 and verify that it functions correctly.

   e. Make a second copy of your model on your disk. Use your copy to answer Lesson 3f.

   f. Find the dot product and cross product of two vectors $\boldsymbol{A}$ and $\boldsymbol{B}$ where $\boldsymbol{A} = \boldsymbol{i} + 3\boldsymbol{j} + 4\boldsymbol{k}$ and $\boldsymbol{B} = 2\boldsymbol{i} + 4\boldsymbol{j} + 3\boldsymbol{k}$.

4. Now add formulas to your model that calculate the components of a two-dimensional vector from its magnitude and angle (Figure 2.19).

5. Now that your model is partially complete you're ready to explore a special, electronic spreadsheet technique that we call *formula*

SPREADSHEET TECHNIQUE

Formula Composition

The properties of the COPY and the MOVE (or COPY and PASTE and CUT and PASTE) commands can easily produce models from simple templates. In formula composition, the important feature of the COPY (or COPY and PASTE) command is that relative variables are transferred relatively. The MOVE (or CUT and PASTE) command, on the other hand, transfers variables absolutely regardless of their status during a copy.

The technique is best shown using Lesson 5 as example: Given the values for the vectors $\boldsymbol{A}_1$, $\boldsymbol{A}_2$, $\boldsymbol{B}_1$, $\boldsymbol{B}_2$, and $\boldsymbol{C}_1$, find

$$[(\boldsymbol{A}_1 + \boldsymbol{A}_2) \times (\boldsymbol{B}_1 + \boldsymbol{B}_2)] \cdot \boldsymbol{C}_1$$

continued.

SPREADSHEET TECHNIQUE (continued)

Here are the steps:

Step 1 Enter the label Vector A1 into a blank region of spreadsheet (e.g., cell A30) and then enter the *x*-, *y*-, and *z*-components of A1 into cells B30, C30, and D30, respectively.

Step 2 Enter the label Vector A2 into cell A31 and then enter the *x*-, *y*-, and *z*-components of A2 into cells B31, C31, and D31, respectively.

Step 3 COPY (or COPY and PASTE) the sum formulas and label from cells A15 - D15 to A32 - D32. Then relabel A32 as "A1 + A2 =". Notice that the formulas copy *relatively,* and the formula now sums the contents of rows 30 and 31.

Steps 4 - 6 Repeat steps 1, 2, 3 for vectors $\boldsymbol{B}_1$ and $\boldsymbol{B}_2$, entering their names and components into cells A33 - D33 and A34 - D34 respectively. COPY (or COPY and PASTE) the sum formulas from cells A15 - D15 (cells A32 - D32 will also work) into cells B35 - D35. Relabel A35 as B1 + B2.

Step 7 MOVE (or CUT and PASTE) cells A32 - D32 to cells A37 - D37. *Notice that in a MOVE (or CUT and PASTE) operation the formulas still refer to cells B30 - D32. This is the crucial difference between COPY and MOVE (or COPY and PASTE and CUT and PASTE).*

Step 8 MOVE (or CUT and PASTE) cells A35 - D35 to cells A38 - D38.

Step 9 COPY (or COPY and PASTE) the cross-product formulas and label from cells E22 - H22 to cells A39 - D39 and relabel cell A39. You may need to widen column A.

Step 10 Enter the label Vector C1 into cell A40 and then enter the *x*-, *y*-, and *z*-components of C1 into cells B40, C40, and D40, respectively.

Step 11 COPY (or COPY and PASTE) the dot-product formula from cell C22 to cell C41 and label cell A41 appropriately.

Cell C41 contains your desired answer, but you may wish to tidy things up for subsequent use using the following steps.

Step 12 MOVE (or CUT and PASTE) the block of cells A33 - D34 to A32 - D33.
MOVE (or CUT and PASTE) cells A40 - D40 to cells A34 - D34.
MOVE (or CUT and PASTE) cells A37 - D39 to cells A35 - D37.
MOVE (or CUT and PASTE) cells A41 - C41 to cells A38 - C38.

Now you have cells for the five input vectors in a block followed by formulas for the intermediate calculations followed by the result. *This calculation might just be a portion of a larger problem.* Perhaps now you would move the input vectors to other locations as dictated by the global layout of the problem.

composition. This technique is extremely useful when you need to prepare a very complicated formula. Formula composition retains the integrity of the original vector formula blocks and provides a sequential record of the operations performed to solve the problem. In general, the technique consists of building and testing the pieces of the formula separately and then combining them by adroitly using your spreadsheets COPY and MOVE commands (COPY and PASTE and CUT and PASTE on some software/hardware combinations). Model 2.6 already contains a set of thoroughly tested "pieces" with which you can compose the solution to $[(\boldsymbol{A}_1 + \boldsymbol{A}_2) \times (\boldsymbol{B}_1 + \boldsymbol{B}_2)] \cdot \boldsymbol{C}_1$ where $\boldsymbol{A}_1 = 5\boldsymbol{i} + \boldsymbol{j} + 3\boldsymbol{k}$, $\boldsymbol{A}_2 = \boldsymbol{i} + \boldsymbol{j} + \boldsymbol{k}$, $\boldsymbol{B}_1 = \boldsymbol{i} + 3\boldsymbol{j} + 4\boldsymbol{k}$, $\boldsymbol{B}_2 = 2\boldsymbol{i} + 4\boldsymbol{j} + 3\boldsymbol{k}$, $\boldsymbol{C}_1 = 3\boldsymbol{i} + \boldsymbol{j} + 3\boldsymbol{k}$. The spreadsheet technique "Formula Composition" describes the exact series of steps required to solve this problem.

6. Use the formula composition technique to produce the formula for $(\boldsymbol{A}_1 + \boldsymbol{A}_2) \cdot [(\boldsymbol{B}_1 + \boldsymbol{B}_2) \times \boldsymbol{C}_1]$. For various values of the five input vectors, compare the results of this formula with the formula in Lesson 5. Your results should verify the important theorem of vector algebra $\boldsymbol{A} \cdot (\boldsymbol{B} \times \boldsymbol{C}) = (\boldsymbol{A} \times \boldsymbol{B}) \cdot \boldsymbol{C}$.

PROBLEMS

1. Forces are treated as vectors in physics. Suppose that two forces $\boldsymbol{F}_A = \boldsymbol{i} + \boldsymbol{j} + \boldsymbol{k}$ and $\boldsymbol{F}_B = 2\boldsymbol{i} - \boldsymbol{j}$ act on a point. Find the following:

   a. The vector components and magnitude of $\boldsymbol{F}_A + \boldsymbol{F}_B$.

   b. The vector components and magnitude of $\boldsymbol{F}_A - \boldsymbol{F}_B$.

   c. The included angle between $\boldsymbol{F}_A$ and $\boldsymbol{F}_B$.

   d. The value of $\boldsymbol{F}_A \cdot \boldsymbol{F}_B$.

   e. The vector components and magnitude of $\boldsymbol{F}_A \times \boldsymbol{F}_B$.

   f. The vector components and magnitude of $\boldsymbol{F}_B \times \boldsymbol{F}_A$.

2. Suppose that two forces $\boldsymbol{F}_A = 3.70\boldsymbol{i} - 4.10\boldsymbol{j} + 7.35\boldsymbol{k}$ and $\boldsymbol{F}_B = -6.05\boldsymbol{i} - 0.15\boldsymbol{j} + 3.41\boldsymbol{k}$ act on a point. Repeat Problem 1a - f for these new values.

3. Two women pull horizontally on ropes attached to a post. The angle between the ropes is 53°. If Sarah exerts a force of 200 Newtons (N) and Maxine exerts a force of 150 N what is the magnitude of the

resultant force? What is the angle that this result force makes with Sarah's direction of pull?

4. A weight (force) of 200 N is supported by two cords A and B. Cord A makes an angle α with the horizontal direction and cord B an angle β.

   a. Draw a sketch. (If you wish, introduce suitable x-, y-values into your spreadsheet and use your graphics commands to make the sketch.)

   b. Consider the symmetric case (both angles equal). Find the tension (force) in each cord when $\alpha = \beta = 45°$. What is the tension when the angles are both 30°? 5°? 0.01°?

   c. Consider the asymmetric case. Find the tension in each cord when

α	β	**Tension**
45°	30°	______
5°	30°	______
0°	90°	______

5. A cruise ship moves southeast from New York with a yeading of 135° from North to the mouth of the Amazon River. The ship maintains a constant velocity with a speed of 40 ft s^{-1} while Gilberto (aboard the ship) walks at a speed of 6.3 ft s^{-1} (with respect to the ship) and at an angle of 48° from the bow direction. What is his resultant velocity with respect to the Earth? Is it possible for him to walk in such a way that his velocity with respect to the Earth is due North at 0.7 ft s^{-1}?

6. A boat traverses the Amazon River at a location where the river is 100 miles wide and the current is 3 mph.

   a. If the boat moves perpendicular from the shore at a speed of 17 mph (with respect to the water) for 10 min, how far downstream and how far across has the boat moved? What is the resultant velocity vector (with respect to the river bank) for this journey?

   b. Suppose that the maximum boat speed with respect to the water is 21 mph and the goal is to reach a small island in the river in 10 min. The island is 4 miles downstream and 1.5 miles from the bank. Try different directions to determine whether this feat can be achieved at a constant velocity.

7. A helium balloon is supported by the surrounding air such that it drifts upward at a constant speed of 2.1 m s^{-1} with respect to the air. A horizontal breeze blowing toward the northeast pushes the balloon at 8.6 mph. What is the balloon's resultant velocity with respect to the Earth? After 1.3 min the balloon breaks and drifts downward at the constant speed of 2.3 m s^{-1} while still being pushed by the breeze. Where does it land with respect to its starting location, and what is the direction of the resultant displacement vector?

8. A fighter pilot moves horizontally at 1800 mph (2640 ft s^{-1}) with respect to the air while climbing at 147 ft s^{-1}. While on a heading of 14° from due north, the pilot's radar identifies another fighter plane traveling 36° from due north at 1500 mph. The plane is 70 miles away and 200 ft higher. The paths are projected to cross. What is their relative velocity magnitude and direction? Will the planes collide if they maintain their presents velocities?

MODELS

VALUES

	A	B	C	D	E	F	G	H
1	Vector Operations							
2								
3	VECTOR COMPONENTS X-Y				VECTOR ANGLE AND MAGNITUDE			
4	Magnitude =		150			i	j	
5	Angle(deg) =		53		Vector =	90	120	
6	Angle(rad) =		0.925					
7					Magnitude =		150	
8		i	j	k	Angle(rad) =		0.9273	
9	Vector =	90.272	119.8	0	Angle(deg) =		53.13	
10								
11	ADDITION				ANGLE BETWEEN A AND B			
12		i	j	k		i	j	k
13	Vector A =	200	0	0	Vector A =	200	10	0
14	Vector B =	90	120	0	Vector B =	90	120	0
15	Sum =	290	120	0	Angle(rad) =		0.8773	
16	Magnitude =		313.85		Angle(deg) =		50.268	
17								
18	DOT PRODUCT				CROSS PRODUCT			
19		i	j	k		i	j	k
20	Vector A =	200	10	0	Vector A =	200	10	0
21	Vector B =	90	120	0	Vector B =	90	120	0
22	Dot Product =		19200		Cross Product =	0	0	23100
23					Magnitude =		23100	

Cell C6 contains =(C5*2*PI())/360
Cell B9 contains =C4*COS(C6)
Cell C9 contains =C4*SIN(C6)
Cell G7 contains =SQRT(F5^2+G5^2)
Cell G8 contains =ATAN2(F5,G5)
Cell G9 contains =(G8*360)/(2*PI())
Cell B15 contains =B13+B14
Cell C15 contains =C13+C14
Cell D15 contains =D13+D14
Cell C16 contains =SQRT((B15*B15)+(C15*C15)+(D15*D15))
Cell G15 contains =ACOS(((F13*F14)+(G13*G14)+(H13*H14))/(SQRT((F13*F13)+(G13*G13)+(H13*H13))*SQRT((F14*F14)+(G14*G14)+(H14*H14))))
Cell G16 contains =(360*G15)/(2*PI())
Cell C22 contains =(B20*B21)+(C20*C21)+(D20*D21)
Cell F22 contains =(G20*H21)-(H20*G21)
Cell G22 contains =(H20*F21)-(F20*H21)
Cell H22 contains =(F20*G21)-(G20*F21)
Cell G23 contains =SQRT((F22*F22)+(G22*G22)+(H22*H22))

Model 2.6

BUILD and verify Model 2.6. The unusual structure of this model requires a somewhat different presentation format. Only VALUES are shown, followed by a list of formula-containing cells and their contents. This model, unlike many of the models in this book, places a number of separate calculations in a single model.

NOTICE that the contents of cell G8 uses the ATAN2 function described in the first spreadsheet technique in this exercise. If your software fails to support ATAN2(), you should replace this cell with the formula =ATAN(G5/F5).

NOTICE also that this spreadsheet is designed specifically for use with the spreadsheet technique "Formula Composition."

EXERCISE **2.4**

NONUNIFORM ACCELERATION

TO GET THE MOST OUT OF THIS EXERCISE, YOU SHOULD ALREADY KNOW

The definitions of the terms:
derivative
definite integral
indefinite integral
velocity
acceleration

How to analytically evaluate simple indefinite integrals.

Exercise 2.1 introduced motion under uniform (constant) acceleration. When acceleration, a, is constant, it is easy to find analytic expressions for position and velocity (Equations 2.1 - 2.4). Suppose, however, that acceleration is nonuniform (Equation 2.29).

$$\frac{dv}{dt} = a(t) \tag{2.29}$$

where v is velocity, t is time, and $a(t)$ is acceleration as a function of time.

Velocity may not be expressible in terms of simple functions you already know but it can always be expressed as an integral (Equation 2.30).

$$v(t) = \int_0^t a(t)dt + v_0 \tag{2.30}$$

A similar situation exists with respect to position (Equations 2.31, 2.32).

$$\frac{ds}{dt} = v(t) \tag{2.31}$$

$$s(t) = \int_0^t v(t)dt + s_0 \tag{2.32}$$

where s is position and $v(t)$ is velocity as a function of time (perhaps obtained from Equation 2.30).

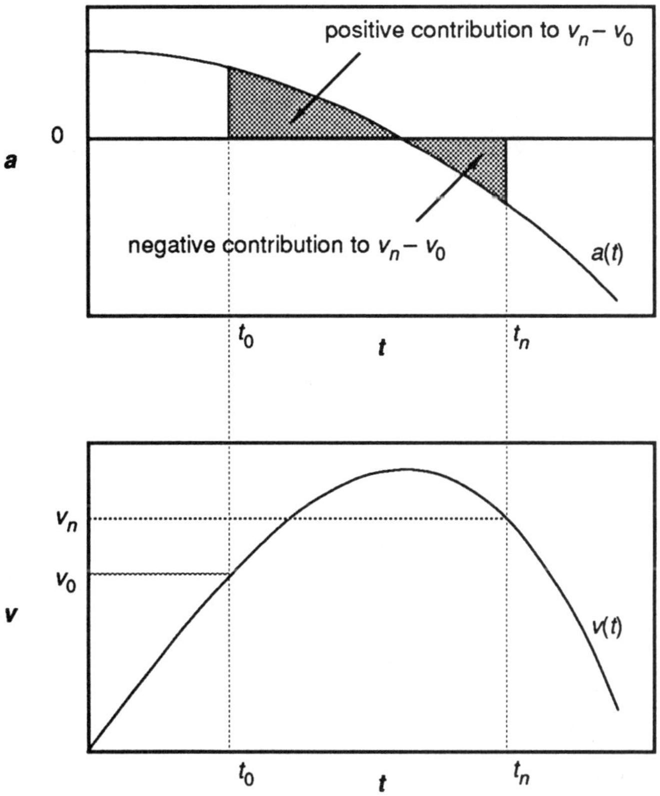

Figure 2.20

The value of the definite integral

$$\int_{t_0}^{t_n} a(t)dt$$

can be interpreted graphically as the area bounded by the t-axis, the vertical lines through t_0 and t_n, and the function $a(t)$. Areas above the t-axis have positive values, and areas below the t-axis have negative values.

In this and the next exercise, you will explore numeric integration methods. Using numeric methods, you can expand your physics explorations to those problems for which the solutions are not simple functions.

Suppose that you wish to know the change in velocity between two times, t_0 and t_n, of an object undergoing a nonuniform acceleration *that is a known function of time* $a(t)$. If v_0 is velocity at t_0 and v_n is velocity at t_n, then

$$v_n - v_0 = \int_{t_0}^{t_n} a(t)dt \qquad (2.33)$$

Equation 2.33 can be represented graphically as the area bounded by the t-axis, the vertical lines passing through t_0 and t_n, and the function $a(t)$ (Figure 2.20). Numeric integration methods take advantage of the fact that this area can be approximated by summing the areas of a large number n of rectangles or other regular shapes that are bounded by the axes and the function (Figure 2.21).

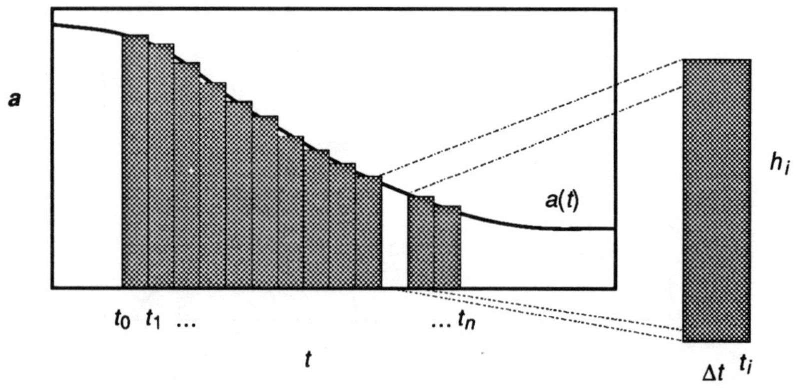

Figure 2.21

The numeric value of a definite integral can be approximated by summing the areas of a large number n of rectangles of known area. If each rectangle has a width Δt and height $a(t)$, then the area of each rectangle is $a(t)\Delta t$, and the value of the definite integral can be approximated by

$$\int_{t_0}^{t_n} a(t)dt \sim \sum_{i=1}^{n} h_i \, \Delta t$$

If rectangular elements of width Δt are used to approximate the area specified by the definite integral, the area of each element is simply the product of Δt and height h of the rectangle. But what height should you use?

For each interval you could take the acceleration at either end (Figure 2.22a, c). You could also take any value of the acceleration within the interval or even some weighted average of several such values. Each such choice yields a different numeric approximation to the integral, and for particular accelerations, one choice is usually better than another.

For the types of smooth functions that usually arise in physics applications, a good choice for height h of the rectangle is the simple, equally weighted average of the functional values at the two ends (Figure 2.22b, Equation 2.34). This rule gives the exact integral if the acceleration is a linear function of time; for that reason, it is called the linear approximation. Equation 2.34 is used to estimate the height of the rectangle in this exercise:

$$h = \frac{a(t - \Delta t) + a(t)}{2} \tag{2.34}$$

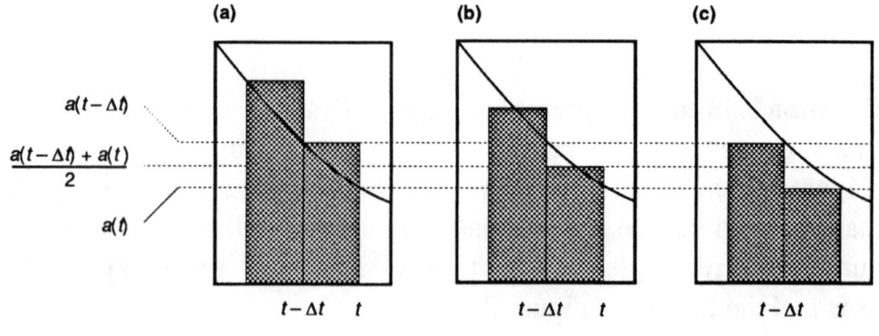

Figure 2.22

The sums of large numbers of rectangular area elements are often used to estimate the value of a definite integral. For a curved function $a(t)$ the height of the rectangle could be taken as any value between $a(t-\Delta t)$ and $a(t)$ or a weighted average of several such values.

(a) If $a(t-\Delta t)$ is chosen as the height for a descending function, then the area under the curve will be consistently overestimated.

(b) The average of $a(t-\Delta t)$ and $a(t)$ is often a good choice for the height of a rectangular area element.

(c) If $a(t)$ is chosen as the height, then the area under the curve will be consistently underestimated.

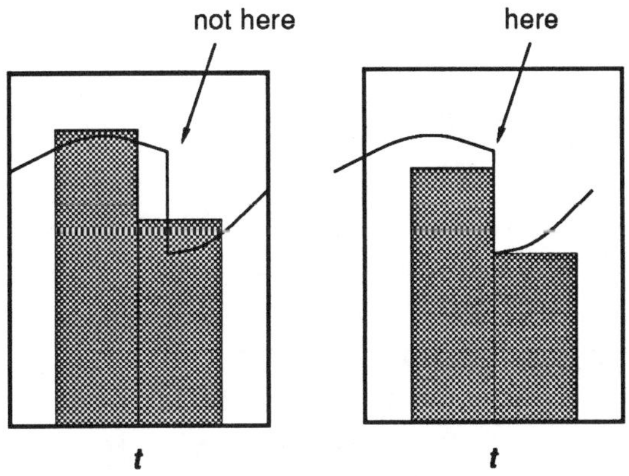

Figure 2.23

There are times when a sudden jump does occur in acceleration (at least on the time scale of the total time interval being considered). The boundaries of the rectangles should be arranged so that the jump occurs on one of them.

Thus, you can estimate the integral in Equation 2.33 by dividing the area under the curve $a(t)$ into n rectangles of width Δt and summing their individual areas (Equation 2.35). The accuracy of the estimate increases as Δt decreases.

$$v_n - v_0 = \int_{t_0}^{t_n} a(t)dt \sim \sum_{i=1}^{n} \frac{a(t_i - \Delta t) + a(t_i)}{2} \Delta t \qquad (2.35)$$

One final note, most approximation methods assume that $a(t)$ is well behaved in the sense that there are no sudden jumps to extremely large positive or large negative values inside one of the rectangles. Accelerations in nature tend to be reasonably accommodating, but sometimes it is a convenient approximation to assume that an acceleration jumps suddenly. In such cases, the boundaries should be chosen so that the jump occurs on one of them (Figure 2.23).

LESSONS

1. Build and verify Model 2.7. This model estimates the velocity at time t of an object subject to the nonuniform acceleration described by Equation 2.36:

$$a(t) = 2 - \frac{t^2}{10} \qquad (2.36)$$

You can, at any time change the function $a(t)$ by entering a new formula into cell C2 and copying it into cells B10 - B35.

Because $a(t)$ can be integrated analytically (Equation 2.37), the difference between the velocity estimate and its actual value is also computed:

Figure 2.24

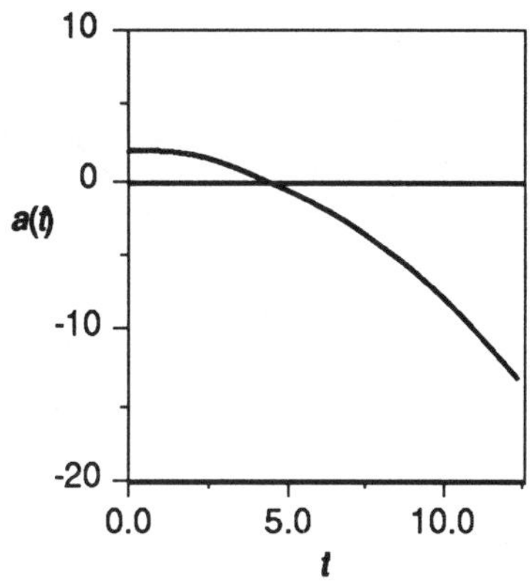

XY-PLOT

Data
Horizontal (x) axis: cells A10 - A35 (t)
Vertical (y) axis: cells B10 - B35 [$a(t)$]

Figure 2.25

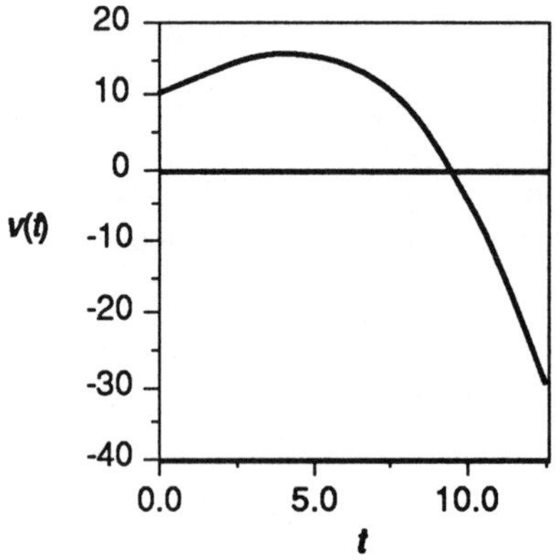

XY-PLOT

Data
Horizontal (*x*) axis: cells A10 - A35 (*t*)
Vertical (*y*) axis: cells D10 - D35 [*v*(*t*)]

Figure 2.26

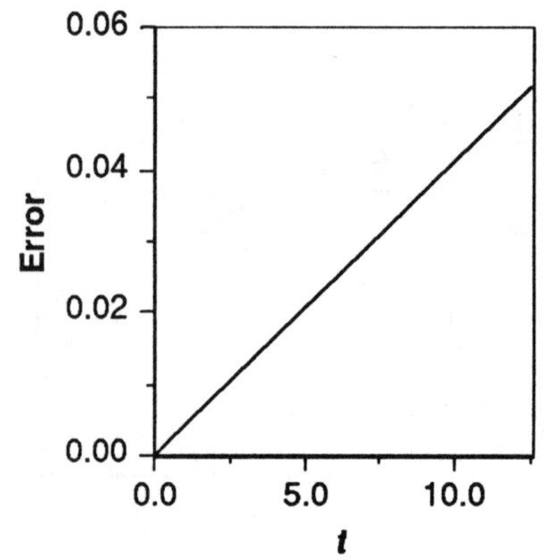

XY-PLOT

Data
Horizontal (*x*) axis: cells A10 - A35 (*t*)
Vertical (*y*) axis: cells E10 - E35 (error)

$$v(t) = \int_0^t a(t)dt + v_0 = 2t - \frac{t^3}{30} + v_0 \qquad (2.37)$$

where v_0 is the initial velocity.

a. Examine the model carefully until you understand its structure and its relationship to Equation 2.35.

b. Plot acceleration as a function of time (Figure 2.24) and velocity as a function of time (Figure 2.25) on-screen and explore different views of $a(t)$ and $v(t)$ by varying Δt. Extend the model beyond row 35 if you wish to explore further.

c. Examine the area beneath $a(t)$ as a function of time and compare it with the value of $v(t)$. What happens to velocity while $a(t)$ is positive, zero, and negative?

2. Confirm that when $\Delta t = 0.5$ and $t = 10$, the error of the numeric estimate (column C) from the analytic value (column D) is –0.04166 m s^{-1} or 1.25%. Column C computes the error using Equation 2.38.

$$\text{Error} = \text{Actual} - \text{Estimate} \qquad (2.38)$$

What happens to the error at $t = 10$ when $\Delta t = 0.2$?

3. Because the estimate of area beneath $a(t)$ is obtained by summing area elements, each of which contributes a small error to the sum, you might expect the overall error to increase as you increase the number of area elements (holding Δt constant, of course). This happens in this and other physics problems. There are cases, however, in which the error hardly grows at all after a certain time. Plot error vs t (Figure 2.26) on-screen for a number of different values of Δt.

Notice the linear increase. What is the slope? How does the error (and its slope) change with Δt?

Many problems in physics do not have solutions in terms of easily calculated functions, and the error of the numeric estimate is difficult to know beforehand. In these cases you can get an estimate of the error by observing how the calculated result changes as you reduce Δt.

PROBLEMS

1. An object moves along a straight path with nonuniform acceleration (Equation 2.39).

$$a(t) = 8 - 12t^2 \tag{2.39}$$

where $a(t)$ is acceleration (m s^{-2}) and t is time (s). Initial velocity, v_0 = 0 m s^{-1}, and initial position, s_0 = 6 m.

a. Modify Model 2.7 to calculate velocity as a function of time numerically. Plot your results on-screen.

b. By visual examination, determine the maximum positive velocity attained and the time that this value is achieved. What is the acceleration at this time?

SPREADSHEET TECHNIQUE

Speeding Up Model Building

Large spreadsheets often take a long time to recalculate, and building or modifying a spreadsheet can be painfully slow when the spreadsheet recalculates with every change in cell contents. There are two solutions.

- Select the spreadsheet option that turns off automatic recalculation. Build your model. Turn on automatic recalculation.

  There are two disadvantages with this technique: (1) You may wish to monitor spreadsheet results while building the model; (2) it is very easy to forget to re-select automatic recalculation once your model is built (and thereby increase your chances for erroneous results).

- If your model is tabular in nature, build only the first few rows of the model before copying the table down to its final row. When modifying a model, erase all but the first few rows of the table before making your changes. Copy the table down to its final row only after your modifications have been entered and verified. (The disadvantage of this method is that the copy operation usually takes considerably longer than a recalculation. You should not copy large groups of formulas too frequently.)

  This procedure also provides a nice method for compressing your worksheet for storage. Erase all but a few rows to save disk storage space.

c. At a fixed value of t, by what factor does the velocity error decrease if you halve Δt?

Δt (s)	**Velocity Error at $t = 1$ s**
0.2	______
0.1	______
0.05	______

d. Modify your model further to calculate position as a function of time (using numeric integration techniques) and plot your results on-screen.

[Hint: you can use the velocity in column C and the elapsed time values in column A to create a new column for the integral of velocity (i.e., position). Don't forget to incorporate "initial position" into your model and to add it to your result.]

e. What is the maximum displacement, when does it occur, and what is the velocity at this time? At what time is the displacement zero?

f. The analytical result for the displacement is

$$s(t) = 6 + 4t^2 - t^4 \tag{2.40}$$

Use this result to calculate the error in displacement calculated numerically. Plot this error vs time. Does it increase linearly? At a fixed t value, by what factor does the position error decrease if you halve Δt?

Δt (s)	**Distance Error at $t = 1$ s**
0.2	______
0.1	______
0.05	______

2. The motion of a particular, springlike device is governed by the nonuniform acceleration described in Equation 2.41.

$$a(t) = a_0 \sin(\omega t) \tag{2.41}$$

where $a_0 = 2\ \text{m s}^{-1}$ and $\omega = 1\ \text{rad s}^{-1}$.

For $v_0 = -2$ m s^{-1} and $x_0 = 0$ m,

a. Plot acceleration vs t from $t = 0$ to $t = 10$ s.

b. Plot velocity vs t. Use several values of Δt to assure that your error is acceptable (e.g., no noticable change on your plotting screen). By observation, determine the value of the period of oscillation.

c. Estimate the time-averaged acceleration, velocity, and displacement — that is, the average value of these quantities over a long period of time.

d. Describe in words, which a bright high school student would understand, the relationship between a plot of acceleration vs time and velocity vs time. How is one derived from the other?

e. Plot position vs t.

f. Describe in words, which a bright high school student would understand, the relationship between a plot of velocity vs time and position vs time. How is one derived from the other?

g. Plot velocity vs position. At what position does velocity have its greatest positive value?

h. Suppose that the initial velocity is zero. What do you estimate the time-averaged acceleration, velocity, and position to be? Describe the general behavior of velocity and position vs time in this case.

i. You may also find it interesting to plot and think about such pairs as velocity vs position, acceleration vs position, and acceleration vs velocity. As you will see later, accelerations in real problems are often known as a function of position (for example) rather than time, and so it is interesting to look at plots such as these.

3. A ball with shallow cuts in the surface is found to fall from rest through a medium such as air with an acceleration expressed by Equation 2.42.

$$a(t) = A\,e^{-bt} \qquad (2.42)$$

where $A = -0.71\ \mathrm{ms}^{-2}$ is a constant that depends on the viscosity of the air and the properties of the ball.

The ball is released from rest 100 m above the surface of the ground.

a. Plot velocity vs t when $b = 0.2\ \mathrm{s}^{-1}$.

b. At what velocity does acceleration become zero (to 1 part in 10,000)? This velocity is called the terminal velocity. What will be the velocity of the ball when it hits the ground if it is dropped from 200 m? 1000 m?

c. Does the ball reach terminal velocity before it has fallen 10 m? What is the smallest value of b (which is related to the viscosity of the medium) permits this result?

4. When a small body falls slowly through a liquid, there is a significant retarding force acting on it due to viscosity. The acceleration of a small sphere under these conditions is

$$a(t) = -(Av_0 - g_{\mathrm{eff}})\ \mathrm{e}^{-At} \tag{2.43}$$

where $A = 6\pi\eta r/m$, $g_{\mathrm{eff}} = [(\rho-\rho_l)/\rho]g$, η is viscosity, r is the radius of the sphere, m is the mass of the sphere, ρ is the average density of the sphere, ρ_l is the density of the liquid, and v_0 is the initial velocity of the sphere.

Consider a small glass bead ($\rho = 2.5\ \mathrm{g\ cm}^{-3}$, $r = 0.1$ cm) falling from rest in heavy oil ($\rho_l = 1.26\ \mathrm{g\ cm}^{-3}$, $\eta = 1\ \mathrm{g\ cm}^{-1}\mathrm{s}^{-1}$).

a. Plot velocity vs t. What is the value of the terminal velocity?

b. How long a time is needed to reach one half of terminal velocity? How long a time is needed to reach 90% of terminal velocity?

c. Now explore the behavior of a small plastic ball in a swimming pool. Take the radius of the ball as 2.5 cm and its average density as $1.3\ \mathrm{g\ cm}^{-3}$. Water has a density of $1\ \mathrm{g\ cm}^{-3}$ and a viscosity of $0.01\ \mathrm{g\ cm}^{-1}\ \mathrm{s}^{-1}$.

d. Finally, explore the behavior of a hollow ball in water. The ball has a radius of 2.5 cm and a mass of 60 g in water.

MODELS

FORMULAS

	A	B	C	D	E
1	Nonuniform Acceleration				
2	a(t) = 2-0.1*t^2	0	=2-0.1*B2*B2		
3					
4	Δt =	0.5	s		
5	v0 =	10	m/s		
6					
7			Num.	Anal.	
8	t	a(t)	v(t)	v(t)	error
9	s	m/s^2	m/s	m/s	m/s
10	0	Formula Set I	=B5	Formula Set III	Formula Set IV
11	=A10+B4		Formula Set II		
12	=A11+B4				
34	=A33+B4				
35	=A34+B4				

VALUES

	A	B	C	D	E
1	Nonuniform Acceleration				
2	a(t) = 2-0.1*t^2	0	2		
3					
4	Δt =	0.5	s		
5	v0 =	10	m/s		
6					
7			Num.	Anal.	
8	t	a(t)	v(t)	v(t)	error
9	s	m/s^2	m/s	m/s	m/s
10	0	2	10	10	0
11	0.5	1.975	10.9938	10.9958	0.00208
12	1	1.9	11.9625	11.9667	0.00417
34	12	-12.4	-23.65	-23.6	0.05
35	12.5	-13.625	-30.156	-30.104	0.05208

Formula Set I Prototype Cell is B10
=2-(0.1*A10*A10)

Formula Set II Prototype Cell is C11
=C10+(**B4** *(B11+B10)/2)

Formula Set III Prototype Cell is D10
=10+(2*A10)-((0.1*A10*A10*A10)/3)

Formula Set IV Prototype Cell is E10
=D10-C10

Model 2.7

BUILD and verify Model 2.7.

ENTER data using cells B4 and B5 (Δ*t* and v_0, respectively). Δ*t* determines the width of each area element and the span of *t* values for which velocity may be calculated. v_0 = velocity at *t* = 0.

READ output from the model at cells B10 - C35.

Cells A10 - A35 (*t*) contain the values of t at which acceleration and velocity are calculated. Cells B10 - B35 [*a*(*t*)] calculate acceleration using Equation 2.36.

Cells C10 - C35 [*v*(*t*)] contain a running total of the area beneath *a*(*t*) as *t* increases. Each cell calculates the area element of height [*a*(*t* – Δ*t*) + *a*(*t*)]/2 and width Δ*t* and adds to it the area computed by the previous cell. Cell C10, of course, contains the initial velocity.

Cells D10 - D35 (velocity) calculate *v* analytically using Equation 2.30 and cells E10 - E35 (error) calculate the difference between the numeric estimate (cells C10 - C35) and the analytic value (cells D10 - D35).

NOTICE that the accuracy of the numeric estimate of velocity increases as Δ*t* decreases. If the range of *t* values becomes too small as Δ*t* decreases, extend the model beyond row 35.

NOTICE that a master copy of the formula defining acceleration is present in cell C2. B2 contains a dummy argument for this formula.

EXERCISE **2.5**

VELOCITY BY SIMPSON'S RULE

TO GET THE MOST OUT OF THIS EXERCISE, YOU SHOULD ALREADY KNOW

The definitions of the terms:
derivative
definite integral
indefinite integral
velocity
acceleration

How to analytically evaluate simple indefinite integrals.

In Exercise 2.4 you integrated a function $a(t)$ between two times, t_0 and t_n, by summing the areas of large numbers of rectangular elements. Although the accuracy of your estimate increases as the number of elements increases, significant error remains possible for "reasonable" numbers of rectangular elements (Figure 2.27a). Simpson's rule is a commonly used device that dramatically increases the accuracy of your estimate of the value of a definite integral without significantly increasing computing overhead. Simpson's rule achieves this increased accuracy by approximating the function $a(t)$ in the range $t - \Delta t$ to t with a quadratic function and then exactly integrating this approximation (Figure 2.27b). In fact, Simpson's rule gives exact answers for integrands $a(t)$ that are constant, linear, quadratic, or cubic polynomial functions. You rarely need to compute definite integrals using higher-order area elements. You should, however, be especially wary of integrands that become infinite in the interval or at its ends — a condition that can arise in some physical problems which you will see later.

Consider a parabola defined by the three points $a(t-\Delta t)$, $a(t-\Delta t/2)$, and $a(t)$. The area beneath this curve, within the interval $t-\Delta t$ and t can be computed exactly using Equation 2.44 (Figure 2.28):

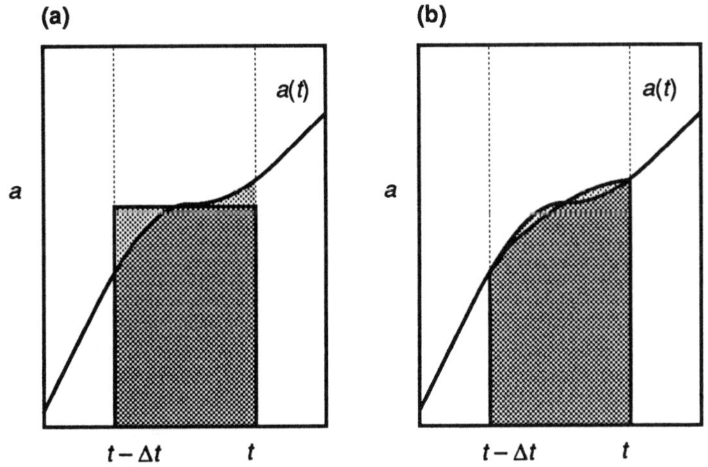

Figure 2.27

(a) Estimations of the area beneath function $a(t)$ with a rectangular shape are prone to error.

(b) Error is reduced when quadratic approximation is used to "fit" the top of the "rectangle" to the function $a(t)$.

$$\text{Area} = \Delta t\left[\frac{a(t - \Delta t) + 4\,a(t - \frac{\Delta t}{2}) + a(t)}{6}\right] \tag{2.44}$$

In this exercise, you will build a model based on Equation 2.44 capable of finding the value of an indefinite integral to a remarkably high degree of accuracy. You will use this model to examine the velocity of objects

SPREADSHEET TECHNIQUE

Naming Cells

Many spreadsheets permit you to *name* cells and ranges of cells. Named cell ranges can make spreadsheet manipulation easier and more straightforward.

For example, you could make the formula in Formula Set I of Model 2.1 easier to read if you named cell B6 "s0," cell By "v0," cell B8 "a," and cell A12 "t." Then, in a spreadsheet program using the symbol $ to indicate an absolute address, the formula would read $s0+$v0*t+0.5*$a*t*t. This formula is nearer to a normal algebraic formula and is easier to examine for errors. Your spreadsheet program may use another device to indicate the absolute device and may have restrictions on allowable names, but the idea is the same and your formulas will be less cryptic.

You can also use named cells or cell ranges to identify the variables within formulas. If, for example, you name the contents of column E "numeric" and the contents of column F "analytic," you can write the formula for error in column G "=analytic-numeric."

If you name cells and cell ranges with names that correspond to their function, you are less likely to make mistakes when building spreadsheets. However, with some spreadsheet programs you just be careful to avoid one very subtle pitfall. *You should never use a name that looks like the name of a cell* (e.g., A1, B32, AX6) or names that contain arithmetic symbols (e.g., +, –, /, *, ^). Confusion can result because some spreadsheets treat such names in special ways.

Figure 2.28

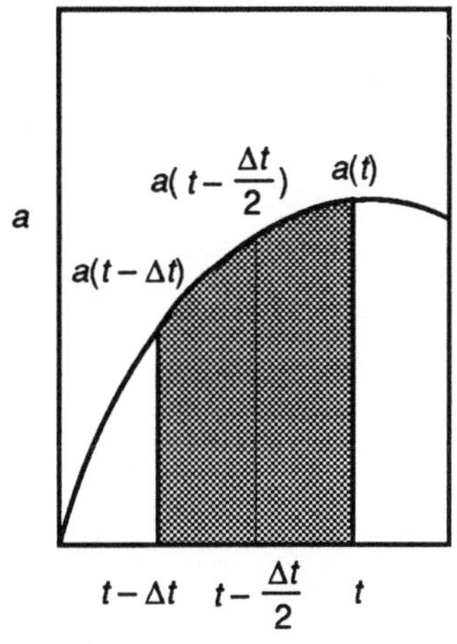

The area beneath a parabola in the interval from $t - \Delta t$ to t can be calculated using Equation 2.44.

subject to nonuniform acceleration just as you did in Exercise 2.4 and to contrast the accuracy of the two numeric integration techniques now available to you.

LESSON

Build and verify Model 2.8. This model closely parallels Model 2.7 and estimates the velocity at time t of an object subject to the nonuniform acceleration described by Equation 2.45. This is the same acceleration function used in the exercises of Exercise 2.4 — faciliting comparison of the accuracy of the two methods of numeric integration.

$$a(t) = 2 - \frac{t^2}{10} \qquad (2.45)$$

You can, at any time change the function $a(t)$ by entering a new formula into cell C2 and copying it into cells B10 - B35 and D10 - D35.

Columns F and G are used to compute the integral analytically and to compare the results of the numeric and analytic method.

a. Examine the model carefully until you understand its structure and its relationship to Model 2.7.

b. Repeat Lessons 1-3 of Experiment 2.4 and confirm the dramatically increased accuracy of Simpson's rule. Indeed, because Equation 2.45 is a quadratic function, Simpson's rule gives precisely the same result as does the analytic method, although computer truncation errors may cause slight inaccuracy in the last significant digit.

PROBLEMS

1. A springlike device is governed by the acceleration function in Equation 2.46:

   $$\mathrm{a}(t) = -a_0 \cos(\omega t) \qquad (2.46)$$

   where $a_0 = 2$ cm s^{-2} and $\omega = 1$ rad s^{-1}.

   Evaluate the integral using Simpson's rule from $t = 0$ s to $t = 10$ s when $v_0 = 2$ (cm s^{-1}). Plot the value of the integral vs position and determine the period.

2. The acceleration of a bullet in a rifle barrel is given by Equation (2.47):

$$a(t) = \begin{cases} 200 - 3 \times 10^{5}t & 0 < t < 1/15 \\ 0 & 1/15 \le t \end{cases} \qquad (2.47)$$

where a is the acceleration (m s^{-2}) and t is the time (s).

a. Plot $a(t)$ vs t.

b. Use Simpson's rule to plot $v(t)$. Equation 2.47 indicates that, when the acceleration becomes zero, it remains zero. Physically this represents what happens after the bullet emerges from the barrel (air resistance ignored). You can automatically impose this condition by using the spreadsheet technique "Conditionals."

c. Compare the velocity of emergence from the barrel obtained via thc numeric and analytic methods. What is the error of the numeric method?

3. A single-stage rocket fired vertically from rest at the Earth's surface burns its fuel in an elapsed time of 30 s. The acceleration function is given by Equation 2.48.

$$a(t) = (48000k - 9.8) \quad \text{for} \quad 0 \le t \le 20$$

$$a(t) = 48000k - 5000k(t - 20) - 9.8 \quad \text{for} \quad 20 \le t \le 30 \qquad (2.48)$$

where acceleration is given in meters per second, time is in seconds, and k is a constant.

SPREADSHEET TECHNIQUE

Conditionals

Often in physics one expression (call it E1) must be used when some condition (call it C) is true, and another expression (E2) otherwise. Spreadsheet programs allow you to write expressions that evaluate as either true or false and so are perfectly designed to express a condition. The expression =(A1=A2), for example, evaluates to true if the contents of cells A1 and A2 are identical and evaluates to false when they are not. In addition, all serious spreadsheet programs include the built-in function IF(). IF() assigns either E1 or E2 as the value of the cell in which it is placed depending on the value of conditional C (Figure 2.29).

As an example, consider Problem 2. You might exploit the IF() statement by writing the conditional into the expression for acceleration (Figure 2.30a). You might also express the conditional in an additional column (Figure 2.30b). When formulas become very complex, (Figure 2.30b) is probably what you'll want to do.

Figure 2.29

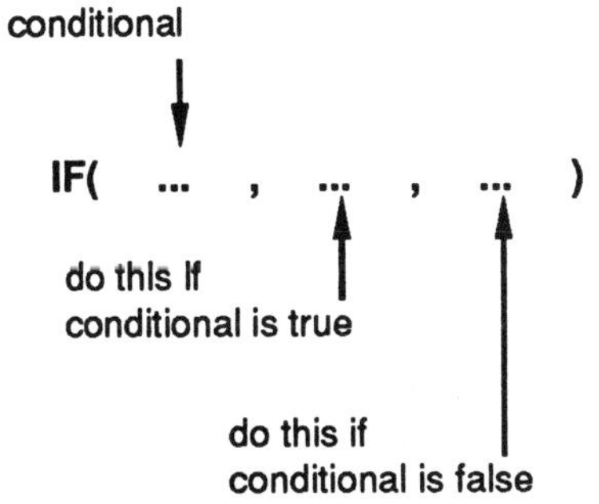

Spreadsheet programs include a built-in function, IF() for making decisions.

Figure 2.30

(a)

	A	B
1	t	a(t)
2	0	=IF(A2<=1/15,20000-30000*A2,0)
3	0.01	=IF(A3<=1/15,20000-30000*A3,0)
4	0.02	=IF(A4<=1/15,20000-30000*A4,0)
5	0.03	=IF(A5<=1/15,20000-30000*A5,0)

(b)

	A	B	C
1	t	a1(t)	a2(t)
2	0	=20000-30000*A2	=IF(A2<=1/15,B2,0)
3	0.01	=20000-30000*A3	=IF(A3<=1/15,B3,0)
4	0.02	=20000-30000*A4	=IF(A4<=1/15,B4,0)

The IF() built-in function can be used in alternate ways depending on the complexity of the expression to be evaluated.

a. What must be the value of k for velocity to equal 8 m s^{-1} when the burn ceases?

b. Plot $a(t)$ vs t and plot $v(t)$ vs t. Find the acceleration 15 s into the flight.

MODELS

Model 2.8

BUILD and verify Model 2.8.

ENTER data using cells B4 and B5 (Δt and v_0, respectively). Δt determines the width of each area element and the span of t values for which velocity may be calculated. v_0 = velocity at t = 0.

Cells B2 - C2 contain the dummy variable and master acceleration formula, respectively. Use the master acceleration formula (cell C2) to copy $a(t)$ into columns B and D. If your spreadsheet supports macros, you may wish to write a macro to automate this procedure.

READ output from the model at cells A10 - G35.

Cells A10 - A35 (t) give the values of t at which acceleration and velocity are calculated. Cells B10 - B35 [$a(t)$] give acceleration at time = t, using Equation 2.45.

Cells C10 - C35 give the values of $t - \Delta t/2$ required for Simpson's rule. Cells D10 - D35, $a[t - (\Delta t/2)]$, give acceleration at time = $t - \Delta t/2$.

Cells E10 - E35 calculate $v(t)$ numerically using Simpson's rule.

Cells F10 - F35 calculate $v(t)$ analytically using Equation 2.37 and cells G10 - G35 (error) give the difference between the numeric estimate (cells E10 - E35) and the analytic value (cells F10 - F35).

NOTICE that the model closely parallels the structure of Model 2.7 — only the algorithm used to calculate the area element has been changed. If the range of t values becomes too small as Δt decreases, extend the model beyond row 35.

FORMULAS

	A	B	C	D	E	F	G
1	Simpson's Rule						
2	a(t) = 2-0.1*t^2	0	=2-0.1*B2*B2				
3							
4	Δt =	0.5	s				
5	v0 =	10	m/s				
6	Δt/2 =	=B4/2					
7					Num.	Anal.	
8	t	a(t)	t - (Δt/2)	a(t-Δt/2)	v(t)	v(t)	Error
9	s	m/s^2	s	m/s^2	m/s	m/s	m/s
10	0	Formula Set I	Formula Set II	Formula Set III	=B5	Formula Set V	Formula Set VI
11	=A10+B4				Formula Set IV		
12	=A11+B4						
34	=A33+B4						
35	=A34+B4						

VALUES

	A	B	C	D	E	F	G
1	Simpson's Rule						
2	a(t) = 2-0.1*t^2	0	2				
3							
4	Δt =	0.5	s				
5	v0 =	10	m/s				
6	Δt/2 =	0.25					
7					Num.	Anal.	
8	t	a(t)	t - (Δt/2)	a(t-Δt/2)	v(t)	v(t)	Error
9	s	m/s^2	s	m/s^2	m/s	m/s	m/s
10	0	2	-0.25	1.99375	10	10	0
11	0.5	1.975	0.25	1.99375	10.9958	10.9958	0
12	1	1.9	0.75	1.94375	11.9667	11.9667	0
34	12	-12.4	11.75	-11.806	-23.6	-23.6	3.6E-15
35	12.5	-13.625	12.25	-13.006	-30.104	-30.104	0

Formula Set I Prototype Cell is B10
=2-0.1*A10*A10

Formula Set II Prototype Cell is C10
=A10-(**B6**)

Formula Set III Prototype Cell is D10
=2-0.1*C10*C10

Formula Set IV Prototype Cell is E11
=E10+(**B4***((B10+(4*D11)+B11)/6))

Formula Set V Prototype Cell is F10
=2*A10-0.1*A10*A10*A10/3+10

Formula Set VI Prototype Cell is G10
=F10-E10

CHAPTER 3

FORCES

By the work one knows the workman — Fables of La Fontaine

The Lion is known by his Claws — Newton, as known by his colleagues

Although Aristotle and Galileo are often proposed as the forebearers of *kinematics* (the study of the motions of objects), it is Sir Isaac Newton (1642 - 1727) who plays the dominant role in the history of *dynamics* (the study of the forces that produce such motions). Indeed the *three laws of motion* that physicists use to study dynamics are often called simply Newton's laws (even though the first law really belongs to Galileo).

- *The Law of Inertia:* An object maintains its state of motion unless acted on by a net external force.

- *The Law of Net Force:* The net external force on a body equals the product of its mass and its acceleration.

- *The Law of Force Pairs:* Forces always come in equal and opposite pairs (i.e., for every action there is a reaction).[1]

While it is Newton's second law of motion (the law of net force) that relates force to acceleration (and thus to velocity and position as well), the concepts of inertia and force pairs are crucial to understanding every problem in dynamics. *All three laws are inseparable and apply to all problems related to the physics of forces!*

Recall from Chapter 2 the mathematical descriptions of the motions of objects. Such motions can be described in terms of either position, velocity, or acceleration, and the functions describing these three properties can be derived each from the other. Velocity as a function of time can be obtained by differentiating position with respect to time (Equation 3.1). Acceleration as a function of time can be obtained by taking either the first derivative of velocity with respect to time or the second derivative of position with respect to time (Equation 3.2):

$$v(t) = \frac{ds(t)}{dt} \tag{3.1}$$

$$a(t) = \frac{dv(t)}{dt} = \frac{d^2s(t)}{dt^2} \tag{3.2}$$

where $s(t)$ is position, $v(t)$ is velocity, and $a(t)$ is acceleration.

Newton's second law of motion offers you the opportunity to determine the acceleration of an object from the net forces acting on it, but the velocity and/or position of the object are often the quantities of primary interest. For this reason, you are often called on to integrate acceleration with respect to time to obtain velocity (Equation 3.3) or to integrate velocity with respect to time to obtain position (Equation 3.4):

$$\frac{ds(t)}{dt} = v(t) = \int a(t)dt \qquad (3.3)$$

$$s(t) = \int v(t)dt \qquad (3.4)$$

Chapter 2 also introduced several different techniques for differentiating and integrating equations of motion, using numeric techniques. You must know explicitly the acceleration or velocity as a function of time in order to use these techniques. In this chapter, you round out your skills by learning how to obtain position from acceleration in cases where the acceleration is not an explicitly known function of time only but is an explicitly known function of velocity and/or position and/or time (Equation 3.5). This in fact is the usual situation in Nature and you will see many examples of it.

$$\frac{d^2s(t)}{dt^2} = a(s, v, t) \qquad (3.5)$$

In this chapter, you consider the ideal case of an object without physical spatial dimensions (i.e., a *point object* whose mass does not change with time). The motions of objects with spatial extent are ignored until Chapter 5.

The laws of motion constitute a benchmark in the intellectual history of the human race. Seventeenth- and eighteenth-century natural philosophers quickly established, in the laboratory and in the heavens, the validity of these laws throughout the observable universe of that time and raised Newton to the stature of the greatest scientist of all time. Discoveries in the last few centuries have not diminished this stature.

Sir Isaac describes his own intellectual adventure:

> *I do not know what I may appear to the world, but to myself I seem to have been only like a boy playing on the seashore, and diverting myself in now and then finding a smoother pebble or a prettier shell than ordinary, whilst the great ocean of truth lay all undiscovered before me.*

EXERCISE **3.1**

THE SECOND LAW OF MOTION: EULER'S METHOD

TO GET THE MOST OUT OF THIS EXERCISE, YOU SHOULD ALREADY KNOW

The three laws of motion

When Isaac Newton described the three laws of motion in his *Principia* (1687), the world became part of a mechanical universe — the knowledge of any object's initial location and movement and all the forces acting on it could be used to calculate any future location and movement of the object.

Although all three of the laws of motion are intimately intertwined, the second law specifies the form of the equations that you must manipulate to solve problems in mechanics. For any given object, the second law describes the relationship between net external force and acceleration. For a point particle in which the mass does not change with time,

$$\Sigma \boldsymbol{F} = m\boldsymbol{a} = m\frac{d^2\boldsymbol{s}}{dt^2} \tag{3.6}$$

where $\Sigma\boldsymbol{F}$ is the vector sum of external forces, m is the mass of the object, $\boldsymbol{a}$ is its acceleration, $\boldsymbol{s}$ is its position (displacement), and t is elapsed time. Note $\boldsymbol{F}$, $\boldsymbol{a}$, and $\boldsymbol{s}$ are vector quantities, while m and t are scalars.

Equation 3.6 can, of course, be resolved into its vector components. For motion directed in the xy-plane, Equation 3.6 is equivalent to Equations 3.7 and 3.8:

$$\Sigma F_x = m\frac{d^2x}{dt^2} \tag{3.7}$$

$$\Sigma F_y = m\frac{d^2y}{dt^2} \tag{3.8}$$

where ΣF_x, ΣF_y are the sum of the external forces along the x- and y-axes, x and y are the position of the object along the x- and y-axes, m is the mass of the object, and t is the elapsed time.

The problems and examples in your physics textbook use analytic techniques to solve the differential Equations 3.7 and 3.8. Recall, for example, Equations 2.1 and 2.2 in Exercise 2.1. Most differential equations, however, are either difficult or impossible to solve in terms of simple, well-known functions. Even the simple problem of the gravitational attraction of two point masses requires a mathematical trick for its solution (Figure 3.1).[1]

In this exercise you will explore a numeric technique, called Euler's method, for solving differential equations. It offers you a general and simple approach for obtaining reasonable approximations to the solutions of second-order differential equations. In the following exercise, you will explore a more complicated but far more accurate technique—the fourth-order Runge-Kutta method for solving the same class of equations. Runge-Kutta methods can accurately handle many mechanics problems.

First-order differential equations such as Equation 3.9 in which the velocity is known explicitly as a function of time can be solved by integration (Equation 3.10) using either analytic or numeric methods such as Simpson's rule.

$$\frac{dx(t)}{dt} = v(t) \tag{3.9}$$

$$x = \int v(t)dt \tag{3.10}$$

In dynamics, however, you are typically faced with a second-order differential equation in which the force (and so the acceleration) is not known explicitly as a function of time (Equation 3.6). In dynamics the force may be an explicit function of the position only (e.g., an approximation of a pendulum swinging in the air), or an explicit function of both position and velocity (as with a pendulum swinging in oil), or even an explicit function of position, velocity, and time (as with a pendulum swinging in oil, with the supporting string gradually being shortened). Other combinations are, of course, possible, but these three are the most common.

Figure 3.1

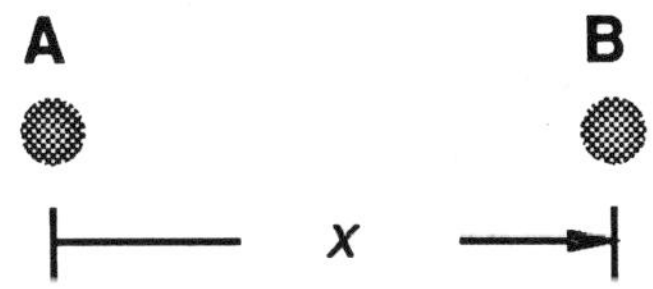

The gravitational force of attraction exerted on point mass *B* by point mass *A* is inversely proportional to the square of the distance *x* that separates them. Thus, the acceleration of *B* toward *A* is given by

$$a_x = \frac{d^2x}{dt^2} = k\,x^{-2}$$

where a_x is the magnitude of the acceleration of *B* toward *A*, *x* is the distance separating *A* and *B*, and *k* is the negative proportionality constant.

This differential equation is difficult to solve until you have discovered the trick. Numeric methods such as Euler's method or the Runge-Kutta method offer a general technique for numerically solving such differential equations.

Figure 3.2

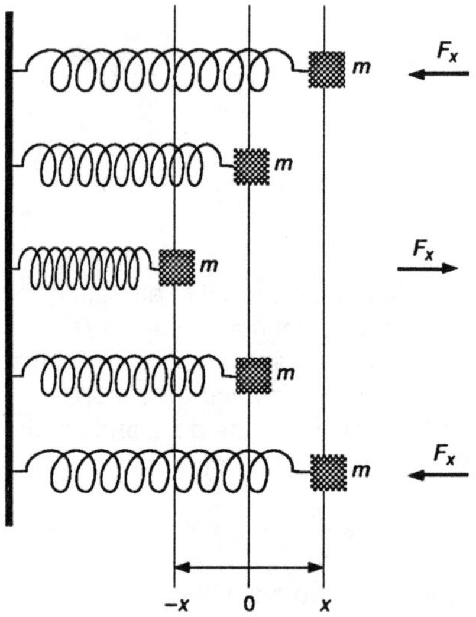

An ideal, horizontal spring with a mass m attached to its free end oscillates freely, displacing the mass a variable distance x with time. The force exerted by the spring on the mass is expressed by Hooke's law (Equation 3.11).

As a simple example in which the force is a function of the position only, consider a mass m on a spring with spring constant k (Figure 3.2). When the mass is displaced a distance x from its equilibrium position, it experiences a force of magnitude kx directed back toward the equilibrium position (Equation 3.11):

$$F_x = -kx \tag{3.11}$$

Thus, from the second law of motion, you get

$$ma_x = -kx \qquad \text{or} \qquad a_x = \frac{dx^2}{dt^2} = -\left(\frac{k}{m}\right)x \tag{3.12}$$

and acceleration is a function of x rather than t.

This situation is far different from that in kinematics because you cannot get $a(t)$ until you know $x(t)$, which you cannot know until you know $a(t)$, etc. This apparently hopeless situation has a subtle solution.

Consider first the simpler case of

$$\frac{dx}{dt} = f(x,t) \tag{3.13}$$

where $f(x,t)$ is a known function of position x and time t. The simplest numeric method for solving this type of equation was proposed by Euler and is similar to the rectangular integration method that you encountered in Exercise 2.4. Recall the approximation for the derivative

$$\frac{dx}{dt} \cong \frac{x(t + \Delta t) - x(t)}{\Delta t} = f(x(t),t) \tag{3.14}$$

Thus

$$x(t + \Delta t) = x(t) + \Delta t\, f(x(t),t) \tag{3.15}$$

Because f *is a known function, you can now calculate the position at time* t + Δt *if you know it at time* t *(Equation 3.15).* Successive application of this rule, easily performed on a spreadsheet, allows you to approximately solve the first-order differential equation. If this is difficult to see now, you will certainly see it after you have built a few models.

This approach can be generalized so that it leads to a solution for differential equations of any order—for our purposes, the second-order differential equations that arise out of the second law (Equation 3.16). Simply break the single, second-order differential equation (Equation

3.16) into two, first-order equations (Equations 3.17 and 3.18) and apply Euler's method to both. Thus, write

$$\frac{d^2x(t)}{dt^2} = a(x,v,t) \tag{3.16}$$

as

$$\frac{dv(t)}{dt} = a(x,v,t) \tag{3.17}$$

$$\frac{dx(t)}{dt} = v(t) \tag{3.18}$$

Notice that Equations 3.16 and 3.17 are differential equations that have been represented in their *general form*. $a(x, v, t)$ declares that a can be expressed as a function of x, v, and t—including the special case where one or two of the independent variables are absent.

Suppose that in Equation 3.17 the acceleration a is explicitly known as a function of its arguments x, v, and t. Euler's method can be applied to approximate $v(t + \Delta t)$, $v(t + 2\Delta t)$, etc. (e.g., Equations 3.19 and 3.20).

$$v(t + \Delta t) \sim v(t) + a(x,v,t)\Delta t \tag{3.19}$$
$$v(t + 2\Delta t) \sim v(t + \Delta t) + a(x(t + \Delta t), v(t + \Delta t), t + \Delta t)\Delta t \tag{3.20}$$

But where do you get $x(t + \Delta t)$? Just as Equations 3.19 and 3.20 can be used to approximate solutions to Equation 3.17, Equations 3.21 and 3.22 can be used to approximate Equation 3.18:

$$x(t + \Delta t) \sim x(t) + v(t)\Delta t \tag{3.21}$$
$$x(t + 2\Delta t) \sim x(t + \Delta t) + v(t + \Delta t)\Delta t \tag{3.22}$$

Where does $v(t + \Delta t)$ come from? Equation 3.19! *The important point to realize is that at each step in your calculations you have just enough*

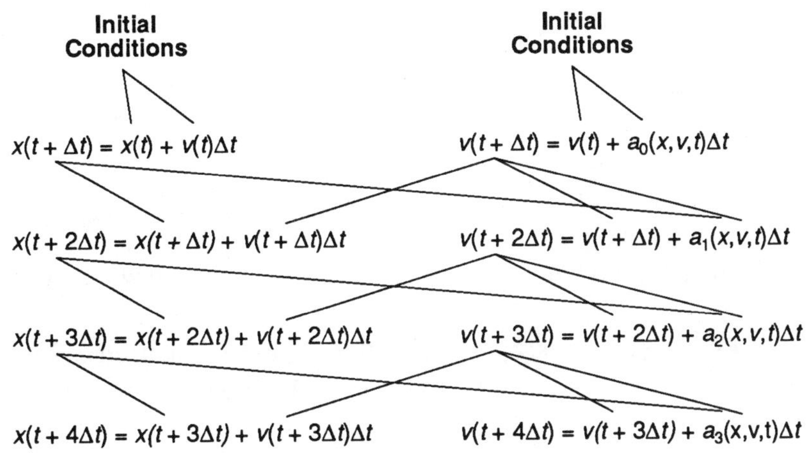

Figure 3.3

Logical progression of calculations for Euler's method. Notice that the acceleration function (a_0, a_1, a_2, etc.) is also driven by the results of the previous calculation or by the initial conditions.

information from the previous step to proceed. The complete logic is indicated in Figure 3.3.

The method just described, Euler's method, has the characteristic that the error in the approximation grows very fast as the number of iterations increases. In fact, in many cases the solution rapidly becomes worthless! Euler's method is important because it is easy to understand and quick and easy to model. Once you clearly understand Euler's method, you will be able to understand the more complex Runge-Kutta method presented in the next exercise. The Runge-Kutta method is a generalization of Simpson's rule for integration and is remarkably accurate for a wide range of physical problems.

LESSONS

1. Consider an ideal, horizontal, oscillating spring with a mass m attached to its free end (Figure 3.2). Hooke's law states that the force F_x exerted by the spring on its attached mass is proportional to the product of the displacement x of the mass from its resting position and a constant k that measures the stiffness of the spring (Equation 3.23):

$$F_x = -kx \qquad (3.23)$$

Applying Newton's second law of motion to Hooke's law yields Equation 3.24:

$$ma_x = -kx \quad \text{or} \quad a_x = \frac{d^2x}{dt^2} = -\left(\frac{k}{m}\right)x \qquad (3.24)$$

where a_x is the acceleration of the mass along the x-axis.

Equation 3.24 can be solved analytically[2] to yield

$$x(t) = A\sin(\omega t + \phi_0) \qquad (3.25)$$

$$v(t) = A\,\omega\cos(\omega t + \phi_0) \qquad (3.26)$$

where $\omega = \sqrt{\frac{k}{m}}$, $A = \sqrt{x_0^2 + \left(\frac{v_0}{\omega}\right)^2}$, $\phi_0 = \tan^{-1}\left(\frac{v_0}{\omega x_0}\right)$

x_0 is the position when t is zero, and v_0 is the velocity when t is zero.

You can verify this by twice differentiating Equation 3.25 with respect to t and then substituting x for $A\sin(\omega t + \phi_0)$ in the result.

You can approximately solve Equation 3.24 numerically by using Euler's method.

Build and verify Model 3.1. Examine the model and its description. Compare it to the flow of values outlined in Figure 3.3. Do not proceed until you fully understand Euler's method and how it is implemented in Model 3.1. The tools developed in this and the following exercise are used throughout the remainder of this book.

2. If Hooke's law (Equation 3.24) had no analytic solution, Model 3.1 would still provide estimates of the velocity and position of the object moving according to Hooke's law. Plot acceleration, velocity, and position as a function of time (Figure 3.4). Use the following values: $\Delta t = 0.1$ s, $x_0 = 3$ cm, $v_0 = 1$ cm s^{-1}, $k = 1$ dyne cm^{-1}, and $m = 4$ g.

   a. Examine the plots of $a(t)$, $v(t)$, and $x(t)$ and use your knowledge of the geometric interpretation of a derivative to explain the shapes of $v(t)$ and $x(t)$ in terms of $a(t)$. For example, is $v(t)$ rising or falling when $a(t)$ is negative; what is $x(t)$ doing when $v(t)$ equals zero?

   b. Picture in your mind the behavior of a mass hung by a spring and the forces acting on it. Why must the mass exhibit oscillatory motion?

   c. Consider a spring for which $k = 1$ dyne cm^{-1} and $m = 1.5$ g. The mass attached to the spring is initially motionless ($v_0 = 0$) but extended 2 cm from its resting position ($x_0 = 2$ cm). Observe the position of the mass with time via $x(t)$ in your plot corresponding to Figure 3.4.

      What is the period of oscillation of the mass? How does the period change as the mass of the attached object increases? How does the period change as the stiffness of the spring decreases (k decreases)?

   d. How does the spring behave if the mass is initially motionless at the spring's resting position ($v_0 = 0$, $x_0 = 0$).

3. The plots of $x(t)$ vs t and $v(t)$ vs t were both obtained using Euler's method. You can use your knowledge of the analytic solution to Equation 3.23 to obtain a measure of the method's accuracy.

Figure 3.4

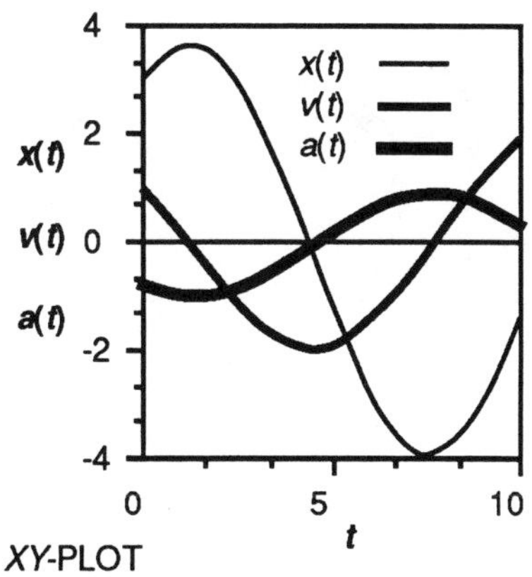

XY-PLOT

Data
Horizontal (*x*) axis: cells A14 - A114 (time)
Vertical (*y*) axis: cells B14 - B114 (position), C14 - C114 (velocity), D14 - D114 (acceleration)

Figure 3.5

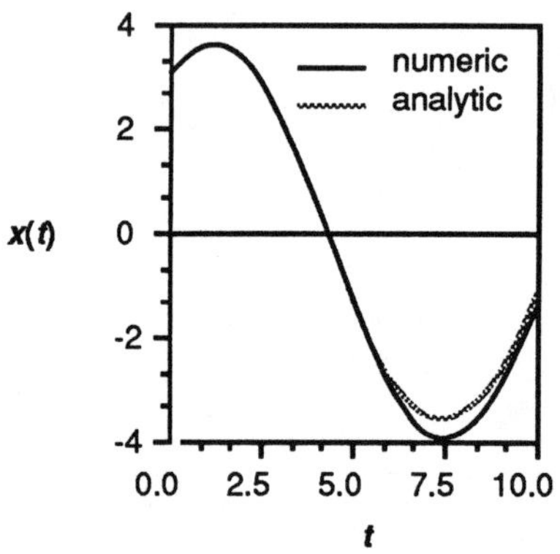

XY-PLOT

Data
Horizontal (*x*) axis: cells A9 - A109 (*t*)
Vertical (*y*) axis: cells B9 - B109 [numeric *x*(*t*)], E9 - E109 [analytic *x*(*t*)]

a. Plot both the numeric and the analytic solutions to $x(t)$ (Figure 3.5) for each of the input values below. From the plots, read the maximum displacement and the maximum error.

Δt (s)	x_0 (cm)	v_0 (cm s^{-1})	k (dyne cm^{-1})	m (g)	Maximum $x - x_0$	Maximum Error
0.1	3	1	1	4	____	____
0.1	3	1	1	1	____	____
0.1	0	2	1	4	____	____
0.1	0	2	1	1	____	____
0.1	0	2	3	1	____	____
0.1	3	0	3	1	____	____

As you can see, there are numerous conditions under which the analytic and numeric solutions diverge rapidly; furthermore, in all cases, the error grows rapidly with time.

b. Quantify the error associated with Euler's method by comparing the values of $x(t)$ obtained by the numeric and analytic methods when $t = 1$. Extend Model 3.1 by adding additional rows to complete the table below for four different values of Δt and when $x_0 = 0$ cm, $v_0 = 2$ cm s^{-1}, $k = 1$ dyne cm^{-1}, and $m = 1$ g.

Δt	No. Steps	$x(t)$ at $t = 1$ Numeric	$x(t)$ at $t = 1$ Analytic	Error
0.1	10	1.76	1.68	0.082
0.01	100	1.691	1.683	0.0084
0.001	____	____	____	____
0.005	____	____	____	____

Notice that the error in the position calculated by Euler's method is approximately proportional to Δt. Thus, reducing Δt by a factor of ten increases the accuracy by one decimal place. *The resulting improvement is not very encouraging because the number of calculations and the calculation time, approximately, also increase by a factor of ten!*

You will find a vast improvement in accuracy for a given number of calculations with the Runge-Kutta method of the next exercise.

PROBLEMS

1. A mass M hangs from a vertically oriented spring (Figure 3.6). The mass of the spring is much less than M and can be safely ignored.

   Newton's second law for this system is represented by Equation 3.27:

$$M\frac{d^2y}{dt^2} = -ky - Mg \tag{3.27}$$

   where k is the force constant of the spring (dyne cm^{-1}), g is the acceleration of gravity (980 cm s^{-2}), y_0 is the initial position (0 cm), and v_0 is the initial velocity (–2 cm s^{-1}).

   a. Use Model 3.1 and an appropriate on-screen graphic to plot the position of the mass with time for a system in which $k = 1000$ dyne cm^{-1} and $M = 500$ g. Apply Euler's method using $\Delta t = 0.2$ s.

      By visually inspecting your plot, determine the period of oscillation T and compare it with its predicted value (Equation 3.28):

$$T = \frac{2\pi}{\omega} \tag{3.28}$$

$$\text{where } \omega = \sqrt{\frac{k}{m}}$$

   b. Plot the velocity of the mass vs time for the same spring as in Lesson 1a. Try $\Delta t = 0.06$ s. What is the maximum velocity?

   c. With the same mass and spring constants as above, what is the motion if the mass is initially at $y_0 = -2$ cm and $v_0 = 0$? Why?

2. Part of a uniform rope hangs over the edge of a frictionless table. If x is the amount of overhang, then the acceleration a of any fixed point on the rope is given by

$$a = \frac{d^2x}{dt^2} = \frac{gx}{L} \quad 0 < x \leq L \tag{3.29}$$

   where $g = 9.8$ m sec^{-2} and L (total rope length) = 2 m.

   Assume that 0.1 m of rope initially overhangs the edge of the table and that the rope is initially motionless ($x_0 = 0.1$ m, $v_0 = 0$ m s^{-1}).

   a. Plot x vs time and determine the moment at which the last bit of rope leaves the table and the rope enters free fall.

Figure 3.6

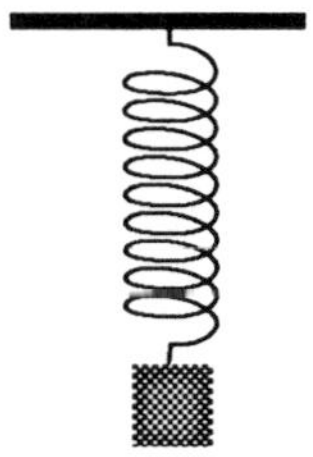

Mass M on an ideal, massless spring.

b. How much time is needed for the rope to enter free fall if the initial length of rope hanging over the edge is 0.2 m? 0.4 m? 1 m?

Overhang (m)	T_{Analytic} (s)	T_{Numeric} (s)
0.2	1.388	____
0.4	1.057	____
1.0	0.603	____

c. If the rope is initially hanging 1 m over the table, can you find an initial velocity value so that the rope ends up with zero velocity and its total length on the table?

3. Just 0.1 m of a 2 m length of uniform rope hangs over the edge of a table. Friction force acts in the opposite direction of the velocity of the sliding rope. The acceleration of the rope can be described by

$$a = \frac{d^2x}{dt^2} = g(1 + \mu)\left(\frac{x}{L}\right) - \mu g \quad 0 < x \le L \tag{3.30}$$

where $\mu = 0.2$ (the coefficient of kinetic friction) and $g = 9.8 \text{ m s}^{-2}$.

a. For what values of initial velocity will the rope slide off the table?

b. If the initial velocity is miniscule but not zero, for what values of x_0 will the rope slide off the table?

c. If the initial velocity is miniscule but not zero and the length of overhanging rope is 0.8 m, how much time will pass before the rope leaves the table entirely?

4. A 1.5-kg mass experiences an impulsive force in Newtons governed by the equation

$$F(t) = A\, t\, e^{-100t} \tag{3.31}$$

where $A = 500$ and t = time (s).

a. Plot x vs t. Is there a maximum x?

b. Plot v vs t. Is there a maximum v?

MODELS

FORMULAS

	A	B	C	D	E	F
1	Simple Harmonic Motion by Euler's Method					
2	DDx = -(k/m) x					
3						
4	Δt =	0.1	s	k =	1	dyne/cm
5	x0 =	0	cm	m =	1	g
6	v0 =	2	cm/s			
7				omega =	=SQRT(E4/E5)	
8				A =	=SQRT((B5*B5)+(B6/E7)^2)	
9				phi0 =	=ATAN2(B6,E7*B5)	
10						
11		Euler's Method Calculations			Anal.	
12	t	x(t)	v(t)	a(t)	x(t)	Error
13	s	cm	cm/s	cm/s^2	cm	cm
14	0	=B5	=B6	Formula Set III	Formula Set IV	Formula Set V
15	=A14+B4	Formula Set I	Formula Set II			
16	=A15+B4					
113	=A112+B4					
114	=A113+B4					

VALUES

	A	B	C	D	E	F
1	Simple Harmonic Motion by Euler's Method					
2	DDx = -(k/m) x					
3						
4	Δt =	0.1	s	k =	1	dyne/cm
5	x0 =	0	cm	m =	1	g
6	v0 =	2	s/cm			
7				omega =	1	
8				A =	2	
9				phi0 =	0	
10						
11		Euler's Method Calculations			Anal.	
12	t	x(t)	v(t)	a(t)	x(t)	Error
13	s	cm	cm/s	cm/s^2	cm	cm
14	0	0	2	0	0	0
15	0.1	0.2	2	-0.2	0.19967	0.00033
16	0.2	0.4	1.98	-0.4	0.39734	0.00266
113	9.9	-1.4012	-2.9578	1.40123	-0.9151	-0.4862
114	10	-1.697	-2.8177	1.69701	-1.088	-0.609

Formula Set I
Prototype Cell is B15

=B14+(**B4***C14)

Formula Set II
Prototype Cell is C15

=C14+(**B4***D14)

Formula Set III
Prototype Cell is D14

=(-**E4**/**E5**)*B14

Formula Set IV
Prototype Cell is E14

=**E8***SIN((**E7***A14)+**E9**)

Formula Set V
Prototype Cell is F14

=B14-E14

Model 3.1

BUILD and verify Model 3.1. The model uses the Euler method to solve second-order differential equations and is initialized with a solution to Hooke's law ($d^2x/dt^2 = -(k/m)x$.

Natural-order recalculation mode is required.

ENTER data using cells B4 - B6 (Δt, initial position, initial velocity) and cells E4 and E5 (spring constant, mass).

ENTER the formula for d^2x/dt^2 into cell D14 and copy into the range D14 - D114. If the differential equation has an analytic solution that you wish to compare with the numeric solution, enter the expression into cell E14 and copy into the range E14 - E114.

READ output from the model at cells A14 - A114 (t), B14 - B114 [$x(t)$], C14 - C114 [$v(t)$]. If an analytical solution for d^2x/dt^2 exists and is placed in cells E14 - E114, the difference between the numeric and analytic result is calculated in cells F14 - F114.

NOTICE that Equations 3.19 - 3.22 are used repeatedly in columns B, C, and D according to the logical steps of Euler's method.

EXERCISE **3.2**

THE SECOND LAW OF MOTION: THE RUNGE-KUTTA METHOD

TO GET THE MOST OUT OF THIS EXERCISE, YOU SHOULD ALREADY KNOW

The three laws of motion
Euler's method

As noted in Exercise 3.1, it is often useful to represent differential equations in their *general form* (e.g., Equation 3.32):

$$\frac{d^2x(t)}{dt^2} = a(x,v,t) \tag{3.32}$$

$a(x,v,t)$ declares that a can be expressed as a function of x, v, and t — including the special case where one or two of the independent variables are absent.

Recall that the first step in solving a second-order differential equation using Euler's method is to change its representation into two first-order differential equations (Equations 3.33 and 3.34):

$$\frac{dv(t)}{dt} = a(x,v,t) \tag{3.33}$$

$$\frac{dx(t)}{dt} = v \tag{3.34}$$

and then to propagate solutions recursively using Equations 3.35 and 3.36.

$$x(t + \Delta t) \sim x(t) + v(t)\Delta t \tag{3.35}$$

$$v(t+\Delta t) \sim v(t) + a(x(t),v(t),t)\Delta t \qquad (3.36)$$

You may wish to review Figure 3.3 of Exercise 3.1 to recall how these equations are used.

The form of the Runge-Kutta method that you will use in this book replaces the terms $v\Delta t$ and $a\Delta t$ with different terms that serve the same purpose (Equations 3.37 and 3.38).[1] Otherwise, the design of your spreadsheet remains very similar to Euler's method of Exercise 3.1.

$$x(t+\Delta t) \sim x(t) + \frac{k_{x0} + 2k_{x1} + 2k_{x2} + k_{x3}}{6} \qquad (3.37)$$

$$v(t+\Delta t) \sim v(t) + \frac{k_{v0} + 2k_{v1} + 2k_{v2} + k_{v3}}{6} \qquad (3.38)$$

where

$$k_{x0} = \Delta t\, v(t) \qquad k_{v0} = \Delta t\, a(x(t), v(t),\ t)$$

$$k_{x1} = \Delta t\left(v(t) + \frac{k_{v0}}{2}\right) \qquad k_{v1} = \Delta t\, a\!\left(x(t) + \frac{k_{x0}}{2}, v(t) + \frac{k_{v0}}{2}, t + \frac{\Delta t}{2}\right)$$

$$k_{x2} = \Delta t\left(v(t) + \frac{k_{v1}}{2}\right) \qquad k_{v2} = \Delta t\, a\!\left(x(t) + \frac{k_{x1}}{2}, v(t) + \frac{k_{v1}}{2}, t + \frac{\Delta t}{2}\right)$$

$$k_{x3} = \Delta t\,(v(t) + k_{v2}) \qquad k_{v3} = \Delta t\, a(x(t) + k_{x2}, v(t) + k_{v2}, t + \Delta t)$$

You may notice a strong similarity between the general form of these equations and Simpson's rule formulas that you learned in Exercise 2.5. In fact, if the acceleration happens to be a function of time only, the Runge-Kutta method becomes exactly the same as Simpson's rule for integration. In a sense then, these Runge-Kutta formulas are a generalization of Simpson's rule.

Spreadsheets built around the Runge-Kutta method can become very large because of the arrays of cells devoted to the calculation of the eight quantities k_{x0} - k_{v3}. It is easy to get lost unless you build your models around a clear organizing principle. In this exercise, we have chosen a design that trades compactness for ease of construction and modification. Notice that the constants k_{v0}, k_{v1}, k_{v2}, and k_{v3} all include expressions for acceleration that contain different values for the arguments x, v, and t (e.g., k_{v0} uses the arguments $x(t)$, $v(t)$, and t while k_{v3} uses the arguments $x(t) + k_{x2}$, $v(t) + k_{v2}$, and $t + \Delta t$). Clearly, there are at least two different ways in which a Runge-Kutta model might be built:

- Prepare four different expressions for acceleration in k_{v0} - k_{v3} where the arguments are modified by formulas contained within the expression.

- Prepare a single expression for acceleration in k_{v0} - k_{v3} choosing arguments that, when copied relatively, point to formulas that calculate different argument values.

The second method occupies more room in your spreadsheet but has the advantage that a single spreadsheet template can be used to solve all (one-dimensional) problems involving Newton's second law. You simply prepare your expression for acceleration in one location and then copy it to all other appropriate locations.

Before following the step-by-step instructions for building a Runge-Kutta model, examine the organizing principles behind the model carefully (Figure 3.7). Shortly thereafter you will discover that this model can be built swiftly through judicious use of your spreadsheet's COPY command.

- Columns A - E constitute a group (group 0) of five columns used to calculate model results — time, position, and velocity (columns A, B, C) — and the Runge-Kutta constants k_{x0} and k_{v0} (columns D and E).

  Column A is self-explanatory.

  Columns B and C represent Equations 3.37 and 3.38.

  Columns D and E (k_{x0} and k_{v0}) retrieve their arguments from cell B4 (Δt) and from the first three columns in the group of five.

- Columns F - J make up a group of five columns (group 1) used to calculate the Runge-Kutta constants k_{x1} and k_{v1} (columns I and J). These constants are calculated from arguments obtained from cell B4 (Δt) and the first three columns of this second group of five.

- Columns K - O constitute group 2 used to calculate k_{x2} and k_{v2} (columns N and O). Again, the argument list consists of cell B4 (Δt) and the first three columns of this group.

- Columns P - T (group 3) calculate k_{x3} and k_{v3} (columns S and T).

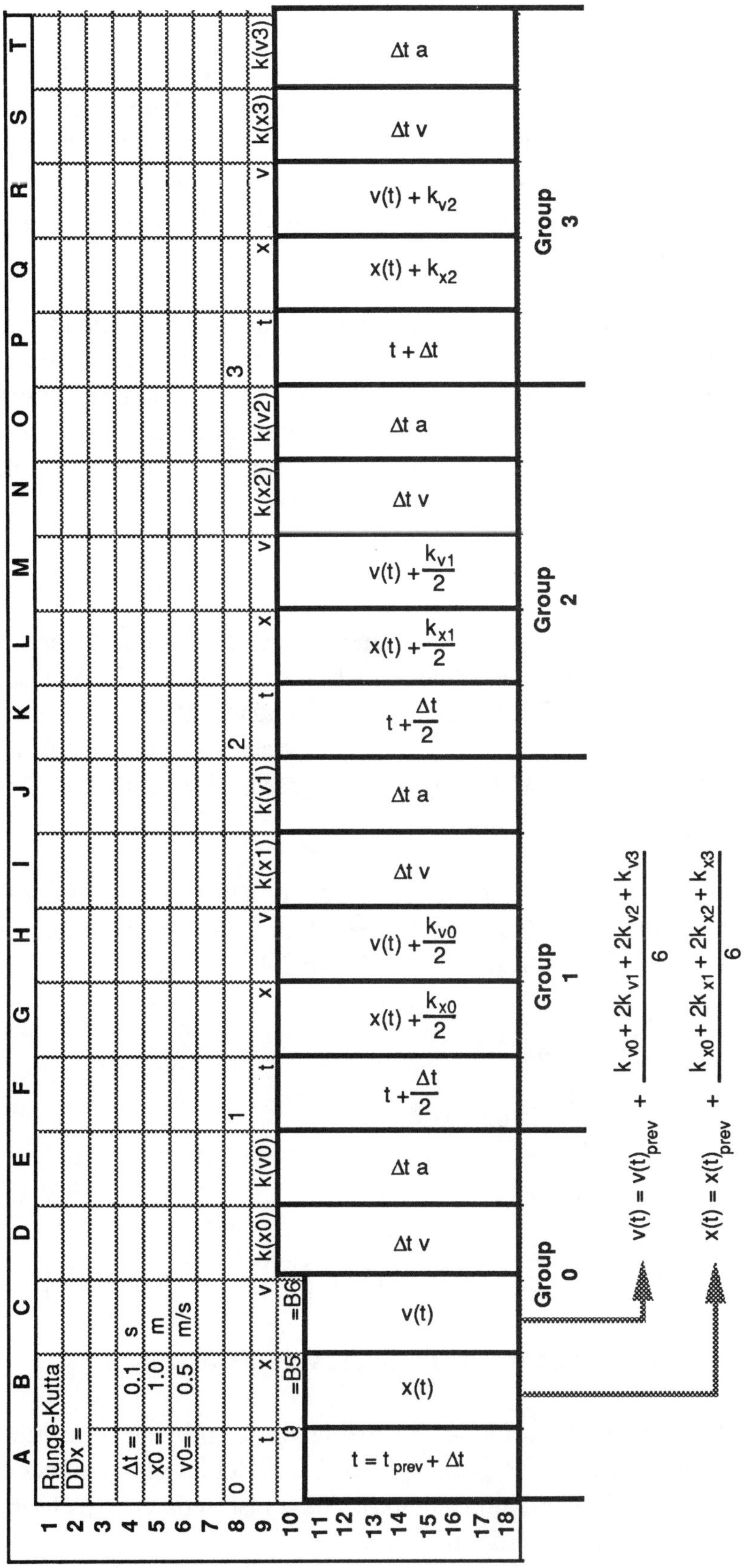

Figure 3.7

The Runge-Kutta method can be implemented on an electronic spreadsheet in a variety of forms. This model trades its relatively large size for ease of maintenance. That is, the formula for acceleration is identical in columns E, J, O, and T and can be changed rapidly to accommodate new problems. A complete description of this model and its design may be found in the main text.

Notice that columns D - T are used only to calculate the quantities (and their arguments) demanded by Equations 3.37 and 3.38 in columns B and C. *The design of this spreadsheet is very similar to that used by Euler's method but contains more arithmetic.* As noted earlier, the formula for acceleration, $a(x,v,t)$, used by k_{v0} - k_{v3} is identical in all columns because of the judicious positioning of argument lists. This feature renders the model portable across a wide range of problems involving differential equations.

Runge-Kutta models are easier to use than to talk about and the payoff for greater complexity is exceptional accuracy; they serve as basic tools throughout your explorations in this book. You should take the time to become thoroughly familiar with their construction. If you wish to understand the mathematical foundation for this method, you should consult any good textbook of numeric methods.

LESSONS

Lesson 1 of Exercise 3.1 considered an ideal, horizontal, oscillating spring with a mass m attached to its free end (Figure 3.2). You solved this problem using Euler's method and found that after just a few cycles of the motion the numeric solutions deviated significantly from the analytic solutions. Now, let's treat the same problem using the Runge-Kutta technique given above.

Recall from that exercise that the application of Newton's second law of motion to Hooke's law yields Equation 3.39, which can be solved analytically to yield Equations 3.40 and 3.41:

$$\frac{d^2x(t)}{dt^2} = -\left(\frac{k}{m}\right)x(t) \qquad (3.39)$$

$$v(t) = A\ \omega \cos(\omega t + \phi_0) \qquad (3.40)$$

$$x(t) = A \sin(\omega t + \phi_0) \qquad (3.41)$$

where $\omega = \sqrt{\frac{k}{m}}$, $A = \sqrt{x_0^2 + \left(\frac{v_0}{\omega}\right)^2}$, $\phi_0 = \tan^{-1}\left(\frac{v_0}{\omega x_0}\right)$

x_0 is the displacement at $t = 0$, and v_0 is the velocity at $t = 0$.

1. Build and verify Model 3.2 using the step-by-step instructions below (Model 3.2 solves Equation 3.39 [Hooke's law] numerically using the Runge-Kutta method). These instructions help clarify the detailed internal order of this model.[2]

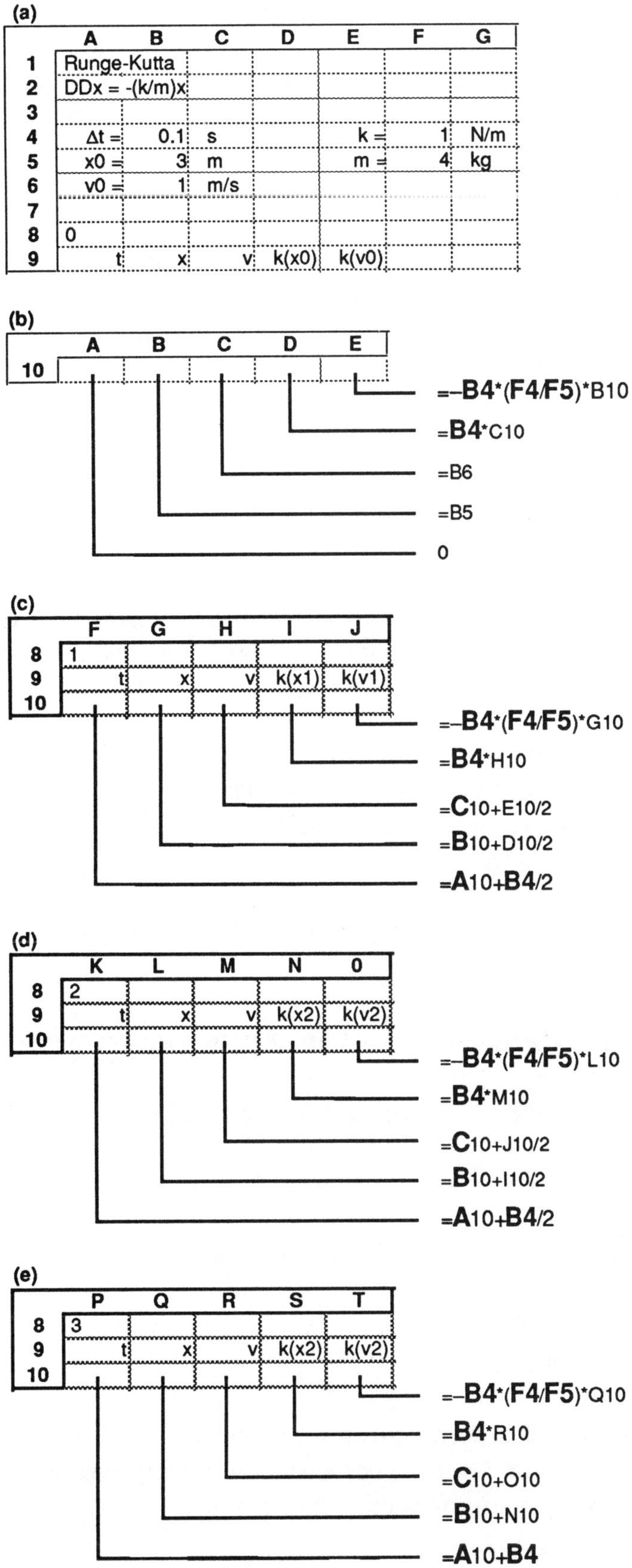

Figure 3.8

The Runge-Kutta model described in Model 3.2 (and Figure 3.7) can be built quickly by using your spreadsheet program's COPY command. See Lesson 1 for details.

a. Enter the labels and values for cells A1 - G9 shown in Figure 3.8a. The zero in cell A8 is used to identify the zeroth block of calculations that will appear below (i.e., the Runge-Kutta constants k_{x0} and k_{v0} are calculated in columns D and E).

b. Enter the values and formulas shown in Figure 3.8b. Take special care with respect to *absolute* and *relative* formula variables. Cell A10 contains the initial time value of zero, and cells B10 and C10 contain x_0 and v_0, respectively. Verify that the formulas in cells D10 and E10 represent the Runge-Kutta quantities k_{x0} and k_{v0}, respectively.

c. COPY the contents of cells A8 - E10 to cells F8 - J10. EDIT cells F8, I9, and J9 from 0 to 1. Cell F8 identifies the beginning of the next block of calculations. Change cells F10, G10, and H10 to read as shown in Figure 3.8c.

d. COPY the contents of cells F8 - J10 to cells K8 - O10. EDIT cell K8, N9, and O9 to contain 2. This region of your model should read as shown in Figure 3.8d.

e. COPY the contents of cells F8 - J10 to cells P8 - T10. EDIT cell P8, S9, and T9 to contain 3. EDIT cells P10, Q10, and R10 to read as shown in Figure 3.8e (i.e. /2 is removed from the end of each expression).

f. Place the formula A10 + **B4** (B4 is an *absolute* variable) in cell A11.

g. Place the formula B10+(D10+2*(I10+N10)+S10)/6 in cell B11.

h. COPY the contents of cell B11 into cell C11.

i. COPY the contents of cells D10 - T10 into D11 - T11.

j. COPY the contents of A11 - T11 down as many rows as desired.

k. Your model is now complete. Verify your values with Model 3.2 (VALUES).

Take the time to thoroughly review this model and the principles of its construction — it is a cornerstone for many exercises to come. You might also consider building a macro that automates the copying process.

You can change the model to accommodate any new problem by preparing a new acceleration formula in cell E10, multiplying it by Δt, and copying it down the E column as far as you wish. Next, copy the E column into the J, O, and T columns (i.e., the k_{v1}, k_{v2}, and k_{v3} columns).[3] Your model is now ready to solve your new force expression.

2. There are two adaptations of Model 3.2 that you may wish to consider in special circumstances.

SPREADSHEET TECHNIQUE

Compact Runge-Kutta Models

Models 3.2 - 3.4 are designed for easy building and easy modification to fit new physical situations. Runge-Kutta models can be made more compact (saving memory and, perhaps, increasing calculation speed) by using a model such as that described in Model 3.5. In Model 3.5,

a. Cells B4 - B6 are used to input Δt and the initial values of velocity (v_0) and position (x_0). Cells F1 and F2 are constants used by the expression for Hooke's law. Cells F4 - F6 are calculations that are needed only if the analytic solution to Hooke's law is desired.

b. Column A (starting at cell A11 and continuing downward) calculates the values of the independent variable t for which values of $x(t)$ and $v(t)$ are desired.

c. Columns B and C reproduce Equations 3.37 and 3.38.

d. Column D represents Hooke's law solved for acceleration (Equation 3.39).

e. Columns E and F are used to compute $x(t)$ analytically and to determine the error between the analytic and the Runge-Kutta solutions. Columns E and F can be deleted from your model if you wish.

f. Columns G - N calculate the Runge-Kutta values k_{x0} - k_{v3}.

   Take particular note of columns J, L, and N. Each column uses a version of the acceleration function with different arguments. Every time you wish to explore a new physical problem, you must change the expression for d^2x/dt^2 in column D, J, L, and N as well. This is the price to pay for the increase in compactness.

This type of model, however, becomes as easy to use as Model 3.2 if your spreadsheet program supports *function macros*. See the spreadsheet technique "Function Macros" later in this exercise if you wish to explore this important feature of recent spreadsheet programs.

Figure 3.9

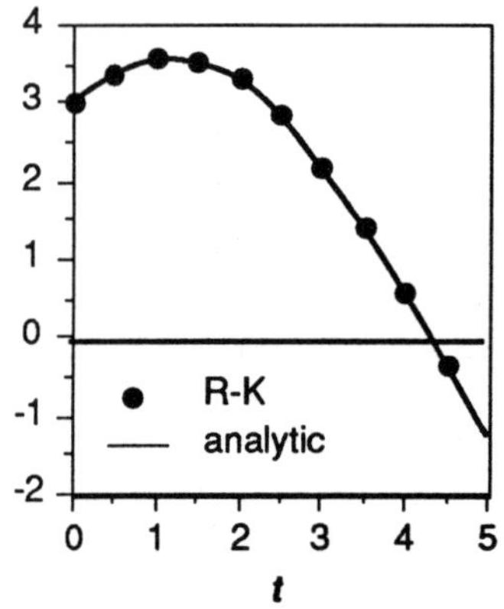

XY-PLOT

Data
Horizontal (*x*) axis: cells A11 - A61 (*t*)
Vertical (*y*) axis: cells B11 - B61 [numeric *x*(*t*)], E11 - E61 [analytic *x*(*t*)]

a. Model 3.2 shows explicitly only position and velocity. Acceleration is implicit in the Runge-Kutta constants but does not appear explicitly.[4]

Use your spreadsheet special commands to insert a new column to the left of the present column A and prepare an appropriate formula for acceleration in this column (Model 3.3). Use this column for examining or plotting acceleration in conjunction with position and velocity.

b. Model 3.2 makes no provision for comparing numeric and analytic solutions to the same differential equation.[5] Starting from Model 3.3, insert an additional two columns to the left of the acceleration column (currently column A). Use one of these columns to build an expression for the analytic solution for position and the other column to prepare an expression for the difference in value between the numeric and analytic solutions. You may wish to simplify the formulas in your analytic solution column by calculating ω, A, and ϕ_0 in convenient locations of your spreadsheet (Model 3.4). This also improves the execution time of your model.

Models 3.2 - 3.4 are really the same model with only minor editing changes to accommodate special situations. Each model is used where appropriate in the lessons and problems that follow.

3. The Runge-Kutta method given here is extremely accurate in many situations. To determine how accurate, plot both the numeric and the analytic solutions to $x(t)$ (Figure 3.9) for each of the input values below. Find the maximum value of x and the maximum error in x for $t < 5s$ in each case.

Δt	x_0	v_0	k	m	Maximum x	Maximum Error
0.1	3	1	1	4	____	____
0.1	3	1	1	1	____	____
0.1	0	2	1	4	____	____
0.1	0	2	1	1	____	____
0.1	0	2	3	1	____	____

a. Examine the error values in column F. How do these errors compare with the errors associated with Euler's method?

b. If you plot the analytic and numeric solutions together, do any of the conditions above cause these plots to diverge perceptibly?

Can you find any conditions where the numeric results are unsatisfactory?

c. Find the error at $t = 1$ s for the values $\Delta t = 0.1, 0.01$, and 0.001 (if possible).

Δt	Error @ $t = 1$ s
0.1	______
0.01	______
0.001	______

Notice that increasing the calculation load by a factor of ten yields four additional decimal places of increased accuracy — a vast improvement over Euler's method.

4. Use Model 3.2 to plot acceleration, velocity, and position as a function of time (Figure 3.10). Use the values with which you verified your model (i.e., $\Delta t = 0.1$ s, $x_0 = 3$ m, $v_0 = 1$ m s^{-1}, $k = 1$ N m^{-1}, $m = 4$ g). Now you can study the behavior of your horizontal spring with confidence.

   a. Examine the plot of $x(t)$ vs time and determine the value of the period of oscillation. (You will have to extend your model or use a larger value for Δt.)

Figure 3.10

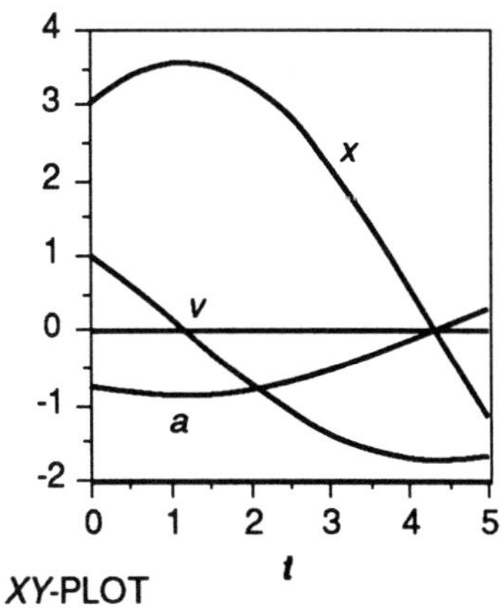

XY-PLOT

Data
Horizontal (*x*) axis: cells A11 - A61 (*t*)
Vertical (*y*) axis: cells B11 - B61 (position), C11 - C61 (velocity), D11 - D61 (acceleration)

SPREADSHEET TECHNIQUE

Function Macros

Compact Runge-Kutta solutions to second-order differential equations require you to write four different expressions for d^2x/dt^2 using four different arguments for the expression. Newer spreadsheets eliminate this nuisance by permitting you to create your own functions (often called *function macros*). For example, Model 3.5 might be written as shown in Model 3.6 where a function M!F(*t*, *x*, arg1, arg2, arg3, arg4) is written as a function macro and called by the model repeatedly. (M!F(...) is a feature of the spreadsheet program Excel that means function F located in file M. Other programs may use a different nomenclature.)

The argument list of the function contains all possible inputs to F. In most instances, only a few of these arguments need be used.

If your spreadsheet supports function macros, you may want to build a model that exploits this feature. You may find that some expressions have unacceptable calculation times using this feature, but it is an easy way to survey a large number of expressions and to verify an "optimized" model where you take the time to write the expression directly.

b. Consider a spring for which $k = 1$ dyne cm^{-1} and $m = 1.5$ g. The mass attached to the spring is initially motionless ($v_0 = 0$) but extended 2 cm from its resting position ($x_0 = 2$ cm).

What is the period of oscillation of the spring? How does the period change as the mass of the attached object increases? How does the period change as the stiffness of the spring decreases (k increases)?

The analytic expression for the period T of a horizontal, oscillating mass is $T = 2\pi/\omega$. Compare your answers with the results obtained using this expression.

c. How does the spring behave if the mass is initially motionless at the spring's resting position ($v_0 = 0$, $x_0 = 0$)?

PROBLEMS

1. The analytic expressions for motion under constant acceleration are well known (Equations 2.1 - 2.4). Modify Model 3.2 to calculate $x(t)$ and $v(t)$ when acceleration is a constant. Compare your Runge-Kutta results with those obtained analytically — the numeric method gives exact results for these expressions, and the errors you see are rounding errors.

2. Part of a uniform rope hangs over the edge of a frictionless table. If x is the amount of overhang, then the acceleration of any fixed point on the rope a is given by

$$a_x = \frac{d^2x}{dt^2} = \frac{gx}{L} \qquad (3.42)$$

where $g = 9.8$ m s^{-2} and L (total rope length) = 2 m.

Assume that 0.1 m of rope initially overhangs the edge of the table and that the rope is motionless ($x_0 = 0.1$ m, $v_0 = 0$ m s^{-1}).

a. Plot x vs time and determine the moment at which the last bit of rope leaves the table and the rope enters free fall.

b. How much time is needed for the rope to enter free fall if the initial length of rope hanging over the edge is 0.2 m? 0.4 m? 1 m?

Overhang (m)	T_{Theor} (s)	T_{Num} (s)
0.2	1.388	____
0.4	1.057	____
1.0	0.603	____

c. If the rope is initially hanging 1 m over the table, can you find an initial velocity value so that the rope ends up with zero velocity and its total length on the table?

3. Just 0.1 m of a 2 m length of uniform rope hangs over the edge of a table. Including friction, the acceleration of the rope can be described by

$$a_x = \frac{d^2x}{dt^2} = g(1+\mu)\left(\frac{x}{L}\right) - \mu g \tag{3.43}$$

where μ is the coefficient of kinetic friction and $g = 9.8$ m s^{-2}.

a. If $\mu = 0.2$, for what values of initial velocity will the rope slide off the table?

b. If $\mu = 0.2$ and the initial velocity is miniscule but not zero (i.e., the coefficient of kinetic friction still applies), for what values of x_0 will the rope slide off the table?

c. If $\mu = 0.2$, the initial velocity is miniscule but not zero, and the length of overhanging rope is 0.8 m, how much time will pass before the rope leaves the table entirely?

4. Two space stations have been assembled in space, each of mass $M = 3 \times 10^7$ kg. They are separated by 1 km in a region of space where the gravitational forces from other bodies are negligible. Assume that the predominant force between them is the attractive gravitational force due to their own masses so that each spacecraft obeys

$$\left(\frac{M}{2}\right)\frac{d^2x}{dt^2} = -\frac{GMM}{x^2} \tag{3.44}$$

where $G = 6.67 \times 10^{-11}$ N m^2 kg^{-2} is Newton's gravitational constant and x is the distance between the two stations.

a. If their initial velocity is zero, plot x vs t. Will the spacecraft collide within a human being's normal lifetime if they are initially at rest with respect to one another?

Figure 3.11

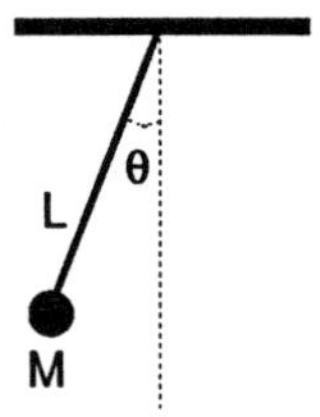

b. What initial velocity in the x-direction is needed to prevent collision?

5. An ideal simple pendulum of length L has an angular acceleration described by

$$\frac{d^2\theta}{dt^2} = -\frac{g}{L}\sin\theta \tag{3.45}$$

where θ is the angle measured from the vertical (Figure 3.11). Often, a small-angle approximation for angles of displacement less than about 0.15 radians is applied so that $\sin\theta$ is well approximated by θ itself. In this angular range, this approximation has a fractional error of less than about 6 parts in 1500. The resulting equation then yields sinusoidal solutions.

a. Make the small-angle approximation and use $g = 9.8$ m s–2, $L = 1$ m, $\theta_0 = 0.1$ radians, and $d\theta/dt = 0$ at $t = 0$. Plot θ vs. t and find the period of oscillation.

b. Change the initial angle to 0.8 radians and use the small angle approximation again. This initial angle is really too large for the approximation. Plot θ vs t and interpret. What is the period?

c. Do not make the small-angle approximation. Instead keep $\sin\theta$ as given. Repeat the calculations for the two initial angles of 0.1 radians and 0.8 radians. What are the periods now? The mass spends a relatively large time at large angles. Why?

d. Investigate the motion when the initial angle is very near 180°, both with and without an initial velocity.

FORMULAS

	A	B	C	D	E	F	G	H	I	J	K	L	M	N	O	P	Q	R	S	T
1	Runge-Kutta																			
2	DDx = -(k/m)x																			
3																				
4	Δt =	0.1	s		k =	1	N/m													
5	x0 =	3	m		m =	4	kg													
6	v0 =	1	m/s																	
7																				
8	0					1					2					3				
9	t	x	v	k(x0)	k(v0)	t	x	v	k(x1)	k(v1)	t	x	v	k(x2)	k(v2)	t	x	v	k(x3)	k(v3)
10	0	=B5	=B6	Form Set IV	Form Set V	Form Set VI	Form Set VII	Form Set VIII	Form Set IX	Form Set X	Form Set XI	Form Set XII	Form Set XIII	Form Set XIV	Form Set XV	Form Set XVI	Form Set XVII	Form Set XVIII	Form Set XIX	Form Set XX
11	Form Set I	Form Set II	Form Set III																	
12																				
13																				

VALUES

	A	B	C	D	E	F	G	H	I	J	K	L	M	N	O	P	Q	R	S	T
1	Runge-Kutta																			
2	DDx = -(k/m)x																			
3																				
4	Δt =	0.1	s		k =	1	N/m													
5	x0 =	3	m		m =	4	kg													
6	v0 =	1	m/s																	
7																				
8	0					1					2					3				
9	t	x	v	k(x0)	k(v0)	t	x	v	k(x1)	k(v1)	t	x	v	k(x2)	k(v2)	t	x	v	k(x3)	k(v3)
10	0	3	1	0.1	-0.08	0.05	3.05	0.963	0.096	-0.08	0.05	3.048	0.962	0.096	-0.08	0.1	3.096	0.924	0.092	-0.08
11	0.1	3.096	0.924	0.092	-0.08	0.15	3.142	0.885	0.089	-0.08	0.15	3.14	0.885	0.088	-0.08	0.2	3.185	0.845	0.085	-0.08
12	0.2	3.185	0.845	0.085	-0.08	0.25	3.227	0.805	0.081	-0.08	0.25	3.225	0.805	0.08	-0.08	0.3	3.265	0.765	0.076	-0.08
13	0.3	3.265	0.765	0.076	-0.08	0.35	3.303	0.724	0.072	-0.08	0.35	3.301	0.723	0.072	-0.08	0.4	3.338	0.682	0.068	-0.08

MODELS

Model 3.2

BUILD and verify Model 3.2.

ENTER data using cells B4 - B6 (Δt, initial position, initial velocity) and cells F4 - F5 (the spring constant required by Hooke's law and mass).

READ output from the model in columns A (time), B (position), and C (velocity).

NOTICE that columns D - T are used to calculate the Runge-Kutta *k*-values. The acceleration function is implicit in the definition of k_{v0}, k_{v1}, k_{v2}, and k_{v3}.

Model 3.2 (continued)

Formula Set I Prototype Cell is A11
=A10+B4

Formula Set II Prototype Cell is B11
=B10+(D10+2*(I10+N10)+S10)/6

Formula Set III Prototype Cell is C11
=C10+(E10+2*(J10+O10)+T10)/6

Formula Set IV Prototype Cell is D10
=B4*C10

Formula Set V Prototype Cell is E10
=-B4*(F4/F5)*B10

Formula Set VI Prototype Cell is F10
=A10+B4/2

Formula Set VII Prototype Cell is G10
=B10+D10/2

Formula Set VIII Prototype Cell is H10
=C10+E10/2

Formula Set IX Prototype Cell is I10
=B4*H10

Formula Set X Prototype Cell is J10
=-B4*(F4/F5)*G10

Formula Set XI Prototype Cell is K10
=A10+B4/2

Formula Set XII Prototype Cell is L10
=B10+I10/2

Formula Set XIII Prototype Cell is M10
=C10+J10/2

Formula Set XIV Prototype Cell is N10
=B4*M10

Formula Set XV Prototype Cell is O10
=-B4*(F4/F5)*L10

Formula Set XVI Prototype Cell is P10
=A10+B4

Formula Set XVII Prototype Cell is Q10
=B10+N10

Formula Set XVIII Prototype Cell is R10
=C10+O10

Formula Set XIX Prototype Cell is S10
=B4*R10

Formula Set XX Prototype Cell is T10
=-B4*(F4/F5)*Q10

FORMULAS

	A	B	C	D	E	F	G	H	I	J	K	L	M	N	O	P	Q	R	S	T	U
1		Runge-Kutta																			
2		DDx = -(k/m)x																			
3																					
4		Δt =	0.1	s		k =	1	N/m													
5		x0 =	3	m		m =	4	kg													
6		v0 =	1	m/s																	
7																					
8		0					1					2					3				
9	a	t	x	v	k(x0)	k(v0)	t	x	v	k(x1)	k(v1)	t	x	v	k(x2)	k(v2)	t	x	v	k(x3)	k(v3)
10	Form Set I	0	=B5	=B6	Form Set V	Form Set VI	Form Set VII	Form Set VIII	Form Set IX	Form Set X	Form Set XI	Form Set XII	Form Set XIII	Form Set XIV	Form Set XV	Form Set XVI	Form Set XVII	Form Set XVIII	Form Set XIX	Form Set XX	Form Set XXI
11		Form Set II	Form Set III	Form Set IV																	
12																					
13																					

VALUES

	A	B	C	D	E	F	G	H	I	J	K	L	M	N	O	P	Q	R	S	T	U
1		Runge-Kutta																			
2		DDx = -(k/m)x																			
3																					
4		Δt =	0.1	s		k =	1	N/m													
5		x0 =	3	m		m =	4	kg													
6		v0 =	1	m/s																	
7																					
8		0					1					2					3				
9	a	t	x	v	k(x0)	k(v0)	t	x	v	k(x1)	k(v1)	t	x	v	k(x2)	k(v2)	t	x	v	k(x3)	k(v3)
10	-0.75	0	3	1	0.1	-0.08	0.05	3.05	0.963	0.096	-0.08	0.05	3.048	0.962	0.096	-0.08	0.1	3.096	0.924	0.092	-0.08
11	-0.77	0.1	3.096	0.924	0.092	-0.08	0.15	3.142	0.885	0.089	-0.08	0.15	3.14	0.885	0.088	-0.08	0.2	3.185	0.845	0.085	-0.08
12	-0.8	0.2	3.185	0.845	0.085	-0.08	0.25	3.227	0.805	0.081	-0.08	0.25	3.225	0.805	0.08	-0.08	0.3	3.265	0.765	0.076	-0.08
13	-0.82	0.3	3.265	0.765	0.076	-0.08	0.35	3.303	0.724	0.072	-0.08	0.35	3.301	0.723	0.072	-0.08	0.4	3.338	0.682	0.068	-0.08

Model 3.3

BUILD and verify Model 3.3.

ENTER data using cells C4 - C6 (Δt, initial position, initial velocity) and cells G4 and G5 (the spring constant required by Hooke's law and mass).

READ output from the model in columns A (acceleration), B (time), C (position), and D (velocity).

NOTICE that Model 3.3 can be created from Model 3.2 by simply inserting a column to the left of column A (in Model 3.2) and preparing and copying a formula for the acceleration function. This acceleration function has a different form from that implicit in the Runge-Kutta *k*-values k_{v0}, k_{v1}, k_{v2}, and k_{v3} and must be prepared separately from them.

Model 3.3 (continued)

Formula Set I	Prototype Cell is A10	=F10/**C4**
Formula Set II	Prototype Cell is B11	=B10+**C4**
Formula Set III	Prototype Cell is C11	=C10+(E10+2*(J10+O10)+T10)/6
Formula Set IV	Prototype Cell is D11	=D10+(F10+2*(K10+P10)+U10)/6
Formula Set V	Prototype Cell is E10	=**C4** *D10
Formula Set VI	Prototype Cell is F10	=-**C4** *(**G4**/**G5**)*C10
Formula Set VII	Prototype Cell is G10	=**B**10+**C4**/2
Formula Set VIII	Prototype Cell is H10	=**C**10+E10/2
Formula Set IX	Prototype Cell is I10	=**D**10+F10/2
Formula Set X	Prototype Cell is J10	=**C4** *I10
Formula Set XI	Prototype Cell is K10	=-**C4** *(**G4**/**G5**)*H10
Formula Set XII	Prototype Cell is L10	=**B**10+**C4**/2
Formula Set XIII	Prototype Cell is M10	=**C**10+J10/2
Formula Set XIV	Prototype Cell is N10	=**D**10+K10/2
Formula Set XV	Prototype Cell is O10	=**C4** *N10
Formula Set XVI	Prototype Cell is P10	=-**C4** *(**G4**/**G5**)*M10
Formula Set XVII	Prototype Cell is Q10	=**B**10+**C4**
Formula Set XVIII	Prototype Cell is R10	=**C**10+O10
Formula Set XIX	Prototype Cell is S10	=**D**10+P10
Formula Set XX	Prototype Cell is T10	=**C4** *S10
Formula Set XXI	Prototype Cell is U10	=-**C4** *(**G4**/**G5**)*R10

FORMULAS

	A	B	C	D	E	F	G	H	I	J	K	L	M	N	O	P	Q	R	S	T	U	V	W
1				Runge-Kutta																			
2				DDx = -(k/m)x																			
3							=SQRT(E5^2+(E6/B4)^2)																
4	w =	=SQRT(I4/I5)		Δt =	0.1	s		=ATAN2(E6,B4*E5)	1	N/m													
5	A =			x0 =	3	m		m =	4	kg													
6	phi0 =			v0 =	1	m/s																	
7																							
8		Anal		0					1					2					3				
9	Error	x	a	t	x	v	k(x0)	k(v0)	t	x	v	k(x1)	k(v1)	t	x	v	k(x2)	k(v2)	t	x	v	k(x3)	k(v3)
10	Form Set I	Form Set II	Form Set III	0	=B5	=B6	Form Set VII	Form Set VIII	Form Set IX	Form Set X	Form Set XI	Form Set XII	Form Set XIII	Form Set XIV	Form Set XV	Form Set XVI	Form Set XVII	Form Set XVIII	Form Set XIX	Form Set XX	Form Set XXI	Form Set XXII	Form Set XXIII
11				Form Set IV	Form Set V	Form Set VI																	
12																							
13																							

VALUES

	A	B	C	D	E	F	G	H	I	J	K	L	M	N	O	P	Q	R	S	T	U	V	W
1				Runge-Kutta																			
2				DDx = -(k/m)x																			
3																							
4	w =	0.5		Δt =	0.1	s		k =	1	N/m													
5	A =	3.606		x0 =	3	m		m =	4	kg													
6	phi0 =	0.983		v0 =	1	m/s																	
7																							
8		Anal		0					1					2					3				
9	Error	x	a	t	x	v	k(x0)	k(v0)	t	x	v	k(x1)	k(v1)	t	x	v	k(x2)	k(v2)	t	x	v	k(x3)	k(v3)
10	0	3	-0.75	0	3	1	0.1	-0.08	0.05	3.05	0.963	0.096	-0.08	0.05	3.048	0.962	0.096	-0.08	0.1	3.096	0.924	0.092	-0.08
11	5E-09	3.096	-0.77	0.1	3.096	0.924	0.092	-0.08	0.15	3.142	0.885	0.089	-0.08	0.15	3.14	0.885	0.088	-0.08	0.2	3.185	0.845	0.085	-0.08
12	9E-09	3.185	-0.8	0.2	3.185	0.845	0.085	-0.08	0.25	3.227	0.805	0.081	-0.08	0.25	3.225	0.805	0.08	-0.08	0.3	3.265	0.765	0.076	-0.08
13	1E-08	3.265	-0.82	0.3	3.265	0.765	0.076	-0.08	0.35	3.303	0.724	0.072	-0.08	0.35	3.301	0.723	0.072	-0.08	0.4	3.338	0.682	0.068	-0.08

Model 3.4

BUILD and verify Model 3.4.

ENTER data using cells E4 - E6 (Δt, initial position, initial velocity) and cells I4 - I5 (the spring constant required by Hooke's law and mass).

READ output from the model in columns C (acceleration), D (time), E (position), and F (velocity). The numeric solution for position (column E) can be compared with the analytic solution (column B) in column A.

NOTICE that Model 3.4 can be created from Model 3.3 by simply inserting two columns to the left of column A (in Model 3.3) and preparing and copying an expression for the analytic solution to the problem and an expression that calculates the difference between the numeric and analytic solutions. For many expressions, you may need to include additional formulas elsewhere on the spreadsheet (e.g., cells B4 - B6).

Model 3.4 (continued)

Formula Set I	Prototype Cell is A10	=B10-E10
Formula Set II	Prototype Cell is B10	=B5*SIN(B4*D10+B6)
Formula Set III	Prototype Cell is C10	=-(I4/I5)*E10
Formula Set IV	Prototype Cell is D11	=D10+E4
Formula Set V	Prototype Cell is E11	=E10+(G10+2*(L10+Q10)+V10)/6
Formula Set VI	Prototype Cell is F11	=F10+(H10+2*(M10+R10)+W10)/6
Formula Set VII	Prototype Cell is G10	=E4*F10
Formula Set VIII	Prototype Cell is H10	=-E4*(I4/I5)*E10
Formula Set IX	Prototype Cell is I10	=D10+E4/2
Formula Set X	Prototype Cell is J10	=E10+G10/2
Formula Set XI	Prototype Cell is K10	=F10+H10/2
Formula Set XII	Prototype Cell is L10	=E4*K10
Formula Set XIII	Prototype Cell is M10	=-E4*(I4/I5)*J10
Formula Set XIV	Prototype Cell is N10	=D10+E4/2
Formula Set XV	Prototype Cell is O10	=E10+L10/2
Formula Set XVI	Prototype Cell is P10	=F10+M10/2
Formula Set XVII	Prototype Cell is Q10	=E4*P10
Formula Set XVIII	Prototype Cell is R10	=-E4*(I4/I5)*O10
Formula Set XIX	Prototype Cell is S10	=D10+E4
Formula Set XX	Prototype Cell is T10	=E10+Q10
Formula Set XXI	Prototype Cell is U10	=F10+R10
Formula Set XXII	Prototype Cell is V10	=E4*U10
Formula Set XXIII	Prototype Cell is W10	=-E4*(I4/I5)*T10

FORMULAS

	A	B	C	D	E	F	G	H	I	J	K	L	M	N
1	Runge-Kutta				k =	1	N/m							
2	DDx = -(k/m) x				m =	4	kg							
3														
4	Δt =	0.1	s		w =	=SQRT(F1/F2)								
5	x0 =	3	m		A =	=SQRT((B5*B5)+(B6/F4)^2)								
6	v0 =	1	m/s		phi0 =	=ATAN2(B6,F4*B5)								
7														
8	Change these columns =>			CHNG	CHNG			CHNG		CHNG		CHNG		CHNG
9	t	x(t)	v(t,x)	a(t,x,v)	Anal	Error	k01	k02	k11	k12	k21	k22	k31	k32
10	s	m	m/s	m/s^2	m	m								
11	0	=B5	=B6											
12–61	Form. Set I	Form. Set II	Form. Set III	Form. Set IV	Form. Set V	Form. Set VI	Form. Set VII	Form. Set VIII	Form. Set IX	Form. Set X	Form. Set XI	Form. Set XII	Form. Set XIII	Form. Set XIV

Model 3.5

BUILD and verify Model 3.5.

ENTER data using cells B4 - B6 and F1 and F2 for the model as initialized for Hooke's law. To change the model for other expressions, create a block of input variables in the region from cell E1 - G6 and change the formulas in cells D11 - D61, J11 - J61, L11 - L61, N11 - N61 (the force expressions required for the Runge-Kutta method), and in cells E11 - E61 (the analytic solution to the force expression, if available).

READ output from the model in cells B11 - B61, C11 - C61 (the numeric solutions), D11 - D61 (the values obtained directly from the force expression), and E11 - E61 (the analytic solution for $x(t)$).

Model 3.5 (continued)

VALUES

	A	B	C	D	E	F	G	H	I	J	K	L	M	N
1	Runge-Kutta				k =	1	N/m							
2	DDx = -(k/m) x				m =	4	kg							
3														
4	Δt =	0.1	s		w =	0.5								
5	x0 =	3	m		A =	3.6056								
6	v0 =	1	m/s		phi0 =	0.9828								
7														
8	Change these columns =>			CHNG	CHNG					CHNG		CHNG		CHNG
9	t	x(t)	v(t,x)	a(t,x,v)	Anal	Error	kx0	kv0	kx1	kv1	kx2	kv2	kx3	kv3
10	s	m	m/s	m/s^2	m	m								
11	0	3	1	-0.75	3	0								
12	0.1	3.0962	0.9238	-0.774	3.0962	5E-09	0.1	-0.075	0.0963	-0.076	0.0962	-0.076	0.0924	-0.077
13	0.2	3.1847	0.8453	-0.796	3.1847	9E-09	0.0924	-0.077	0.0885	-0.079	0.0885	-0.079	0.0845	-0.08
14	0.3	3.2652	0.7646	-0.816	3.2652	1E-08	0.0845	-0.08	0.0805	-0.081	0.0805	-0.081	0.0765	-0.082

	A	B	C	D	E	F	G	H	I	J	K	L	M	N
60	4.9	-1.035	-1.727	0.2588	-1.035	-4E-07	-0.175	0.0215	-0.174	0.0237	-0.174	0.0237	-0.173	0.0259
61	5	-1.206	-1.699	0.3016	-1.206	-4E-07	-0.173	0.0259	-0.171	0.028	-0.171	0.028	-0.17	0.0302

Model 3.5 (continued)

Formula Set I Prototype Cell is A12
=A11+B4

Formula Set II Prototype Cell is B12
=B11+(G12+(2*I12)+(2*K12)+M12)/6

Formula Set III Prototype Cell is C12
=C11+(H12+(2*J12)+(2*L12)+N12)/6

Formula Set IV Prototype Cell is D11
=(-F1/F2)*B11

Formula Set V Prototype Cell is E11
=F5*SIN((F4*A11)+F6)

Formula Set VI Prototype Cell is F11
=E11-B11

Formula Set VII Prototype Cell is G12
=B4*C11

Formula Set VIII Prototype Cell is H12
=B4*D11

Formula Set IX Prototype Cell is I12
=B4*(C11+(H12/2))

Formula Set X Prototype Cell is J12
=B4*(-F1/F2)*(B11+(G12/2))

Formula Set XI Prototype Cell is K12
=B4*(C11+(J12/2))

Formula Set XII Prototype Cell is L12
=B4*(-F1/F2)*(B11+(I12/2))

Formula Set XIII Prototype Cell is M12
=B4*(C11+L12)

Formula Set XIV Prototype Cell is N12
=B4*(-F1/F2)*(B11+K12)

Model 3.6

BUILD and verify Model 3.6. Model 3.6 uses function macros to simplify model changes. You will find this model useful only if your spreadsheet program supports function macros and your computer has sufficient processing power (i.e., speed).

ENTER data using cells B4 - B6 and F4 - F7. The model is initialized for Hooke's law and only two of the four cells reserved for arguments (cells F4 and F5) are used. The functions in cells D11 - D61, J11 - J61, L11 - L61, and N11 - N61 are defined by a function macro that you must create. The function macro for Hooke's law (for the spreadsheet program Excel) is shown in the spreadsheet technique "Function Macros."

READ output from the model in cells B11 - B61 [$x(t)$], C11 - C61 [$v(t)$], and D11 - D61 [$a(t)$].

NOTICE that the function is called via a function M!F(). M!F is what function F of file M would be called in the spreadsheet Excel. Other spreadsheets may use a different convention.

FORMULAS

	A	B	C	D	E	F	G	H	I	J	K	L
1	Runge-Kutta											
2	DDx =											
3												
4	Δt =	0.1	s		arg1 =	1	(k)					
5	x0 =	3	m		arg2 =	4	(m)					
6	v0 =	1	m/s		arg3 =	0						
7					arg3 =	0						
8												
9	t	x(t)	v(t,x)	a(t,x,v)	k01	k02	k11	k12	k21	k22	k31	k32
10	s	m	m/s	m/s^2								
11	0	=B5	=B6	Form. Set IV								
12–71	Form. Set I	Form. Set II	Form. Set III		Form. Set V	Form. Set VI	Form. Set VII	Form. Set VIII	Form. Set IX	Form. Set X	Form. Set XI	Form. Set XII

VALUES

	A	B	C	D	E	F	G	H	I	J	K	L
1	Runge-Kutta											
2	DDx =											
3												
4	Δt =	0.1	s		arg1 =	1	(k)					
5	x0 =	3	m		arg2 =	4	(m)					
6	v0 =	1	m/s		arg3 =	0						
7					arg3 =	0						
8												
9	t	x(t)	v(t,x)	a(t,x,v)	k01	k02	k11	k12	k21	k22	k31	k32
10	s	m	m/s	m/s^2								
11	0	3	1	-0.75								
12	0.1	3.0962	0.9238	-0.774	0.1	-0.075	0.0963	-0.076	0.0962	-0.076	0.0924	-0.077
13	0.2	3.1847	0.8453	-0.796	0.0924	-0.077	0.0885	-0.079	0.0885	-0.079	0.0845	-0.08
14	0.3	3.2652	0.7646	-0.816	0.0845	-0.08	0.0805	-0.081	0.0805	-0.081	0.0765	-0.082
70	5.9	-2.564	-1.267	0.6411	-0.133	0.0609	-0.13	0.0625	-0.13	0.0625	-0.127	0.0641
71	6	-2.688	-1.202	0.6719	-0.127	0.0641	-0.124	0.0657	-0.123	0.0657	-0.12	0.0672

Formula Set I
Prototype Cell is A12

=A11+**B4**

Formula Set II
Prototype Cell is B12

=B11+(E12+(2*G12)+(2*I12)+K12)/6

Formula Set III
Prototype Cell is C12

=C11+(F12+(2*H12)+(2*J12)+L12)/6

Formula Set IV
Prototype Cell is D11

=MIF(A11,B11, **F4**,**F5**,**F6**,**F7**)

Formula Set V
Prototype Cell is E12

=**B4***C11

Formula Set VI
Prototype Cell is F12

=**B4***D11

Formula Set VII
Prototype Cell is G12

=**B4***(C11+(F12/2))

Formula Set VIII
Prototype Cell is H12

=**B4***MIF(A11+**B4**/2,B11+(E12/2), **F4**,**F5**,**F6**,**F7**)

Formula Set IX
Prototype Cell is I12

=**B4***(C11+(H12/2))

Formula Set X
Prototype Cell is J12

=**B4***MIF(A11+**B4**/2,B11+(G12/2), **F4**,**F5**,**F6**,**F7**)

Formula Set XI
Prototype Cell is K12

=**B4***(C11+J12)

Formula Set XII
Prototype Cell is L12

=**B4***MIF(A11+**B4**,B11+I12, **F4**,**F5**,**F6**,**F7**)

EXERCISE **3.3**

TRAJECTORIES

TO GET THE MOST OUT OF THIS EXERCISE, YOU SHOULD ALREADY KNOW

How to solve differential equations using the Runge-Kutta method.

Exercise 3.2 was designed to introduce you to a general plan for solving the differential equations that arise out of Newton's second law of motion. All problems were associated with motion in a single dimension. In this and the next exercise, you extend your tools to include motion in two and three dimensions.

Recall that Newton's second law of motion states:

$$\Sigma \boldsymbol{F} = m\boldsymbol{a} = m\frac{d\boldsymbol{v}}{dt} = m\frac{d^2\boldsymbol{s}}{dt^2} \tag{3.46}$$

where $\Sigma \boldsymbol{F}$ is the vector sum of the forces acting on an object, m is the mass of the object, $\boldsymbol{a}$ is the acceleration of the object, $\boldsymbol{v}$ is the velocity of the object, $\boldsymbol{s}$ is the position of the object.

Recall further that $\Sigma \boldsymbol{F}$, $\boldsymbol{a}$, $\boldsymbol{v}$, and $\boldsymbol{s}$ are vector quantities that can be resolved into components along the x-, y-, and z-axes:

$$\Sigma F_x = ma_x = m\frac{dv_x}{dt} = m\frac{d^2x}{dt^2} \tag{3.47}$$

$$\Sigma F_y = ma_y = m\frac{dv_y}{dt} = m\frac{d^2y}{dt^2} \tag{3.48}$$

$$\Sigma F_z = ma_z = m\frac{dv_z}{dt} = m\frac{d^2z}{dt^2} \tag{3.49}$$

These three second-order differential equations can be represented as three pairs of first-order differential equations (Equations 3.50 - 3.52):

$$v_x = \frac{dx}{dt}, \qquad \frac{F_x}{m} = \frac{dv_x}{dt} = a_x \tag{3.50}$$

$$v_y = \frac{dy}{dt}, \qquad \frac{F_y}{m} = \frac{dv_y}{dt} = a_y \tag{3.51}$$

$$v_z = \frac{dz}{dt}, \qquad \frac{F_z}{m} = \frac{dv_z}{dt} = a_z \tag{3.52}$$

and solved using Runge-Kutta methods (Equations 3.53 - 3.58).

$$x(t + \Delta t) \sim x(t) + \frac{k_{sx0} + 2k_{sx1} + 2k_{sx2} + k_{sx3}}{6} \tag{3.53}$$

$$v_x(t + \Delta t) \sim v_x(t) + \frac{k_{vx0} + 2k_{vx1} + 2k_{vx2} + k_{vx3}}{6} \tag{3.54}$$

$$y(t + \Delta t) \sim y(t) + \frac{k_{sy0} + 2k_{sy1} + 2k_{sy2} + k_{sy3}}{6} \tag{3.55}$$

$$v_y(t + \Delta t) \sim v_y(t) + \frac{k_{vy0} + 2k_{vy1} + 2k_{vy2} + k_{vy3}}{6} \tag{3.56}$$

$$z(t + \Delta t) \sim z(t) + \frac{k_{sz0} + 2k_{sz1} + 2k_{sz2} + k_{sz3}}{6} \tag{3.57}$$

$$v_z(t + \Delta t) \sim v_z(t) + \frac{k_{vz0} + 2k_{vz1} + 2k_{vz2} + k_{vz3}}{6} \tag{3.58}$$

where

$$k_{sx0} = \Delta t\, v_x(t) \qquad k_{vx0} = \Delta t\, a_x(t, x(t), v_x(t))$$

$$k_{sx1} = \Delta t \left(v_x(t) + \frac{k_{vx0}}{2} \right) \qquad k_{vx1} = \Delta t\, a_x\left(t + \frac{\Delta t}{2}, x(t) + \frac{k_{sx0}}{2}, v_x(t) + \frac{k_{vx0}}{2} \right)$$

$$k_{sx2} = \Delta t \left(v_x(t) + \frac{k_{vx1}}{2} \right) \qquad k_{vx2} = \Delta t\, a_x\left(t + \frac{\Delta t}{2}, x(t) + \frac{k_{sx1}}{2}, v_x(t) + \frac{k_{vx1}}{2} \right)$$

$$k_{sx3} = \Delta t\, (v_x(t) + k_{vx2}) \qquad k_{vx3} = \Delta t\, a_x(t + \Delta t, x(t) + k_{sx2}, v_x(t) + k_{vx2})$$

$$k_{sy0} = \Delta t\, v_y(t) \qquad k_{vy0} = \Delta t\, a_y(t, y(t), v_y(t))$$

etc.

Figure 3.12

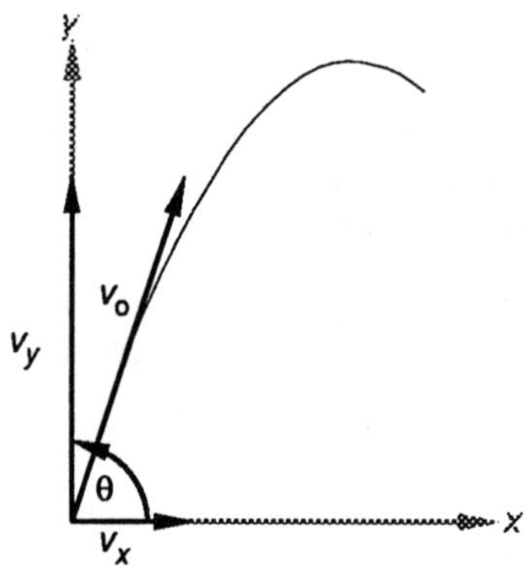

A ball is thrown into the air with an initial speed v_0 and at an angle θ from the horizontal.

The horizontal component of initial velocity is $v_{0x} = v_0 \cos\theta$ and the vertical component of initial velocity is $v_{0y} = v_0 \sin\theta$. $v_z = 0$. Subsequent values of v_x and v_y can be calculated using the Runge-Kutta method. v_z remains zero throughout the ball's motion.

The vertical component of acceleration is simply the downward acceration due to gravity $-g$. No other forces act on the ball once it has been thrown.

LESSONS

A simple example of a three-dimensional trajectory is a ball thrown into a crosswind.

In the absence of a crosswind, the ball describes a two-dimensional path defined by motions in the horizontal (x) and vertical (y) directions (if there is also no air resistance, the motion is described by a parabolic path near the Earth's surface). The ball is thrown into the air with an initial speed v_0 and at an angle θ from the horizontal (Figure 3.12). The three pairs of first-order differential equations that describe this motion are

$$v_{0x} = \frac{dx}{dt} = v_0 \cos\theta, \qquad a_x = \frac{F_x}{m} = \frac{dv_x}{dt} = 0 \tag{3.59}$$

$$v_{0y} = \frac{dy}{dt} = v_0 \sin\theta, \qquad a_y = \frac{F_y}{m} = \frac{dv_y}{dt} = -g \tag{3.60}$$

$$v_{0z} = \frac{dz}{dt} = 0, \qquad a_z = \frac{F_z}{m} = \frac{dv_z}{dt} = 0 \tag{3.61}$$

Because both the initial velocity and acceleration in the z-direction are zero, the z-coordinate remains constant in time and can be ignored.

If a crosswind acts on the ball in a direction normal to the flight of the ball (i.e., the z-direction) the expression for F_z/m in Equation 3.61 must be modified to account for this force. In the lesson that follows, you explore a crosswind that increases with the height of the ball such that Equation 3.61 must be modified to read

$$v_{0z} = \frac{dz}{dt} = 0, \quad a_z = \frac{F_z}{m} = \frac{dv_z}{dt} = \frac{F_0\, e^{y/y_0}}{m} \tag{3.62}$$

where $y_0 = 100$ m and F_0 is a constant equal to the force of the wind at $y = 0$.

1. Build and verify Model 3.7. The model calculates the three components of the position vector s (x,y,z) and the three components of the velocity vector v (v_x,v_y,v_z) as a function of elapsed time. The model is a straightforward extension of the Runge-Kutta methods developed in Exercise 3.2 and represents an example of the compact model described there.

   Study the model and verify that you understand how it is constructed.

2. Begin by setting the force of the crosswind to zero ($F_0 = 0$). Extend the model through enough rows so that the ball strikes the ground, using $\Delta t = .07$ s, $v_0 = 25$ m s^{-1}, $\theta = 30°$, $F_0 = 0$ N, $m = 1$ kg. Generate

SPREADSHEET TECHNIQUE

Correct Geometric Pictures

Most spreadsheet programs are designed to draw pictures without regard to geometric relations in the real world. Sometimes, however, you will want to draw pictures using the same scale for both axes. Here's how.

Measure the lengths of the *x*- and *y*-axes (w_x and w_y, respectively) on your computer graphic display by holding a ruler against the screen. Assuming that your screen is like most screens (i.e., $w_x > w_y$) the following treatment is valid:

- Find a convenient location on your spreadsheet model and prepare formulas that find the largest absolute value of your *x*- and *y*-data. For example, if your *x*-data are arrayed in cells A20 - A120 and your *y*-data are arrayed in cells B20 - B120, then prepare the following formulas:

  x_{max} is =MAX(A20:A120)
  x_{min} is =MIN(A20:A120)
  y_{max} is =MAX(B20:B120)
  y_{min} is =MIN(B20:B120)

  Many spreadsheets permit you to include empty cells in your formulas without incurring additional computational overhead (e.g., =MAX(A20:A7000)). If your spreadsheet permits it, you may wish to use such formulas in this instance to expedite an increase in the number of working rows in your models.

- Write a formula to calculate the largest "spread" of data immediately below the four formulas above.

$$\Delta = \text{MAX}(x_{max} - x_{min}, y_{max} - y_{min})$$

- Above and separate from the *x*-data (but still in column B), prepare the formula

$$\text{Scale-}x = \frac{x_{max} + x_{min}}{2} - \frac{w_x}{2}\frac{\Delta}{w_y}$$

- Above and separate from the y-data (but still in column D), prepare the formula

$$\text{Scale-}y = \frac{y_{max} + y_{min}}{2} - \frac{\Delta}{2}$$

(continued.)

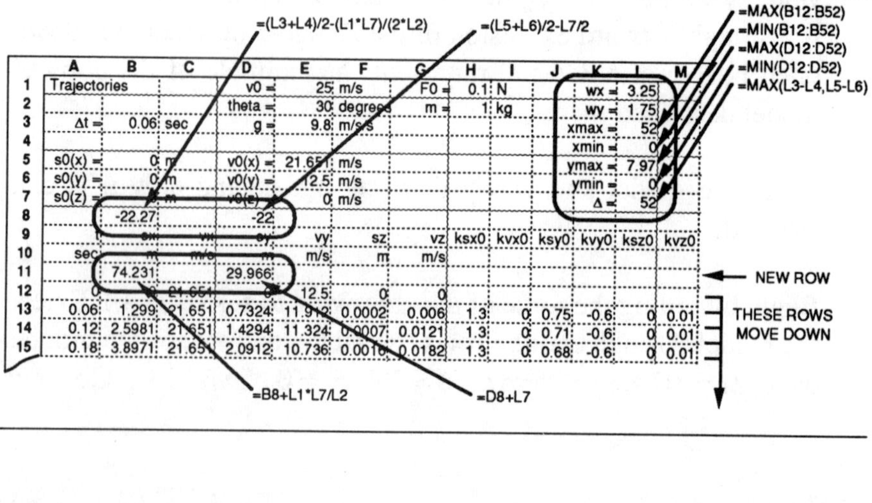

Figure 3.13

The graphics capabilities of electronic spreadsheet programs can be used to display correct geometric pictures of the motions of objects, provided that the same scale is used for both axes. Model 3.7 can be modified to generate such axes by using the methods described in the spreadsheet technique "Correct Geometric Pictures."

SPREADSHEET TECHNIQUE (continued)

- At the head of and included within your *x*-data prepare the formula

$$\text{Head-}x \quad = \quad \text{Scale-}x + \left(\frac{w_x}{w_y}\right)\Delta$$

- At the head of and included within your *y*-data prepare the formula

$$\text{Head-}y \quad = \quad \text{Scale-}y + \Delta$$

When head-*x* and head-*y* are included in your graphics range, your picture will be geometrically correct no matter what your data are. The automatic scaling of your spreadsheet graphics routines may still cause a small amount of distortion, but this technique generally works quite well.

An example of how to use this technique to correctly scale the *x*- and *y*- (but not *z*-) data of Model 3.7 is shown in Figure 3.13.

a plot to follow the movement of the ball in the *x*-*y* and *x*-*z* planes (Figure 3.14). Notice that there is no movement in the *z*-direction.

Explore the behavior of the thrown ball at a number of different angles θ.

a. How far does a ball thrown at 25 m s^{-1} at an angle of 30° travel horizontally before hitting the ground? Can you find another angle at which the ball travels the same distance? (You may need to extend your model by adding rows.)

For angles other than 45°, there are always two trajectory angles that travel the same horizontal distance before striking the ground.

b. Complete and graph (Figure 3.15) the table below to explore the maximum horizontal distance that a ball can travel when thrown at an initial velocity v_0 of 25 m s^{-1} but at different angles.

θ (deg)	s_x (max) (m)
0	______
15	______
30	55.2
45	______
60	55.2
75	______
90	______

Figure 3.14

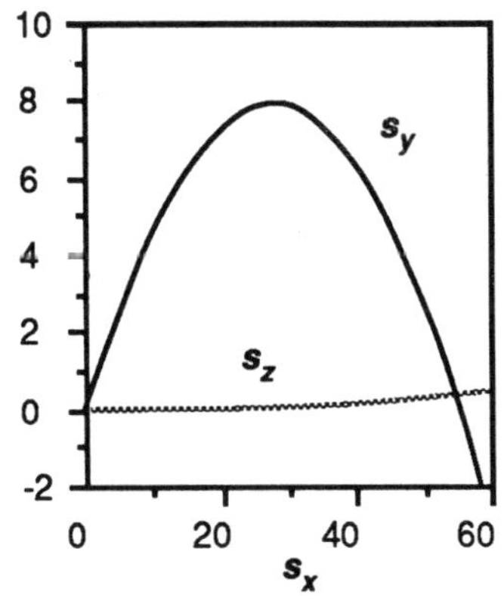

XY-PLOT

Data
Horizontal (*x*) axis: cells B11 - B57 (s_x)
Vertical (*y*) axis: cells D11 - D57 (s_y) and F11 - F57 (s_z)

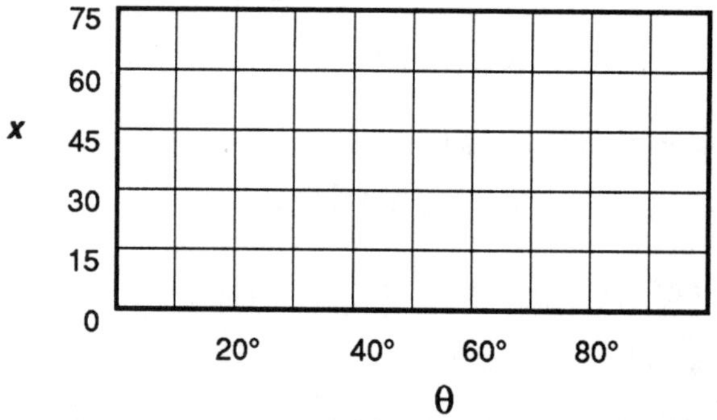

Figure 3.15

Horizontal distance x vs initial angle θ for $v_0 = 25$ m s⁻¹.

c. Explore the influence of initial velocity v_0 on the ball's trajectory by determining the initial conditions that produce a maximum height of 1000 m for the throwing angles of 20°, 40°, and 70° from the horizontal.

3. Introduce a small crosswind into your model by setting $F_0 = 0.1$. What are the initial conditions that will cause a 1-kg sphere to land $s_x = +50$ m straight ahead and $s_z = +10$ m to the right of the x-axis.

PROBLEMS

1. When an object moves through air, it experiences a drag force opposing its motion. Thus, the drag force acts in a direction opposite to the object's vector velocity. For a roughly spherical object a few centimeters or more in radius, a ball for example, the magnitude of this drag force is approximately

$$F_d = \varepsilon\, \rho_{air}\, v^2 A \tag{3.63}$$

where ε is a dimensionless factor that depends on the ball's shape, ρ_{air} is the density of the air, v is the relative velocity beween the ball and the air, and A is the frontal area the ball presents to the air. The factor $\varepsilon \cong 0.5$ for a ball, $A = \pi r^2$ (where r is the ball's radius), and $\rho_{air} = 0.0011$ g cm^{-3} (1.1 kg m^{-3}).

If there is no wind and the ball's velocity with respect to the Earth is $\boldsymbol{v} = v_x\boldsymbol{i} + v_y\boldsymbol{j}$ then the drag force is

$$\boldsymbol{F}_{drag}^{no\ wind} = -F_d\left(\frac{v_x\boldsymbol{i} + v_y\boldsymbol{j}}{v}\right)$$

$$= -\varepsilon\, \rho_{air} \sqrt{v_x^2 + v_y^2} \cdot A\, (v_x\boldsymbol{i} + v_y\boldsymbol{j}) \tag{3.64}$$

On the other hand, if there is a wind opposing the motion (in two dimensions)with a velocity relative to the Earth of $v_{wind} = v_x^w \boldsymbol{i} + v_y^w \boldsymbol{j}$, then the drag on the ball becomes

$$\boldsymbol{F}_{drag}^{wind} = -\varepsilon \rho_{air} \sqrt{(v_x - v_x^w)^2 + (v_y - v_y^w)^2}\, A\,((v_x - v_x^w)\boldsymbol{i} + (v_y - v_y^w)\boldsymbol{j}) \tag{3.65}$$

a. Build a two-dimensional Runge-Kutta spreadsheet model that includes the drag force given by Equation 3.64 with no wind. Use the values for a baseball ($m = 0.15$ kg, $r = 0.035$ m) thrown with an initial velocity $v_0 = 25$ m s^{-1}. Complete the baseball column of the table below.

	Maximum Horizontal Distance			
θ	**Baseball**	**Golf Ball**	**Lead Ball**	**No Air**
10	______	______	______	______
15	______	______	______	______
30	______	______	______	______
45	______	______	______	______
60	______	______	______	______
75	______	______	______	______
90	______	______	______	______

b. Now consider a thrown golf ball of $m = 0.06$ kg and $r = 0.0175$ m. Complete the golf ball column of the table.

c. A lead ball of the same radius as the golf ball ($r = 0.0175$ m) has a mass of 0.26 kg. Complete the table for the lead ball.

d. With no air density (a vacuum), complete the table in the no-air column for a baseball.

2. Modify Problem 1 above for the different objects such that they experience

a. Head winds of 10 km h^{-1} and 40 km h^{-1}.

b. Tail winds of 10 km h^{-1} and 40 km h^{-1}.

3. A ball is launched horizontally from a tall tower at an initial velocity of 5 m s^{-1}. What are the x-, y-, and z-coordinates of the ball during its first 10 s of motion. Assume that the tower is located at (0,0,0) and there is no drag force and no wind. Set $\Delta t = 0.1$ s.

a. Plot y vs x.

b. Suppose that you wish to launch the ball horizontally such that the ball drops at only 3×10^{-5} m for each meter of horizontal movement. What is the initial velocity of the ball?

4. A quarterback throws a football ($m = 0.2$ kg) into the air with an initial speed of 20 m s^{-1}. It arrives at the receiver 30 m away at the same height above the ground. Assume no air effects initially.

   a. At what angle to the horizontal did the quarterback throw the football?

   b. What range of angles will deliver the ball to within 0.5 m of the receiver along the direct line from quarterback to receiver?

   c. Suppose that air friction opposes the motion of the football with a force

$$f_a = 0.02\, v \tag{3.66}$$

   where v is the velocity in m s^{-1}. Now what range of angles will deliver the ball to within 0.5 m of the receiver?

5. A water droplet of mass 10^{-6} kg departs a lawn sprinkler with an initial velocity of 30 m s^{-1} at an angle of 65° from the horizontal.

   a. Where does it hit the ground (assuming that air effects can be neglected)?

   b. The droplet has an initial radius r of 0.1 cm and evaporates at the rate

$$\frac{dr}{dt} = 3 \times 10^{-6}\, v \tag{3.67}$$

   where v is the velocity in m s^{-1}. Does the droplet reach its final destination?

   c. Suppose that the frictional resistance of the air retards the motion of the droplet with a force

$$f_R = 6\pi\eta_{air}\, r\, v \tag{3.68}$$

where $\eta_{air} = 190 \times 10^{-6}$ g cm^{-1} s^{-1}, r = droplet radius (cm) and v = velocity (m s^{-1}) and that the droplet evaporates according to Equation 3.67. How long will the droplet last, and where will it be when it disappears?

d. What is the maximum horizontal distance attainable for the water droplet given air resistance and evaporation?

MODEL

Model 3.7

BUILD and verify Model 3.7.

ENTER data using cells E1 - E3 (initial velocity, initial angle of throw, acceleration of gravity), H1 and H2 (minimum force of crosswind, mass of projectile), B3 (Δt), and B5 - B7 (x-, y-, and z-coordinates of the projectile's starting position).

READ output from the model from columns A (time) and B - G (three-dimensional position of projectile with time).

NOTICE that this model is designed in compact form to save space.

FORMULAS

	A	B	C	D	E	F	G	H	I	J	K	L	M	N
1	Trajectories			v0 =	25	m/s	F0 =	0.1	N					
2				theta =	30	degrees	m =	1	kg					
3	Δt =	0.06	sec	g =	9.8	m/s/s								
4														
5	s0(x) =	0	m	v0(x) =	=E1*COS(E2*(PI()/180))									
6	s0(y) =	0	m	v0(y) =	=E1*SIN(E2*(PI()/180))									
7	s0(z) =	0	m	v0(z) =	0	m/s								
8														
9	t	sx	vx	sy	vy	sz	vz	ksx0	kvx0	ksy0	kvy0	ksz0	kvz0	ksx1
10	sec	m	m/s	m	m/s	m	m/s							
11	0	=B5	=E5	=B6	=E6	=B7	=E7							
12	Form Set I	Form Set II	Form Set III	Form Set IV	Form Set V	Form Set VI	Form Set VII	Form Set VIII	0	Form Set IX	Form Set X	Form Set XI	Form Set XII	Form Set XIII
13									0					
14									0					

FORMULAS (Contd.)

	O	P	Q	R	S	T	U	V	W	X	Y	Z	AA	AB	AC	AD	AE
9	kvx1	ksy1	kvy1	ksz1	kvz1	ksx2	kvx2	ksy2	kvy2	ksz2	kvz2	ksx3	kvx3	ksy3	kvy3	ksz3	kvz3
10																	
11																	
12	0	Form Set XIV	Form Set XV	Form Set XVI	Form Set XVII	Form Set XVIII	0	Form Set XIX	Form Set XX	Form Set XXI	Form Set XXII	Form Set XXIII	0	Form Set XXIV	Form Set XXV	Form Set XXVI	Form Set XXVII
13	0						0						0				
14	0						0						0				

VALUES

	A	B	C	D	E	F	G	H	I	J	K	L	M	N
1	Trajectories			v0 =	25	m/s	F0 =	0.1	N					
2				theta =	30	degrees	m =	1	kg					
3	Δt =	0.06	sec	g =	9.8	m/s/s								
4														
5	s0(x) =	0	m	v0(x) =	21.651	m/s								
6	s0(y) =	0	m	v0(y) =	12.5	m/s								
7	s0(z) =	0	m	v0(z) =	0	m/s								
8														
9	t	sx	vx	sy	vy	sz	vz	ksx0	kvx0	ksy0	kvy0	ksz0	kvz0	ksx1
10	sec	m	m/s	m	m/s	m	m/s							
11	0	0	21.651	0	12.5	0	0							
12	0.06	1.299	21.651	0.7324	11.912	0.0002	0.006	1.3	0	0.75	-0.6	0	0.01	1.3
13	0.12	2.5981	21.651	1.4294	11.324	0.0007	0.0121	1.3	0	0.71	-0.6	0	0.01	1.3
14	0.18	3.8971	21.651	2.0912	10.736	0.0016	0.0182	1.3	0	0.68	-0.6	0	0.01	1.3

VALUES (Contd.)

	O	P	Q	R	S	T	U	V	W	X	Y	Z	AA	AB	AC	AD	AE
9	kvx1	ksy1	kvy1	ksz1	kvz1	ksx2	kvx2	ksy2	kvy2	ksz2	kvz2	ksx3	kvx3	ksy3	kvy3	ksz3	kvz3
10																	
11																	
12	0	0.73	-0.6	0	0.01	1.3	0	0.73	-0.6	0	0.01	1.3	0	0.71	-0.6	0	0.01
13	0	0.7	-0.6	0	0.01	1.3	0	0.7	-0.6	0	0.01	1.3	0	0.68	-0.6	0	0.01
14	0	0.66	-0.6	0	0.01	1.3	0	0.66	-0.6	0	0.01	1.3	0	0.64	-0.6	0	0.01

Model 3.7 (continued)

Formula Set I	Prototype Cell is A12	=A11+**B3**
Formula Set II	Prototype Cell is B12	=B11+(H12+2*N12+2*T12+Z12)/6
Formula Set III	Prototype Cell is C12	=C11+(I12+2*O12+2*U12+AA12)/6
Formula Set IV	Prototype Cell is D12	=D11+(J12+2*P12+2*V12+AB12)/6
Formula Set V	Prototype Cell is E12	=E11+(K12+2*Q12+2*W12+AC12)/6
Formula Set VI	Prototype Cell is F12	=F11+(L12+2*R12+2*X12+AD12)/6
Formula Set VII	Prototype Cell is G12	=G11+(M12+2*S12+2*Y12+AE12)/6
Formula Set VIII	Prototype Cell is H12	=**B3***C11
Formula Set IX	Prototype Cell is J12	=**B3***E11
Formula Set X	Prototype Cell is K12	=**B3***(-**E3**)
Formula Set XI	Prototype Cell is L12	=**B3***G11
Formula Set XII	Prototype Cell is M12	=**B3***(**H1***EXP(D11/100))/**H2**
Formula Set XIII	Prototype Cell is N12	=**B3***(C11+I12/2)
Formula Set XIV	Prototype Cell is P12	=**B3***(E11+K12/2)
Formula Set XV	Prototype Cell is Q12	=**B3***(-**E3**)
Formula Set XVI	Prototype Cell is R12	=**B3***(G11+M12/2)
Formula Set XVII	Prototype Cell is S12	=**B3***(**H1***EXP((D11+J12/2)/100))/**H2**
Formula Set XVIII	Prototype Cell is T12	=**B3***(C11+O12/2)
Formula Set XIX	Prototype Cell is V12	=**B3***(E11+Q12/2)
Formula Set XX	Prototype Cell is W12	=**B3***(-**E3**)
Formula Set XXI	Prototype Cell is X12	=**B3***(G11+S12/2)
Formula Set XXII	Prototype Cell is Y12	=**B3***(**H1***EXP((D11+P12/2)/100))/**H2**
Formula Set XXIII	Prototype Cell is Z12	=**B3***(C11+U12)
Formula Set XXIV	Prototype Cell is AB12	=**B3***(E11+W12)
Formula Set XXV	Prototype Cell is AC12	=**B3***(-**E3**)
Formula Set XXVI	Prototype Cell is AD12	=**B3***(G11+Y12)
Formula Set XXVII	Prototype Cell is AE12	=**B3***(**H1***EXP((D11+V12)/100))/**H2**

EXERCISE **3.4**

ORBITS

In Exercise 3.3, you explored the motion of an object subject to a constant gravitational force (i.e., constant both in terms of magnitude and direction) and discovered that the trajectory of such an object is a parabola when there is no air resistance to the motion (Figure 3.16) or a simple, straight line. Although the assumption of a constant gravitational force is valid for objects traveling short distances close to the Earth's surface, the assumption is invalid once the size of the trajectory approaches that of the radius of the Earth. For such large trajectories,

- The direction of the gravitational force is always toward the center of the Earth.

- The magnitude of the gravitational force decreases with the square of the distance from the center of the Earth.

Under these conditions, the object's path may be a parabola, hyperbola, ellipse, straight line, or circle (Figure 3.16).

It is convenient to study two bodies (of mass m_1 and m_2) in orbit about each other in terms of two new vectors $\boldsymbol{r}$ and $\boldsymbol{R}$ (Figure 3.17, Equations 3.69 and 3.70).

TO GET THE MOST OUT OF THIS EXERCISE, YOU SHOULD ALREADY KNOW

Runge-Kutta solutions for three-dimensional motion (Exercise 3.3)

Figure 3.16

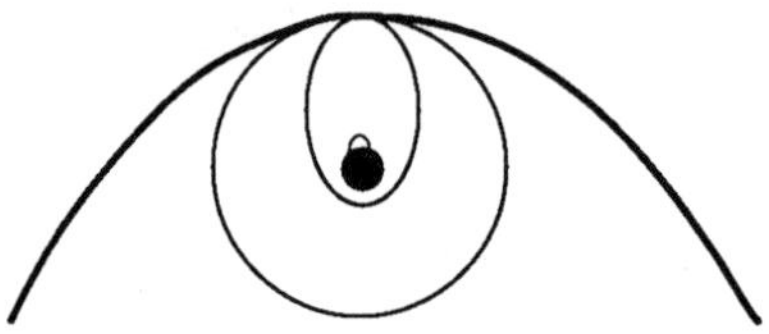

When a trajectory approaches the size of the radius of the Earth, the force of gravity is no longer a constant and a number of nonparabolic paths are possible.

$$r = r_2 - r_1 \tag{3.69}$$

$$R = \frac{m_2 r_2 + m_1 r_1}{m_1 + m_2} \tag{3.70}$$

The vector $\boldsymbol{r}$ is the relative position of mass m_2 with respect to mass m_1, and is called the *relative coordinate vector*. The vector $\boldsymbol{R}$ is called the *center of mass vector* and it extends from an inertial coordinate system origin to a point on the line joining the two masses. This point is called the center of mass of the system; it always lies nearer the larger of the two masses. You can express the coordinates of the two masses in terms of $\boldsymbol{R}$ and $\boldsymbol{r}$ as

$$r_1 = R - \left(\frac{m_1}{m_1 + m_2}\right) r \tag{3.71}$$

$$r_2 = R - \left(\frac{m_2}{m_1 + m_2}\right) r \tag{3.72}$$

When the only forces acting on these bodies are their equal and opposite mutual gravitational attractions, it can be shown that these new vectors satisfy the equations of motion

$$M \frac{d^2 R}{dt^2} = 0 \tag{3.73}$$

$$\mu \frac{d^2 r}{dt^2} = F = \frac{-G m_1 m_2}{r^3} r \tag{3.74}$$

where $\mu = \frac{m_1 m_2}{m_1 + m_2}$, called the *reduced mass*.

$\boldsymbol{F}$ is the force of m_1 on m_2 along the line connecting their centers (the vector $\boldsymbol{r}$), and $G = 6.670 \times 10^{-11}$ N m^2 kg^{-2} (the gravitational constant).

Figure 3.17

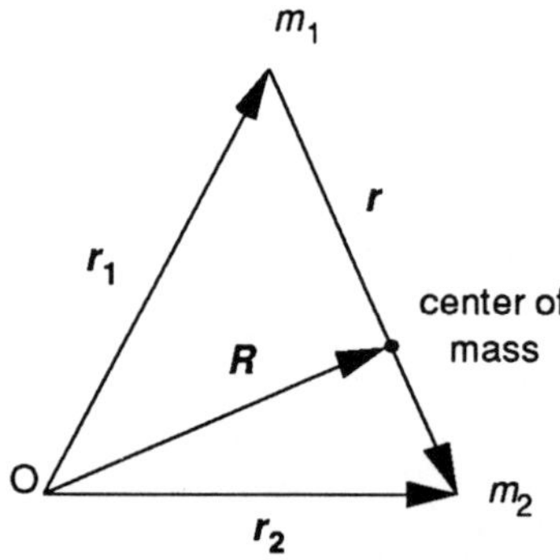

Two bodies of mass m_1 and m_2 in orbit about each other can be studied by introducing a change in coordinates such that two new vectors $\boldsymbol{r}$ and $\boldsymbol{R}$ replace r_1 and r_2. Prudence suggests that the mass m_1 be the most massive (i.e., the fixed point in the system).

You see that the center of mass of the system moves as a point mass M with no forces acting on it. Therefore it moves with a constant velocity relative to an inertial coordinate system and in fact you can always choose to use the coordinate system in which it is at rest as your inertial reference system. For convenience, we will do that in this exercise, and furthermore, we will take the center of mass to be the origin. In this coordinate system, you have

$$R = 0 \tag{3.75}$$

$$r_1 = \left(\frac{-m_2}{M}\right) r \tag{3.76}$$

$$r_2 = \left(\frac{-m_1}{M}\right) r \tag{3.77}$$

When m_1 is very much larger than m_2 (m_1 could be the Earth, for example, and m_2 the Moon), r_2 is almost equal to r, and r_1 is very small compared to r. Thus you can think of m_2 orbiting around m_1. In fact, of course, both masses always orbit around their common center of mass.

Thus, the equation of motion of the relative coordinate vector $\boldsymbol{r}$ can be written

$$\frac{d^2\boldsymbol{r}}{dt^2} = \frac{\boldsymbol{F}}{\mu} = \frac{-GM}{(x^2 + y^2 + z^2)^{3/2}}\boldsymbol{r} \tag{3.78}$$

where $M = m_1 + m_2$, the total mass of the system.

Equation 3.78 can be resolved into its vector components via any number of suitable coordinate systems and for analytical work an

SPREADSHEET TECHNIQUE

Self-Adjusting Time Intervals

In some orbits, an object may move rapidly sometimes and slowly at other times — highly elliptic orbits, for example. Runge-Kutta methods that use constant values of Δt are often inefficient in this situation because Δt values that generate enough points to accurately map rapid motion will hardly show any motion at all when the object is moving very slowly. You can sometimes dramatically increase the efficiency of your Runge-Kutta models (e.g., decrease the number of rows required to fully describe an orbit) by adding a Δt column to your model and computing Δt such that $\Delta t = \Delta s/|v|$, for a constant value of Δs.

- For a two-dimensional Runge-Kutta model, compute Δt by using the expression

$$\Delta t = \frac{\Delta s}{\sqrt{v_x^2 + v_y^2}}$$

- For a three-dimensional Runge-Kutta model, compute Δt by using the expression

$$\Delta t = \frac{\Delta s}{\sqrt{v_x^2 + v_y^2 + v_z^2}}$$

	A	B	C	D
1		Orbits		
2				
3		Δs =	20000000	
4				
5		s0x =	4E+08	m
6		s0y =	0	m
7		s0z =	0	m
8				
9	Δt	t	sx	vx
10		sec	m	m/s
11	33333.33	0	4E+08	0
12	33123.98	33123.98	3.99E+08	-84.2086
13	32517.18	65641.17	3.94E+08	-168.435
14	31577.44	97218.6	3.88E+08	-252.19

$$\frac{\Delta s}{\sqrt{x^2 + y^2 + z^2}}$$

You must also be certain to write suitable formulas within your model — the Δt within the Runge-Kutta quantities, for example.

Model 3.9 describes the Runga-Kutta solution for a two-dimensional orbit in which both the time intervals and the aspect ratios used for plotting the orbit (see spreadsheet technique "Correct Geometric Pictures" in Exercise 3.3) are corrected.

enlightened choice of coordinate system can dramatically simplify your work. The simplest choice is a Cartesian coordinate system in which the force components are resolved along the familiar *x*-, *y*-, and *z*-axes (Equations 3.79 - 3.81).

$$\frac{d^2x}{dt^2} = \frac{-\,G\,M\,x}{(x^2 + y^2 + z^2)^{3/2}} \qquad (3.79)$$

$$\frac{d^2y}{dt^2} = \frac{-\,G\,M\,y}{(x^2 + y^2 + z^2)^{3/2}} \qquad (3.80)$$

$$\frac{d^2z}{dt^2} = \frac{-\,G\,M\,z}{(x^2 + y^2 + z^2)^{3/2}} \qquad (3.81)$$

For orbital motions you can choose to ignore the *z*-component. This simplification is very reasonable for two orbiting bodies because, for any given initial value of ***r*** and ***v***, the *z*-axis can be chosen such that it is perpendicular to the plane defined by ***r*** and ***v***. There is then neither an initial force nor an initial velocity in the *z*-direction and so the *z*-coordinate stays constant (see Equation 3.81).

LESSONS

1. Modify and verify Model 3.7 of Exercise 3.3 such that the *x*-, *y*-, and *z*-components of acceleration are represented by Equations 3.79 - 3.81. Notice that the *z*-component of acceleration is zero and may be eliminated if you wish. These modifications are reflected in Model 3.8. As an alternative, build Model 3.9. Model 3.9 conserves space by ignoring the *z*-axis and also corrects aspect ratios (see spreadsheet technique, "Correct Geometrical Pictures," Exercise 3.3) and uses self-adjusting time intervals (See spreadsheet technique "Self-Adjusting Time Intervals," this exercise).

   Begin by exploring the orbit of the Moon about the Earth at a distance of 4×10^8 m. The moon's tangential velocity is 1000 m s^{-1} (the easiest way to do this is to set $v_y = 1000$ m s^{-1} when $x = 4 \times 10^8$ m and $y = z = 0$). Note that, unless the aspect ratio of your on-screen graphic display is 1:1, the precise shape of your orbit will be incorrect. See the spreadsheet technique in Exercise 3.3 for a correction to this problem.

   Plot the *x*- and *y*- coordinates of Model 3.8 or 3.9 (Figure 3.18).

   a. Extend your model to include enough rows so that at least a complete, single orbit is possible. Does the orbit close?

   b. It is not at all obvious from Equations 3.79 - 3.81 that an inverse square law for attractive force should yield a closed orbit for the

Figure 3.18

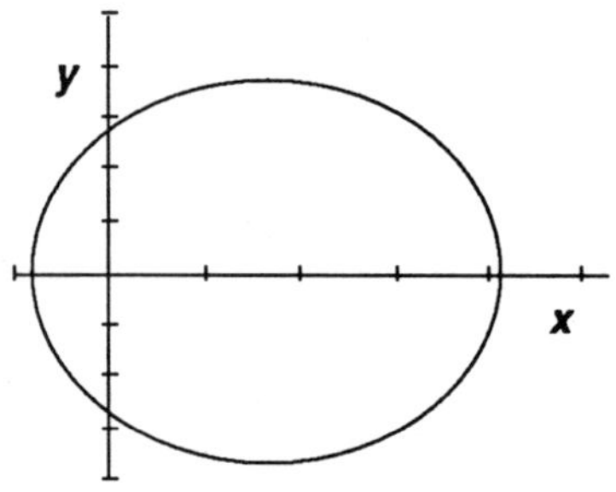

XY-PLOT

Data (using Model 3.8)
Horizontal (*x*) axis: cells B11 - B110
Vertical (*y*) axis: cells D11 - D110

Data (using Model 3.9)
Horizontal (*x*) axis: cells C11 - C110
Vertical (*y*) axis: cells E11 - E110

given initial conditions.[1] Nevertheless, it is a fact that under the conditions given the orbit of the Moon should loop back upon itself precisely.

Although Runge-Kutta methods are extremely accurate, you will find that thc initialization values of Model 3.8 or 3.9 cause the orbit to be slightly displaced after a single cycle. Modify the value of Δt until the model approximates a closed loop and returns to its starting position to at least 1 part in a 1000.

2. Your model was initialized to represent the Moon (m_2) orbiting about the Earth such that m_2 begins its orbit at the x-, y-, and z-coordinates (4×10^8 m, 0 m, 0 m) with an initial speed of 1000 m s^{-1}. Given the actual Earth-Moon distance of 384,000 km, you could easily calculate a good approximation of the Moon's actual speed on your spreadsheet. Remember that it takes the Moon about 28 days to orbit the Earth.

    a. What changes do you notice in the orbit if speed is decreased to 800 m s^{-1}. Now reduce the speed to 400 m s^{-1} and notice that you must reduce Δt to get a sensible orbit. The fast motion around day five is the source of the problem.

    b. The Moon's mass m_2 is about 1% of the Earth's mass. Spacecraft and satellites have miniscule masses compared to these two giants. If the Moon were actually a spacecraft of 50,000 kg traveling at a speed of 800 m s^{-1}, what would happen to the orbit? Why?

3. Explore the orbit of a satellite at a distance of 500 km above the surface of the Earth. Ignore the effect of the Moon. Because the radius of the Earth equals 6.38×10^3 km, this represents a distance of 6.88×10^3 km from the Earth's center of mass. Use $\Delta t = 120$ s (2 min).

    a. Determine by trial and error the initial velocity required to place the satellite in a circular orbit. To help, you should know that Sputnik had a period of about 100 min (6000 s).

    b. What happens to the orbit if the spacecraft were subjected to an impulsive rocket boost inward along the radial direction such that the initial value of $v_x = -5000$ m s^{-1}. Explore other initial radial velocities. Make a table of periods.

PROBLEMS

1. A satellite with a mass of 2000 kg travels with a tangential speed of 1000 m s^{-1} at a distance of 0.384×10^9 m (the distance from the Earth's center to the moon). $G = 6.670 \times 10^{-11}$ N m^2 kg^{-2}, and the mass of the Earth = 5.98×10^{24} kg.

    a. Is the orbit circular?

    You can establish circularity by calculating radial distance as a function of time or by noticing the values of x_{max}, x_{min}, y_{max}, and y_{min}. Alternatively, you can calculate the radial velocity, $\boldsymbol{v} \cdot \boldsymbol{r}/r$, which is zero for a circular orbit. If the orbit is not circular, change the initial tangential speed to make it so.

    b. What is the orbital period (in days)?

2. The satellite in Problem 1 must be brought back to Earth. Beginning at the orbital radius of the Moon (0.384×10^9 m), you wish to bring it to the Earth's radius (6.37×10^6 m).

    a. Give the satellite a *forward* or *backward* boost such that it will approach within 5 Earth radii on the next pass. The boost can be accomplished by suddenly changing the velocity. Assume that the satellite mass changes by a negligible amount.

    How long does it take to get a radius of five Earth radii, and what is the velocity then?

    b. Give the satellite an *inward* boost such that it will collide with the planet.

3. A satellite in need of repair is in a circular orbit 500 km above the Earth's surface. A shuttle of mass 50,000 kg is now in a circular orbit 450 km above the surface and will attempt to rendezvous with the satellite at *x*-, *y*-, and *z*-coordinates (6.884×10^3 km, 0 km, 0 km) by firing its engines. The thrust can be mimicked by changing the velocity suddenly. Several bursts may be necessary.

    Pilot your craft to the rendezvous. Assume that insignificant amounts of mass will be expended during engine firing.

4. Place a spacecraft of mass 50,000 kg into any elliptic orbit with a closest approach of 7.0×10^6 km from the Earth's center.

a. Near apogee (farthest distance), an engine burn accelerates the spacecraft tangential to the orbit for a very small time interval. Compare the new orbit to the old orbit.

b. Near perigee (closest distance), an engine burn decelerates the spacecraft tangential to the orbit. Describe the new orbit.

c. Maneuver the spacecraft such that it escapes Earth orbit.

5. Launch a spacecraft with an initial mass of $m_0 = 10^5$ kg from a nonspinning Earth and attempt to achieve an orbit about 100 miles above its surface. Your spacecraft has both a main engine and an auxiliary engine. The mass loss due to firing the auxiliary engine is negligible but when the main engine is firing, the mass of the spacecraft changes significantly.

   If you wish, assume that the main engine is used only for the initial vertical ascent. The equation of motion during this phase of the flight is

$$m(t)\frac{d^2r}{dt^2} = -m(t)\,g\left(\frac{R_E}{r}\right)^2 + v_g\left(\frac{dm}{dt}\right) \tag{3.82}$$

   where $m(t) = m_0 - (dm/dt)t$ is the decreasing mass of the spacecraft, $dm/dt = 10^3$ kg s^{-1} is the constant rate of mass loss, $v_g = 2400$ m s^{-1} is the propulsion gas exhaust velocity (relative to the spacecraft), $R_E = (2 \times 10^7)/\pi$ m is the radius of the Earth, and $g = 9.8$ m s^{-2} is the acceleration of gravity at the surface of the Earth. Let the main engine burn until the final mass of the spacecraft is $0.1\,m_0$.

   After the burn of the main engine, assume that the auxiliary engine may be pointed in any direction and can produce any constant thrust for as long as you wish (an unrealistic assumption, of course). Design the use of your auxiliary engine to minimze the product of its thrust times its burning time. You may wish to redirect the thrust of the main engine [the $v_g\,(dm/dt)$ term] during the ascent to reduce the requirements on the auxiliary engine.

6. Find a closed orbit for a mass subject to the radial force:

$$F = -\,k_1 r - \frac{k_2}{r^2} \tag{3.83}$$

   where $k_1, k_2 > 0$.

   Initially choose $k_1 = 10$ N m^{-1} and $k_2 = 8$ N m^{-2}. Adjust the values k_1 and k_2 until a closed orbit is achieved.

Model 3.8

BUILD and verify Model 3.8.

ENTER data using cells B3 (Δt), B5 - B7 (x-, y-, z-initial position coordinates), E5 - E7 (x-, y-, z-initial velocities), and E1 - E3 (gravitational constant and the masses of the two bodies).

READ output from the model from columns A (time) and B - G (relative coordinate vector ***r*** with time).

NOTICE that this model is designed in compact form to save space.

MODELS

FORMULAS

	A	B	C	D	E	F	G	H	I	J	K	L	M	N	O
1	Orbits			G =	6.67E-11	Nm^2/kg^2									
2				m1 =	5.98E+24	kg	M =	=E2+E3							
3	Δt =	20000	sec	m2 =	7.34E+22	kg	-GM =	=-E1*H2							
4															
5	s0x =	4E+08	m	v0x =	0	m/s									
6	s0y =	0	m	v0y =	600	m/s									
7	s0z =	0	m	v0z =	0	m/s									
8															
9	t	sx	vx	sy	vy	sz	vz	ksx0	kvx0	ksy0	kvy0	ksz0	kvz0	ksx1	kvx1
10	sec	m	m/s	m	m/s	m	m/s								
11	0	=B5	=E5	=B6	=E6	=B7	=E7								
12–14	Form Set I	Form Set II	Form Set III	Form Set IV	Form Set V	Form Set VI	Form Set VII	Form Set VIII	Form Set IX	Form Set X	Form Set XI	Form Set XII	Form Set XIII	Form Set XIV	Form Set XV

FORMULAS (Contd.)

	P	Q	R	S	T	U	V	W	X	Y	Z	AA	AB	AC	AD	AE
9	ksy1	kvy1	ksz1	kvz1	ksx2	kvx2	ksy2	kvy2	ksz2	kvz2	ksx3	kvx3	ksy3	kvy3	ksz3	kvz3
10																
11																
12–14	Form Set XVI	Form Set XVII	Form Set XVIII	Form Set XIX	Form Set XX	Form Set XXI	Form Set XXII	Form Set XXIII	Form Set XXIV	Form Set XXV	Form Set XXVI	Form Set XXVII	Form Set XXVIII	Form Set XXIX	Form Set XXX	Form Set XXXI

Model 3.8 (continued)

VALUES

	A	B	C	D	E	F	G	H	I	J	K	L	M	N	O
1	Orbits			G =	6.67E-11	Nm^2/kg^2									
2				m1 =	5.98E+24	kg	M =	6E+24							
3	Δt =	20000	sec	m2 =	7.34E+22	kg	−GM =	-4E+14							
4															
5	s0x =	4E+08	m	v0x =	0	m/s									
6	s0y =	0	m	v0y =	600	m/s									
7	s0z =	0	m	v0z =	0	m/s									
8															
9	t	sx	vx	sy	vy	sz	vz	ksx0	kvx0	ksy0	kvy0	ksz0	kvz0	ksx1	kvx1
10	sec	m	m/s	m	m/s	m	m/s								
11	0	4E+08	0	0	600	0	0								
12	20000	3.99E+08	-50.49	11994950	599.2422	0	0	0	-50.47	1E+07	0	0	0	-5E+05	-50.45
13	40000	3.98E+08	-101.099	23959526	596.9595	0	0	-1E+06	-50.53	1E+07	-1.517	0	0	-2E+06	-50.57
14	60000	3.95E+08	-151.946	35862984	593.1242	0	0	-2E+06	-50.71	1E+07	-3.053	0	0	-3E+06	-50.81

VALUES (Contd.)

	P	Q	R	S	T	U	V	W	X	Y	Z	AA	AB	AC	AD	AE
9	ksy1	kvy1	ksz1	kvz1	ksx2	kvx2	ksy2	kvy2	ksz2	kvz2	ksx3	kvx3	ksy3	kvy3	ksz3	kvz3
10																
11																
12	1E+07	-0.757	0	0	-5E+05	-50.52	1E+07	-0.758	0	0	-1E+06	-50.53	1E+07	-1.517	0	0
13	1E+07	-2.28	0	0	-2E+06	-50.64	1E+07	-2.283	0	0	-2E+06	-50.71	1E+07	-3.052	0	0
14	1E+07	-3.831	0	0	-3E+06	-50.87	1E+07	-3.836	0	0	-3E+06	-51.01	1E+07	-4.626	0	0

Model 3.8 (continued)

Formula Set I
Prototype Cell is A12
=A11+**B3**

Formula Set II
Prototype Cell is B12
=B11+(H12+2*N12+2*T12+Z12)/6

Formula Set III
Prototype Cell is C12
=C11+(I12+2*O12+2*U12+AA12)/6

Formula Set IV
Prototype Cell is D12
=D11+(J12+2*P12+2*V12+AB12)/6

Formula Set V
Prototype Cell is E12
=E11+(K12+2*Q12+2*W12+AC12)/6

Formula Set VI
Prototype Cell is F12
=F11+(L12+2*R12+2*X12+AD12)/6

Formula Set VII
Prototype Cell is G12
=G11+(M12+2*S12+2*Y12+AE12)/6

Formula Set VIII
Prototype Cell is H12
=**B3***C11

Formula Set XVI
Prototype Cell is P12
=**B3***(E11+K12/2)

Formula Set XVII
Prototype Cell is Q12
=**B3***H3***(D11+J12/2)/((B11+H12/2)^2+(D11+J12/2)^2+(F11+L12/2)^2)^1.5

Formula Set XVIII
Prototype Cell is R12
=**B3***(G11+M12/2)

Formula Set XIX
Prototype Cell is S12
=**B3***H3***(F11+L12/2)/((B11+H12/2)^2+(D11+J12/2)^2+(F11+L12/2)^2)^1.5

Formula Set XX
Prototype Cell is T12
=**B3***(C11+O12/2)

Formula Set XXI
Prototype Cell is U12
=**B3***H3***(B11+N12/2)/((B11+N12/2)^2+(D11+P12/2)^2+(F11+R12/2)^2)^1.5

Formula Set XXII
Prototype Cell is V12
=**B3***(E11+Q12/2)

Formula Set XXIII
Prototype Cell is W12
=**B3***H3***(D11+P12/2)/((B11+N12/2)^2+(D11+P12/2)^2+(F11+R12/2)^2)^1.5

Model 3.8 (continued)

Formula Set IX Prototype Cell is I12
=**B3*****H3***B11/(B11^2+D11^2+F11^2)^1.5

Formula Set X Prototype Cell is J12
=**B3***E11

Formula Set XI Prototype Cell is K12
=**B3*****H3***D11/(B11^2+D11^2+F11^2)^1.5

Formula Set XII Prototype Cell is L12
=**B3***G11

Formula Set XIII Prototype Cell is M12
=**B3*****H3***F11/(B11^2+D11^2+F11^2)^1.5

Formula Set XIV Prototype Cell is N12
=**B3***(C11+I12/2)

Formula Set XV Prototype Cell is O12
=**B3*****H3***(B11+H12/2)/((B11+H12/2)^2+(D11+J12/2)^2+(F11+L12/2)^2)^1.5

Formula Set XIV Prototype Cell is X12
=**B3***(G11+S12/2)

Formula Set XXV Prototype Cell is Y12
=**B3*****H3***(F11+R12/2)/((B11+N12/2)^2+(D11+P12/2)^2+(F11+R12/2)^2)^1.5

Formula Set XXVI Prototype Cell is Z12
=**B3***(C11+U12)

Formula Set XXVII Prototype Cell is AA12
=**B3*****H3***(B11+T12)/((B11+T12)^2+(D11+V12)^2+(F11+X12)^2)^1.5

Formula Set XXVIII Prototype Cell is AB12
=**B3***(E11+W12)

Formula Set XXIX Prototype Cell is AC12
=**B3*****H3***(D11+V12)/((B11+T12)^2+(D11+V12)^2+(F11+X12)^2)^1.5

Formula Set XXX Prototype Cell is AD12
=**B3***(G11+Y12)

Formula Set XXXI Prototype Cell is AE12
=**B3*****H3***(F11+X12)/((B11+T12)^2+(D11+V12)^2+(F11+X12)^2)^1.5

Model 3.9

BUILD and verify Model 3.9.

ENTER data using cells C3 (Δt), C5 - C7 (x-, y-, z-initial position coordinates), F5 - F7 (x-, y-, z-initial velocities), and F1 - F3 (gravitational constant and the masses of the two bodies). Prudence suggests that m_1 should always be greater than m_2. If you wish to generate xy-plots in the correct geometric perspective: Measure the length of the x-axis on your screen and enter the value into cell L1; measure the length of the y-axis on your screen and enter the value into cell L2. Be sure to include row 11 in your plotting range.

READ output from the model from columns A (time) and B - G (three-dimensional position of projectile with time).

NOTICE that this model is designed in compact form to save space.

FORMULAS

	A	B	C	D	E	F	G	H	I	J	K	L
1		Orbits (2D)			G =	6.67E-11	Nm^2/kg^2				wx =	3.25
2					m1 =	5.98E+24	kg	M =	=F2+F3		wy =	1.75
3		Δs =	2000000		m2 =	7.34E+22	kg	-GM =	=-F1*I2		xmax =	=MAX(C12:C81)
4											xmin =	=MIN(C12:C81)
5		s0x =	4E+08	m	v0x =	0	m/s				ymax =	=MAX(E12:E81)
6		s0y =	0	m	v0y =	1000	m/s				ymin =	=MIN(E12:E81)
7											Δ =	=MAX(L3-L4,L5-L6)
8			=(L3+L4)/2-(L1*L7)/(2*L2)		=(L5+L6)/2-L7/2							
9	Δt	t	sx	vx	sy	vy	ksx0	kvx0	ksy0	kvy0	ksx1	kvx1
10		sec	m	m/s	m	m/s						
11			=C8+L1*L7/L2		=E8+L7							
12		0	=C5	=F5	=C6	=F6						
13–15	Formula Set I	Formula Set II	Formula Set III	Formula Set IV	Formula Set V	Formula Set VI	Form Set VII	Form Set VIII	Form Set IX	Form Set X	Form Set XI	Form Set XII

FORMULAS (Contd.)

	M	N	O	P	Q	R	S	T	U	V
9	ksy1	kvy1	ksx2	kvx2	ksy2	kvy2	ksx3	kvx3	ksy3	kvy3
10										
11										
12										
13–15	Form Set XIII	Form Set XIV	Form Set XV	Form Set XVI	Form Set XVII	Form Set XVIII	Form Set XIX	Form Set XX	Form Set XXI	Form Set XXII

Model 3.9 (continued)

VALUES

	A	B	C	D	E	F	G	H	I	J	K	L
1		Orbits (2D)			G =	6.67E-11	Nm^2/kg^2				wx =	3.25
2					m1 =	5.98E+24	kg	M =	6E+24		wy =	1.75
3		Δs =	20000000		m2 =	7.34E+22	kg	−GM =	-4E+14		xmax =	4E+08
4											xmin =	-4E+08
5		s0x =	4E+08	m	v0x =	0	m/s				ymax =	4E+08
6		s0y =	0	m	v0y =	1000	m/s				ymin =	-1E+08
7											Δ =	8E+08
8			-7.3E+08		-2.6E+08							
9	Δt	t	sx	vx	sy	vy	ksx0	kvx0	ksy0	kvy0	ksx1	kvx1
10		sec	m	m/s	m	m/s						
11			7.4E+08		5.28E+08							
12	20000	0	4E+08	0	0	1000						
13	19999.76	19999.76	3.99E+08	-50.4496	19991588	998.7385	0	-50.47	2E+07	0	-5E+05	-50.42
14	19999.05	39998.81	3.98E+08	-100.775	39932499	994.9569	-1E+06	-50.41	2E+07	-2.523	-2E+06	-50.3
15	19997.87	59996.68	3.95E+08	-150.851	59771945	988.6643	-2E+06	-50.22	2E+07	-5.039	-3E+06	-50.05

VALUES (Contd.)

	M	N	O	P	Q	R	S	T	U	V
9	ksy1	kvy1	ksx2	kvx2	ksy2	kvy2	ksx3	kvx3	ksy3	kvy3
10										
11										
12										
13	2E+07	-1.261	-5E+05	-50.49	2E+07	-1.263	-1E+06	-50.41	2E+07	-2.522
14	2E+07	-3.779	-2E+06	-50.36	2E+07	-3.785	-2E+06	-50.22	2E+07	-5.039
15	2E+07	-6.289	-3E+06	-50.11	2E+07	-6.298	-3E+06	-49.91	2E+07	-7.543

Model 3.9 (continued)

Formula Set I
Prototype Cell is A12

=**C3** /SQRT(D12^2+F12^2)

Formula Set II
Prototype Cell is B13

=B12+A13

Formula Set III
Prototype Cell is C13

=C12+(G13+2*K13+2*O13+S13)/6

Formula Set IV
Prototype Cell is D13

=D12+(H13+2*L13+2*P13+T13)/6

Formula Set V
Prototype Cell is E13

=E12+(I13+2*M13+2*Q13+U13)/6

Formula Set VI
Prototype Cell is F13

=F12+(J13+2*N13+2*R13+V13)/6

Formula Set VII
Prototype Cell is G13

=A12*D12

Formula Set VIII
Prototype Cell is H13

=A12* **I3** *C12/(C12^2+E12^2)^1.5

Formula Set IX
Prototype Cell is I13

=A12*F12

Formula Set X
Prototype Cell is J13

=A12* **I3** *E12/(C12^2+E12^2)^1.5

Formula Set XI
Prototype Cell is K13

=A12*(D12+H13/2)

Formula Set XII
Prototype Cell is L13

=A12* **I3** *(C12+G13/2)/((C12+G13/2)^2+(E12+I13/2)^2)^1.5

Formula Set XIII
Prototype Cell is M13

=A12*(F12+J13/2)

Formula Set XIV
Prototype Cell is N13

=A12* **I3** *(E12+I13/2)/((C12+G13/2)^2+(E12+I13/2)^2)^1.5

Formula Set XV
Prototype Cell is O13

=A12*(D12+L13/2)

Formula Set XVI
Prototype Cell is P13

=A12* **I3** *(C12+K13/2)/((C12+K13/2)^2+(E12+M13/2)^2)^1.5

Formula Set XVII
Prototype Cell is Q13

=A12*(F12+N13/2)

Formula Set XVIII
Prototype Cell is R13

=A12* **I3** *(E12+M13/2)/((C12+K13/2)^2+(E12+M13/2)^2)^1.5

Formula Set XIX
Prototype Cell is S13

=A12*(D12+P13)

Formula Set XX
Prototype Cell is T13

=A12* **I3** *(C12+O13)/((C12+O13)^2+(E12+Q13)^2)^1.5

Formula Set XXI
Prototype Cell is U13

=A12*(F12+R13)

Formula Set XXII
Prototype Cell is V13

=A12* **I3** *(E12+Q13)/((C12+O13)^2+(E12+Q13)^2)^1.5

CHAPTER 4

WORK, ENERGY, AND MOMENTUM

If Newton's Laws are the salt of the Earth, then the conservation laws are manna from heaven. — Anon

Figure 4.1

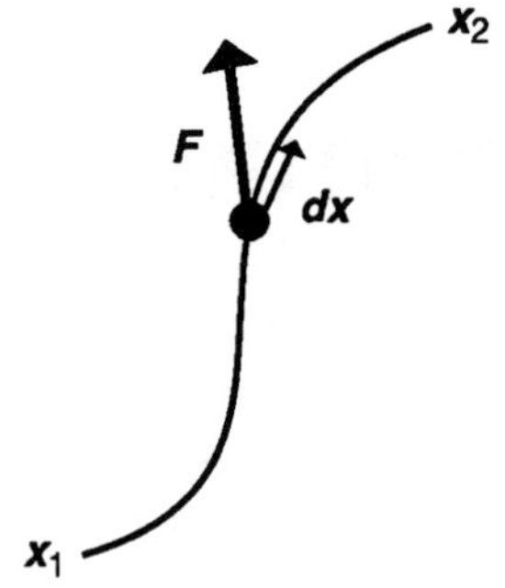

The work done on an object in going from x_1 to x_2 along the actual path it takes is obtained by integrating the dot product $\boldsymbol{F} \cdot \boldsymbol{dx}$ along the path. Notice that, in general, you need to know the actual motion before you can calculate the work done on the object.

It is Newton's three laws of motion that build bridges between movements and the forces that cause them. While all three laws are inseparable and apply to all problems related to the physics of forces, it is Newton's second law of motion (Equation 4.1) that is often seen explicitly because it relates force to acceleration (and thus to velocity and position as well).[1]

$$\Sigma F = ma \tag{4.1}$$

In Chapter 3, you learned how to solve Equation 4.1 using numeric techniques — a particularly important skill because most differential equations can be solved no other way!

The concept of force can always be used to determine the motion of an object, but deep insight into the character of a motion can often be obtained by exploiting the conservation of certain other dynamical quantities. In this chapter, you explore the consequences of *the conservation of mechanical energy* and *the conservation of momentum,* and you explore the circumstances under which they are applicable. In some cases these concepts are even sufficient to determine a motion.

Let's begin with energy. Energy is a useful concept in part because it often gives you considerable qualitative insight into a motion. This insight results from the facts that

- Energy is a scalar quantity.
- Energy is conserved although its distribution in various forms usually changes during a motion.[2]

Force and energy are related via the concept of the *work* done by a force. In the simplest case of a constant force, work (W) is a scalar quantity defined by the dot product of two vectors: $\boldsymbol{F}$ the constant, external force applied to an object and $\boldsymbol{s}$ the actual displacement of the object (Equation 4.2). The dimension of work is the Joule (i.e., N m).

$$W = \boldsymbol{F} \cdot \boldsymbol{s} \tag{4.2}$$

For a nonconstant force, the actual path of an object is ordinarily curved, and the work done by the force between t_1 and t_2 is defined by an integral (Equation 4.3).

$$W = \int_{t_1}^{t_2} \boldsymbol{F} \cdot \boldsymbol{v}\ dt = \int_{x_1}^{x_2} \boldsymbol{F} \cdot \boldsymbol{dx} \qquad (4.3)^{3}$$

where $\mathbf{v}$ is the velocity of the object, and $\mathbf{x}_1$, $\mathbf{x}_2$ are the actual locations of the object at time t_1 and t_2. In the second integral, $\mathbf{dx}$ is the differential displacement of the object along its actual path of motion. The integral is usually called a *path integral* because it is to be taken along the path of the particle (Figure 4.1).

It is important to notice that, in general, you cannot calculate the amount of work that a force does on an object *until after you know the motion.* There are, however, many important exceptions where this is not necessary. In these cases the work offers insight and simplified calculations. Two important examples in which you need not know the motion are:

- The work performed by gravity upon a falling object of mass m is given by Equation 4.4 (Figure 4.2):

$$W_{\text{grav}} = \int_{y_1}^{y_2} (-mg)dy = mg(y_1 - y_2) \tag{4.4}$$

  where W_{grav} is the work performed by gravity over the distance from y_2 to y_1 and g is the acceleration of gravity (9.8 m s^{-2}).

- The work performed by the force used to stretch an ideal spring is given by Equation 4.5 (Figure 4.3):

$$W_{\text{elastic}} = \int_{x_1}^{x_2} -(kx)dx = \frac{k(x_1^2 - x_2^2)}{2} \tag{4.5}$$

  where W_{elastic} is the work performed by an ideal spring as it moves a mass over the distance from x_2 to x_1 and k is the spring modulus.

Consider next the concept of kinetic energy K (Equation 4.6).

$$K = \frac{m\mathbf{v} \cdot \mathbf{v}}{2} \tag{4.6}$$

It is a general consequence of Newton's second law of motion that

$$K(t_2) - K(t_1) = \int_{t_1}^{t_2} \mathbf{F} \cdot \mathbf{v} \ \ dt = W \tag{4.7}$$

Thus, *the amount of work done on an object is equal to the increase in its kinetic energy.*

A second form of energy, *potential energy,* can be defined in special cases in such a way that it has the nice property

$$W = -\Delta U = -[U(\mathbf{x}_2) - U(\mathbf{x}_1)] \qquad (4.8)[4]$$

Figure 4.2

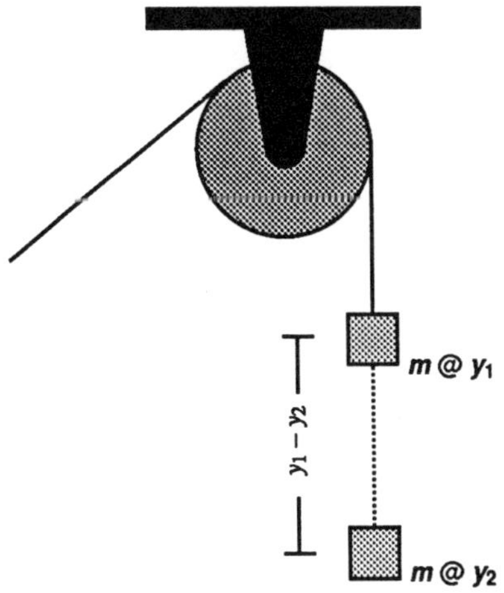

A mass m is attached to a frictionless pulley so that it can be raised and lowered with respect to the ground. The work done as the mass moves is given by Equation 4.4.

Figure 4.3

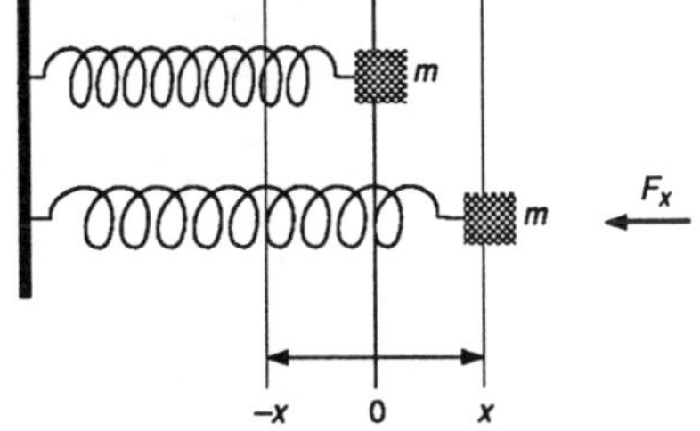

A mass m is attached to a horizontal spring and moved by the tension in the spring. The work done by the spring as the mass moves is given by Equation 4.5.

where ΔU is the change in potential energy as the object moves from $\boldsymbol{x}_1$ (the position of the object at t_1) to $\boldsymbol{x}_2$ (the position of the object at t_2).

In one dimension, work can be calculated without prior knowledge of the motion (and a potential energy function defined) when the force is a function of position only—velocity-dependent forces such as friction and time-dependent forces clearly do not satisfy this condition and so do not have associated potential energy functions. For example, the potential energy of an object in a gravitational field near the Earth's surface is the simple function

$$U(y) = mgy \tag{4.9}$$

where y is the height of the object above some arbitrary reference (e.g., the Earth's surface, a table top, some point of support). By comparing Equations 4.4, 4.8, and 4.9 you can see that this is a suitable definition of the potential energy in this case. Similarly for the case of a spring, Equation 4.5, the potential energy is

$$U(x) = \frac{kx^2}{2} \tag{4.10}$$

In two and three dimensions, you cannot, in general, define a potential energy function even if the force is a function of position alone! It is certainly necessary that the force depends only on position (thus ruling out friction and external forces), but this is not sufficient. There are additional conditions that must be satisfied. You will see what these are later when you have the opportunity to explore forces that depend only on position but for which no potential energy function can be defined.

Returning to Equations 4.7 and 4.8, they can be joined to yield the famous *law of conservation of mechanical energy.*[5]

$$K(t_2) - K(t_1) = -[U(\boldsymbol{x}_2) - U(\boldsymbol{x}_1)] \tag{4.11}$$

or

$$K(t_2) + U(\boldsymbol{x}_2) = U(\boldsymbol{x}_1) + K(t_1) \tag{4.12}$$

It is the sum $E_0 = K + U$ that is conserved although both K and U usually change during a motion. E_0 is called the *total mechanical energy.*

You will verify the truth of Equation 4.12 in Exercises 4.1 and 4.2, and you will see how this equation can be used to calculate the motions of objects in Exercise 4.4. Exercise 4.3 introduces forces that are not functions of position (e.g., friction) and where total mechanical energy is not conserved.

Another important quantity that is sometimes conserved in mechanical systems is momentum. The linear momentum $\boldsymbol{p}$ of a single object is defined as the product of its mass m and velocity $\boldsymbol{v}$ (Equation 4.13).

$$\boldsymbol{p} = m\boldsymbol{v} \tag{4.13}$$

The linear momentum of several objects is the vector sum of the linear momentum of each of them, and when this sum is conserved (i.e., when the sum of all of the forces acting on several objects vanishes), the very important *law of conservation of linear momentum* holds.

Notice that momentum is a vector quantity and so when it is conserved, each of its three components is separately conserved. Thus, the law of conservation of linear momentum is really three separate conservation laws.

The conservation of momentum is a particularly useful tool for studying collisions (first noted by René Descartes in the early 1600s). For example, suppose two particles a and b collide. Before the collision the particles have known masses (m_{a1}, m_{b1}) and velocities ($\boldsymbol{v}_{a1}$, $\boldsymbol{v}_{b1}$). The after-collision masses and velocities (m_{a2}, m_{b2}, $\boldsymbol{v}_{a2}$, $\boldsymbol{v}_{b2}$) are restrained by the conservation of momentum to satisfy Equation 4.14:

$$m_{a1}\boldsymbol{v}_{a1} + m_{b1}\boldsymbol{v}_{b1} = m_{a2}\boldsymbol{v}_{a2} + m_{b2}\boldsymbol{v}_{b2} \tag{4.14}$$

You will explore the conservation of momentum and its role in determining the behavior of collisions in Exercise 4.5.

We'll finish this introduction with a bit of historical perspective.

Although Newton's laws of motion were first used in the late 1600s, Descartes had noticed that momentum was a conserved quantity somewhat earlier. Other contemporary natural philosophers noticed that, in collisions, a quantity that they called *vis viva* (living force) was also conserved. *Vis visa* was the original name of what we now call kinetic energy (from two Greek words meaning "of motion" and "in work," respectively). Its relation to work in mechanics was widely appreciated after Newton's time, and in fact Newton himself knew of what we now call the conservation of mechanical energy.

The concept of energy forms the basis for other formulations of mechanics due to Joseph Louis Lagrange (1736 - 1813) and Sir William

Rowan Hamilton (1805 - 1865). These formulations are computationally more efficient than Newton's laws and are the platforms from which much of modern theoretical physics has evolved.

It was suspected even in pre-Newtonian times that heat was due to the microscopic motions of parts of a body, but it was only in the 1840s when James Prescott Joule (1818 - 1889) showed that a definite amount of heat always appears when friction causes a definite amount of mechanical energy to disappear. Joule's observations showed clearly that heat was just a form of energy, and many other forms of energy were subsequently identified. Such observations ultimately led to a principal cornerstone of modern science — the general law of the conservation of energy.

EXERCISE **4.1**

THE CONSERVATION OF ENERGY

TO GET THE MOST OUT OF THIS EXERCISE, YOU SHOULD ALREADY KNOW

The definitions of the terms:
potential energy
kinetic energy
total mechanical energy

How to solve second-order differential equations using Runge-Kutta methods.

In all mechanical systems, kinetic energy K is a dynamical quantity that measures the ability of the sytem to do work on other systems. It is a very important concept. For a single particle of mass m and velocity $\boldsymbol{v}$, K is defined as

$$K = \frac{m\boldsymbol{v}\cdot\boldsymbol{v}}{2} = \frac{mv^2}{2} \qquad (4.15)$$

and for several particles, K is the sum of the kinetic energies of each of them.

K arises directly out of Newton's second law of motion via a clever examination of the properties of the quantity $\boldsymbol{F}\cdot\boldsymbol{v}$ (Equations 4.16 - 4.18).

$$\boldsymbol{F} = m\boldsymbol{a} = m\frac{d\boldsymbol{v}}{dt} = \frac{d(m\boldsymbol{v})}{dt}$$

$$\boldsymbol{F}\cdot\boldsymbol{v} = \frac{d(m\boldsymbol{v})}{dt}\cdot\boldsymbol{v} = \frac{d}{dt}\left(\frac{mv^2}{2}\right) \qquad (4.16)$$

Because $\boldsymbol{v}\,dt = \boldsymbol{dx}$, a differential part of the particle's path, it follows that

$$\boldsymbol{F}\cdot\boldsymbol{v}\ dt = \boldsymbol{F}\cdot\boldsymbol{dx} = d\left(\frac{mv^2}{2}\right) \qquad (4.17)$$

or

$$dW = dK \tag{4.18}$$

where dW is the differential of work done by the force on the particle and dK is the differential change in kinetic energy as the force F moves the particle from $\boldsymbol{x}$ to $\boldsymbol{x} + \boldsymbol{dx}$. Equation 4.18 simply states that *the amount of work done on an object is equal to the change in its kinetic energy.*

You can always, of course, integrate the right side of Equation 4.17 between two times to obtain

$$\int_{t_1}^{t_2} d\left(\frac{mv^2}{2}\right) = \left(\frac{mv^2}{2}\right)_{t_2} - \left(\frac{mv^2}{2}\right)_{t_1} = \Delta K \tag{4.19}$$

where ΔK is the increase in kinetic energy of the particle between t_1 and t_2.

Integrating the work side of Equation 4.17 is less easy because you must always know not only the force acting on the particle *but also the path that the particle actually took during the time interval* (t_1, t_2). It is a beautiful feature of the fundamental forces of gravity and electricity that the work integral can be calculated without knowing the actual path that a particle takes. Forces that have this very special and important property are called *conservative forces,* and there exists for them a law of energy conservation involving a potential energy function $U(\boldsymbol{x})$. Furthermore, the force itself can be calculated from $U(\boldsymbol{x})$.

In this exercise you explore motion in one dimension and discover that it is very easy to see which forces are conservative.

If an object of mass m is constrained to move along the x-axis only *and it is acted upon by a force* F(x) *that depends on position only*, then you can integrate the work side of Equation 4.17. Furthermore, it is convenient to define a potential energy function $U(x)$ such that

$$U(x) = -\int_{x_0}^{x} F(x)dx \quad + \quad \text{(any constant)} \tag{4.20}$$

where x_0 is any convenient position and "any constant" is any convenient value (usually zero).

Note that, in effect, the potential energy function $U(x)$ is an *indefinite integral* of the force. If you take the constant to be zero, you get $U(x_0) = 0$.

There are two very important properties of $U(x)$. First, Equation 4.17 can be integrated to give

$$\Delta K = -\Delta U \qquad (4.21)$$

where ΔK is the increase in kinetic energy of a particle between any two times and ΔU is the increase in the potential energy of the particle over the same time interval. It should be apparent that the sum $E_0 = K + U$ is a constant. E_0 is called the total mechanical energy of the particle, and the fact that E_0 remains unchanged in these situations is called the *law of conservation of mechanical energy.*

Second, the force itself can be calculated from the potential energy by differentiating Equation 4.20 (Equation 4.22):

$$F(x) = -\frac{dU}{dx} \qquad (4.22)$$

You should also realize that forces depending explicitly on velocity (such as friction) and forces depending explicitly on time (such as a person pushing on a mass) are not conservative, and so when they must be taken into account, the mechanical energy of the particle will not be conserved.

LESSONS

1. A ball thrown vertically upward is subject only to the force of gravity when the retarding effects of the atmosphere are ignored. The change in potential energy during the movement of the ball can be calculated using Equation 4.23:

$$W = -\Delta U = \int_{y_0}^{y} -(mg)dy = mg(y_0 - y) \qquad (4.23)$$

where y_0, y is the starting and ending vertical displacement of the ball measured vertically upward, m is the mass of the ball, and g is the acceleration of gravity (9.8 m s^{-2}).

If the starting position of the ball is taken at the Earth's surface ($y_0 = 0$) and the potential energy of the ball is taken to be zero when resting on the ground, then the potential energy of the ball is a simple function of its height above the ground (Equation 4.24):

$$U = mgy \qquad (4.24)$$

a. Build and verify Model 4.1. Model 4.1 calculates the velocity and position of a ball thrown vertically upward using a one-

Figure 4.4

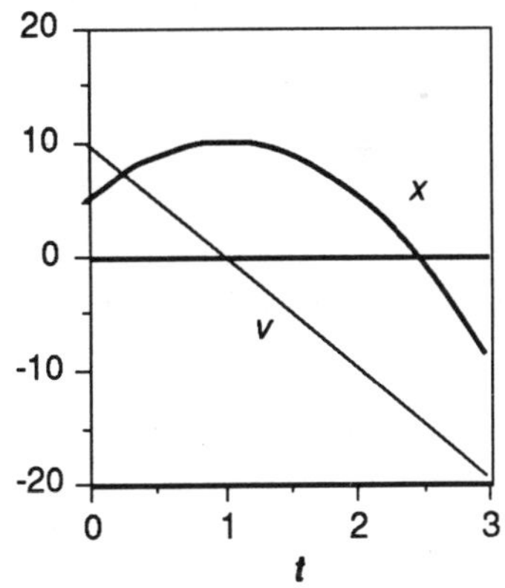

XY-PLOT

Data
Horizontal (*x*) axis: cells E11 - E60 (*t*)
Vertical (*y*) axis: cells F11 - F60 (*x*), G11 - G60 (*v*)

Figure 4.5

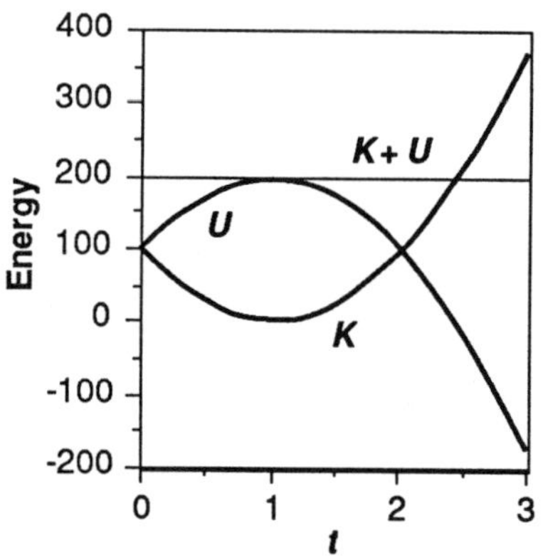

XY-PLOT

Data
Horizontal (*x*) axis: cells E11 - E60 (t)
Vertical (*y*) axis: cells A11 - A60 (*K*+*U*), B11 - B60 (*K*), C11 - C60 (*U*)

dimensional Runge-Kutta model similar to Model 3.3 in the previous chapter. Three additional columns have been added to the extreme left of the model in order to track K (kinetic energy), U (potential energy), and $K+U$ (total mechanical energy).

b. Build on-screen graphics for your model that display the ball's position and velocity (Figure 4.4) and the partition of kinetic and potential energy for the ball during its flight (Figure 4.5).

c. Explore your model by throwing the ball into the air with a number of different initial velocities and from a number of different initial heights.

d. As the ball increases in height above the ground what happens to the velocity and kinetic energy (K) of the ball? What happens to the potential energy (U)? What happens to the total mechanical energy ($K+U$)?

e. What are the values for K, U, and $K+U$ when the ball is at its highest point above the ground? At the moment just before it strikes the ground?

f. What are the potential and kinetic energies of a ball resting in your hand 0.5 m above the ground. Suppose you open your hand and the ball falls to the ground. Describe in simple language what happens to K, U, and $K+U$ during the ball's travels.

g. Your model is not built to deal with events that occur after the ball hits the ground. In reality, what happens?[1]

Figure 4.6

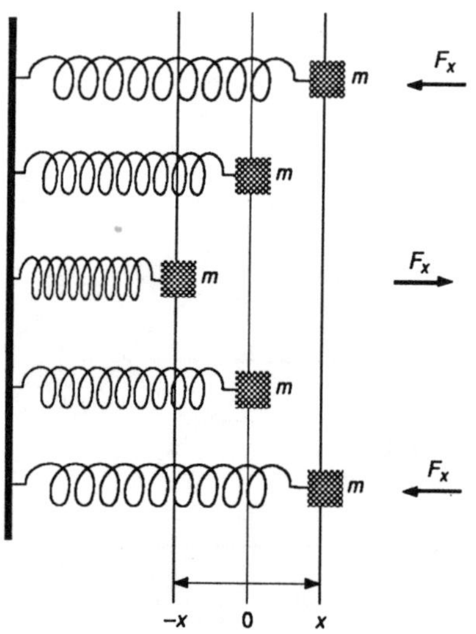

A massless spring oscillates back and forth with a mass attached to one end. If *x* measures the displacement of the spring from its resting position, then changes in the potential energy of the spring are given by Equation 4.25.

2. A massless spring rests on a frictionless surface with a mass m attached to one end (Figure 4.6). An external force is used to start the spring oscillating by pulling it a distance x from its resting position and releasing it (with or without an initial velocity). The change in potential energy during the oscillation of the spring can be calculated using Equation 4.25:

$$W = -\Delta U = \int_{x_0}^{x} (-kx)dx = -k\frac{x^2}{2} + k\frac{x_0^2}{2} \qquad (4.25)$$

where x is the displacement of the spring from its resting position and k is the spring constant. You can take the potential energy to be zero when the spring is at $x = 0$, its resting position, and so write

$$U = k\frac{x^2}{2} \tag{4.26}$$

a. Build and verify Model 4.2. Model 4.2 calculates the velocity and position of the mass attached to the spring using a one-dimensional Runge-Kutta model similar to Model 3.3 in the previous chapter. Three additional columns have been added to the extreme left of the model to track K (kinetic energy), U (potential energy), and $K+U$ (total mechanical energy).

b. Build on-screen graphics for your model that display the position and velocity of the mass (Figure 4.7) and the partition of kinetic and potential energy for the spring system as it oscillates (Figure 4.8).

c. Explore your model by initializing the spring system with a number of different initial displacements and velocities.

d. As the mass departs from its resting position, what happens to the velocity and kinetic energy (K) of the mass? What happens to the potential energy (U)? What happens to the total mechanical energy ($K+U$)?

e. Once the system is set in motion, what are the values for K, U, and $K+U$ when the mass is farthest from its resting position? As it passes through its resting position during each oscillation?

f. Describe in simple language the kinetic and potential energies of your spring system at each stage of its repetitive oscillation.

g. What is the total mechanical energy of a spring system at rest (nonoscillating)?

Figure 4.7

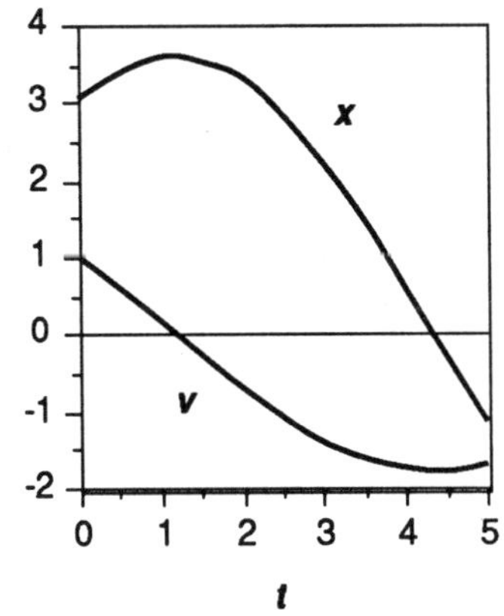

XY-PLOT

Data
Horizontal (*x*) axis: cells E11 - E60 (*t*)
Vertical (*y*) axis: cells F11 - F60 (*x*), G11 - G60 (*v*)

Figure 4.8

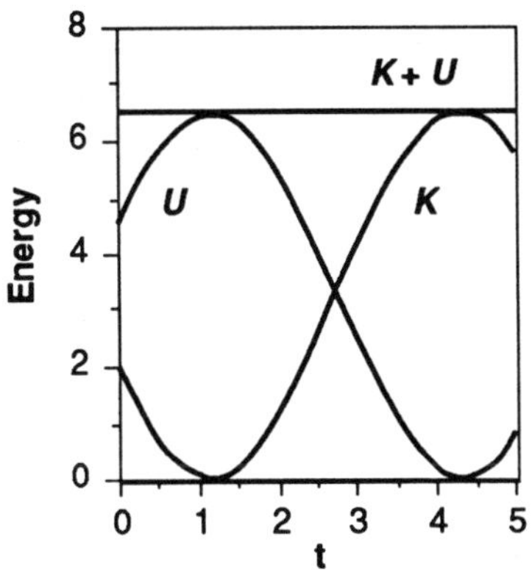

XY-PLOT

Data
Horizontal (*x*) axis: cells E11 - E60 (t)
Vertical (*y*) axis: cells A11 - A60 (*K+U*), B11 - B60 (*K*), C11 - C60 (*U*)

PROBLEMS

1. Inhabitants of planetoid Gbar live within a depression in the surface and experience a gravitational force

$$F = -mg\left(1 - \frac{y}{R}\right) \tag{4.27}$$

where y is the distance below the surrounding surface. Suppose an inhabitant is released from position $y = 548$ m and falls to the bottom of the depression located at $y = 1432$ m. Given a g of 2.45 m s^{-2} and a planetoid radius R of 4570 m, plot the position of the inhabitant vs time. What is the value of the total work done by gravity for the complete journey?

2. Consider the spring which exerts the restoring force F_x (Equation 4.28) on a mass m of 121 g:

$$F_x = -(2x^2 - 3x + 4)x \tag{4.28}$$

Plot of the position of the mass vs time if the mass has an initial velocity v of 23 cm s^{-1} at its equilibrium point ($x = 0$). Plot kinetic energy vs position.

3. An archer pulls back on an arrow set in a compound bow. Until the bow's mechanism aids her effort, the draw requires a force that increases linearly as a function of the distance pulled (i.e., it mimics an ideal spring). Suppose a draw length of 0.7 m requires a force of 253 N. How fast is an arrow of 0.21 kg moving at the moment it loses contact with the bow string? Plot the kinetic energy of the arrow vs position in the bow.

4. An acrobat weighing 134 lbs jumps from a platform located 13.2 ft above a trampoline. Assuming the trampoline has an effective restoring-force constant $k = 87$ lb ft^{-1}, how far below the plane of the resting trampoline will the acrobat travel before beginning his upward journey anew? Plot kinetic energy vs time. Plot the acrobat's kinetic energy vs position. What is the potential energy of the acrobat vs time while in contact with the trampoline?

5. A spider hangs on its self-manufactured thread while a local disturbance gives the thread an extra stretch of 0.42 cm beyond its equilibrium position. If the spider mass is 1.4 g and the equivalent spring constant $k = 1.6$ dynes cm^{-1}, what is the maximum speed experienced by the spider as it rides the thread? Suppose the spider extends its legs (as a kind of air break) to dampen the motion whenever its acceleration exceeds 1.5 cm s^{-2}. When will the spider extend its legs?

MODELS

Model 4.1

BUILD and verify Model 4.1.

ENTER data using cells F4 - F6 (Δt, initial position, initial velocity), J4 and J5 (acceleration of gravity and mass of object).

READ output from the model from columns A - G (total mechanical energy, kinetic energy, potential energy, acceleration, time, position, and velocity).

NOTICE that the model is very similar to the one-dimensional Runge-Kutta models that you built in Chapter 3.

FORMULAS

	A	B	C	D	E	F	G	H	I	J	K	L	M	N
1					Runge-Kutta									
2					DDx =	-g								
3														
4					Δt =	0.06	s		g =	9.8	m/s^2			
5					x0 =	5	m		m =	2	kg			
6					v0 =	10	m/s							
7														
8					0					1				
9	K+U	K	U	a	t	x	v	k(x0)	k(v0)	t	x	v	k(x1)	k(v1)
10	J	J	J	m/s^2	s	m	m/s							
11					0	=F5	=F6							
12–14	Form Set I	Form Set II	Form Set III	Form Set IV	Form Set V	Form Set VI	Form Set VII	Form Set VIII	Form Set IX	Form Set X	Form Set XI	Form Set XII	Form Set XIII	Form Set XIV

FORMULAS (Contd.)

	O	P	Q	R	S	T	U	V	W	X
8	2					3				
9	t	x	v	k(x2)	k(v2)	t	x	v	k(x3)	k(v3)
10										
11–14	Form Set XV	Form Set XVI	Form Set XVII	Form Set XVIII	Form Set XIX	Form Set XX	Form Set XXI	Form Set XXII	Form Set XXIII	Form Set XXIV

VALUES

	A	B	C	D	E	F	G	H	I	J	K	L	M	N
1					Runge-Kutta									
2					DDx =	-g								
3														
4					Δt =	0.06	s		g =	9.8	m/s^2			
5					x0 =	5	m		m =	2	kg			
6					v0 =	10	m/s							
7														
8					0					1				
9	K+U	K	U	a	t	x	v	k(x0)	k(v0)	t	x	v	k(x1)	k(v1)
10	J	J	J	m/s^2	s	m	m/s							
11	198	100	98	-9.8	0	5	10	0.6	-0.59	0.03	5.3	9.706	0.582	-0.59
12	198	88.59	109.4	-9.8	0.06	5.582	9.412	0.565	-0.59	0.09	5.865	9.118	0.547	-0.59
13	198	77.86	120.1	-9.8	0.12	6.129	8.824	0.529	-0.59	0.15	6.394	8.53	0.512	-0.59
14	198	67.83	130.2	-9.8	0.18	6.641	8.236	0.494	-0.59	0.21	6.888	7.942	0.477	-0.59

VALUES (Contd.)

	O	P	Q	R	S	T	U	V	W	X
8	2					3				
9	t	x	v	k(x2)	k(v2)	t	x	v	k(x3)	k(v3)
10										
11	0.03	5.291	9.706	0.582	-0.59	0.06	5.582	9.412	0.565	-0.59
12	0.09	5.856	9.118	0.547	-0.59	0.12	6.129	8.824	0.529	-0.59
13	0.15	6.385	8.53	0.512	-0.59	0.18	6.641	8.236	0.494	-0.59
14	0.21	6.88	7.942	0.477	-0.59	0.24	7.118	7.648	0.459	-0.59

Model 4.1 (continued)

Formula Set I Prototype Cell is A11
=B11+C11

Formula Set II Prototype Cell is B11
=J5*G11^2/2

Formula Set III Prototype Cell is C11
= J5*J4*F11

Formula Set IV Prototype Cell is D11
=-J4

Formula Set V Prototype Cell is E12
=E11+F4

Formula Set VI Prototype Cell is F12
=F11+(H11+2*(M11+R11)+W11)/6

Formula Set VII Prototype Cell is G12
=G11+(I11+2*(N11+S11)+X11)/6

Formula Set VIII Prototype Cell is H11
=F4*G11

Formula Set IX Prototype Cell is I11
=-F4*J4

Formula Set X Prototype Cell is J11
=E11+F4/2

Formula Set XI Prototype Cell is K11
=F11+H11/2

Formula Set XII Prototype Cell is L11
=G11+I11/2

Formula Set XIII Prototype Cell is M11
=F4*L11

Formula Set XIV Prototype Cell is N11
=-F4*J4

Formula Set XV Prototype Cell is O11
=E11+F4/2

Formula Set XVI Prototype Cell is P11
=F11+M11/2

Formula Set XVII Prototype Cell is Q11
=G11+N11/2

Formula Set XVIII Prototype Cell is R11
=F4*Q11

Formula Set XIX Prototype Cell is S11
=-F4*J4

Formula Set XX Prototype Cell is T11
=E11+F4

Formula Set XXI Prototype Cell is U11
=F11+R11

Formula Set XXII Prototype Cell is V11
=G11+S11

Formula Set XXIII Prototype Cell is W11
=F4*V11

Formula Set XXIV Prototype Cell is X11
=-F4*J4

FORMULAS

	A	B	C	D	E	F	G	H	I	J	K	L	M	N
1					Runge-Kutta									
2					DDx = -(k/m)x									
3														
4					Δt =	0.1	s		k =	1	N/m			
5					x0 =	3	m		m =	4	kg			
6					v0 =	1	m/s							
7														
8					0					1				
9	K+U	K	U	a	t	x	v	k(x0)	k(v0)	t	x	v	k(x1)	k(v1)
10	J	J	J	m/s^2	s	m	m/v							
11					0	=F5	=F6							
12	Form	Form	Form	Form	Form	Form	Form	Form	Form	Form	Form	Form	Form	Form
13	Set	Set	Set	Set	Set	Set	Set	Set	Set	Set	Set	Set	Set	Set
14	I	II	III	IV	V	VI	VII	VIII	IX	X	XI	XII	XIII	XIV

FORMULAS (Contd.)

	O	P	Q	R	S	T	U	V	W	X
8	2					3				
9	t	x	v	k(x2)	k(v2)	t	x	v	k(x3)	k(v3)
10										
11										
12	Form	Form	Form	Form	Form	Form	Form	Form	Form	Form
13	Set	Set	Set	Set	Set	Set	Set	Set	Set	Set
14	XV	XVI	XVII	XVIII	XIX	XX	XXI	XXII	XXIII	XXIV

VALUES

	A	B	C	D	E	F	G	H	I	J	K	L	M	N
1					Runge-Kutta									
2					DDx = -(k/m)x									
3														
4					Δt =	0.1	s		k =	1	N/m			
5					x0 =	3	m		m =	4	kg			
6					v0 =	1	m/s							
7														
8					0					1				
9	K+U	K	U	a	t	x	v	k(x0)	k(v0)	t	x	v	k(x1)	k(v1)
10	J	J	J				m/s^2	s	m	m/v				
11	6.5	2	4.5	-0.75	0	3	1	0.1	-0.08	0.05	3.05	0.963	0.096	-0.08
12	6.5	1.707	4.793	-0.77	0.1	3.096	0.924	0.092	-0.08	0.15	3.142	0.885	0.089	-0.08
13	6.5	1.429	5.071	-0.8	0.2	3.185	0.845	0.085	-0.08	0.25	3.227	0.805	0.081	-0.08
14	6.5	1.169	5.331	-0.82	0.3	3.265	0.765	0.076	-0.08	0.35	3.303	0.724	0.072	-0.08

VALUES (Contd.)

	O	P	Q	R	S	T	U	V	W	X
8	2					3				
9	t	x	v	k(x2)	k(v2)	t	x	v	k(x3)	k(v3)
10										
11	0.05	3.048	0.962	0.096	-0.08	0.1	3.096	0.924	0.092	-0.08
12	0.15	3.14	0.885	0.088	-0.08	0.2	3.185	0.845	0.085	-0.08
13	0.25	3.225	0.805	0.08	-0.08	0.3	3.265	0.765	0.076	-0.08
14	0.35	3.301	0.723	0.072	-0.08	0.4	3.338	0.682	0.068	-0.08

Model 4.2

BUILD and verify Model 4.2.

ENTER data using cells F4 - F6 (Δt, initial position, initial velocity), J4 and J5 (spring modulus and mass of attached object).

READ output from the model from columns A - G (total mechanical energy, kinetic energy, potential energy, acceleration, time, position, and velocity).

NOTICE that the model is very similar to the one-dimensional Runge-Kutta models that you built in Chapter 3.

Model 4.2 (continued)

Formula Set I	Prototype Cell is A11	=B11+C11
Formula Set II	Prototype Cell is B11	=**J5** *G11^2/2
Formula Set III	Prototype Cell is C11	=**J4** *F11^2/2
Formula Set IV	Prototype Cell is D11	=-(**J4** /**J5**)*F11
Formula Set V	Prototype Cell is E12	=E11+ **F4**
Formula Set VI	Prototype Cell is F12	=F11+(H11+2*(M11+R11)+W11)/6
Formula Set VII	Prototype Cell is G12	=G11+(I11+2*(N11+S11)+X11)/6
Formula Set VIII	Prototype Cell is H11	=**F4** *G11
Formula Set IX	Prototype Cell is I11	=-**F4** *(**J4** /**J5**)*F11
Formula Set X	Prototype Cell is J11	=**E**11+**F4** /2
Formula Set XI	Prototype Cell is K11	=**F**11+H11/2
Formula Set XII	Prototype Cell is L11	=**G**11+I11/2
Formula Set XIII	Prototype Cell is M11	=**F4** *L11
Formula Set XIV	Prototype Cell is N11	=-**F4** *(**J4** /**J5**)*K11
Formula Set XV	Prototype Cell is O11	=**E**11+**F4** /2
Formula Set XVI	Prototype Cell is P11	=**F**11+M11/2
Formula Set XVII	Prototype Cell is Q11	=**G**11+N11/2
Formula Set XVIII	Prototype Cell is R11	=**F4** *Q11
Formula Set XIX	Prototype Cell is S11	=-**F4** *(**J4** /**J5**)*P11
Formula Set XX	Prototype Cell is T11	=**E**11+**F4**
Formula Set XXI	Prototype Cell is U11	=**F**11+R11
Formula Set XXII	Prototype Cell is V11	=**G**11+S11
Formula Set XXIII	Prototype Cell is W11	=**F4** *V11
Formula Set XIV	Prototype Cell is X11	=-**F4** *(**J4** /**J5**)*U11

THE POTENTIAL ENERGY FUNCTION: ENERGY CONSERVATION IN TWO DIMENSIONS

In the previous exercise, you learned that total mechanical energy E_0 is conserved in one-dimensional mechanical systems in which the force is a function of position only. You also explored how this energy is partitioned between kinetic energy K and potential energy U such that their sum E_0 remains constant (Equation 4.29):

$$E_0 = K + U \tag{4.29}$$

This equation remains valid in two or three dimensions for some forces, including many fundamental forces of Nature. As in one dimension, these forces must be functions of position only, but in contrast to the situation in one dimension, this is not sufficient. *Forces can exist that are functions of position only but for which potential energy functions cannot be written!*[1]

For the two-dimensional case, $U(x,y)$ exists (and the system is conservative) when the force $F(x,y)$ is a function of position only and when the components of $F(x,y)$ satisfy Equation 4.30:

$$F_x = -\frac{\partial U}{\partial x}, \quad F_y = -\frac{\partial U}{\partial y} \tag{4.30}$$

TO GET THE MOST OUT OF THIS EXERCISE, YOU SHOULD ALREADY KNOW

The definitions of the terms:
potential energy
kinetic energy
total mechanical energy
partial derivative

How to solve second-order differential equations using Runge-Kutta methods.

How to calculate partial derivatives.

where F_x is the component of F in the x-direction and F_y is the component of F in the y-direction. The three-dimensional case is the obvious extension.

For any differentiable function $U(x,y)$, it is an identity that

$$\frac{\partial}{\partial x}\left(\frac{\partial U}{\partial y}\right)=\frac{\partial}{\partial y}\left(\frac{\partial U}{\partial x}\right) \qquad (4.31)$$

Therefore, if the force components satisfy Equation 4.30, you get from Equation 4.31

$$\frac{\partial F_y}{\partial x}=\frac{\partial F_x}{\partial y} \qquad (4.32)$$

It can be shown mathematically that the converse of this logic is also true: If the components of force satisfy Equation 4.32, these components can be derived from a potential energy function $U(x,y)$ using Equation 4.30.[2]

You will explore a force in Problem 2 that does not meet the restrictions imposed by Equation 4.32. For the moment, let's examine a force that does meet these conditions. Recall Newton's law of gravitation for a single dimension. According to the law of gravitation, the force of gravity is given by Equation 4.33:

$$F_y=-G\frac{Mm}{y^2} \qquad (4.33)$$

where F is the force of gravity exerted by the Earth on an object; G is the 6.670×10^{-11} N m^2 kg^{-2} (the gravitational constant); M, m are the mass of the Earth and the object, respectively; and y is the distance of the object from the center of the Earth to the object.

Figure 4.9

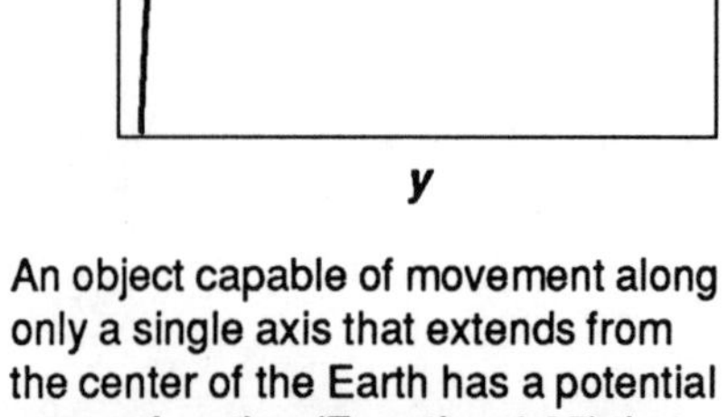

An object capable of movement along only a single axis that extends from the center of the Earth has a potential energy function (Equation 4.35) that can be obtained from Newton's law of gravitation (Equation 4.33).

The change in potential energy of such an object as it departs the Earth can be given by

$$\Delta U=-\int_{y_0}^{y} F\ dy=\int_{y_0}^{y} G\frac{Mm}{y^2}\ dy=-GMm\left(\frac{1}{y}-\frac{1}{y_0}\right) \qquad (4.34)$$

where y_0 and y are the starting and ending heights of the object above the center of the Earth, respectively.

If y_0 is taken as infinity and if the potential energy of the object is chosen as zero when y_0 equals infinity, then Equation 4.34 evaluates to

$$U = U(y) = -G\frac{Mm}{y} \quad (4.35)$$

The shape of $U(y)$ is displayed in Figure 4.9.

Although we have called the one dimension in this calculation y, it could be any direction away from the center of the gravitational force because of gravity's symmetry. Perhaps you can guess that, in two dimensions, the potential energy function for gravity is

$$U = U(x,y) = -G\frac{Mm}{\sqrt{x^2+y^2}} \quad (4.36)$$

The shape of $U(x,y)$ is shown in Figure 4.10.

You can check that this potential energy function yields Newton's law of gravity in two dimensions

$$F_x = -\frac{\partial U}{\partial x} = -\frac{GMmx}{(x^2+y^2)^{3/2}} \quad (4.37a)$$

$$F_y = -\frac{\partial U}{\partial y} = -\frac{GMmy}{(x^2+y^2)^{3/2}} \quad (4.37b)$$

You can also verify that the conditions set forth by Equation 4.32 are satisfied.

Pause for a moment and consider how kinetic energy K must depend on position in these two cases to assure that total mechanical energy remains constant. The exercises that follow are intended to reinforce your understanding of this and other issues.

Figure 4.10

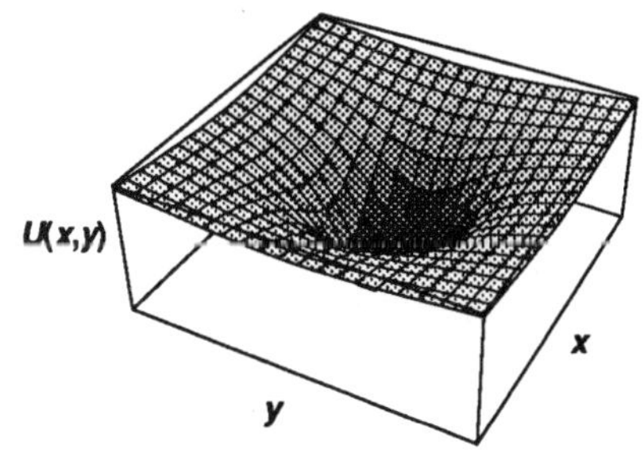

An object in orbit about the Earth moves in a plane defined by the x- and y-axes. The potential energy function for this object (Equation 4.36) can be obtained from Newton's law of gravitation (Equation 4.33).

LESSONS

1. Consider an object that moves on a trajectory such that the force of gravity can be taken as a constant (near the Earth, for example). If the only force acting on this object is the force of gravity, then

$$F_x = 0 \quad (4.38a)$$

$$F_y = -mg \quad (4.38b)$$

where $g = 9.8\ \mathrm{m\,s^{-2}}$, the acceleration of gravity at the Earth's surface.

A potential energy function U from which the force in Equations 4.38a - 4.38b can be obtained is

Figure 4.11

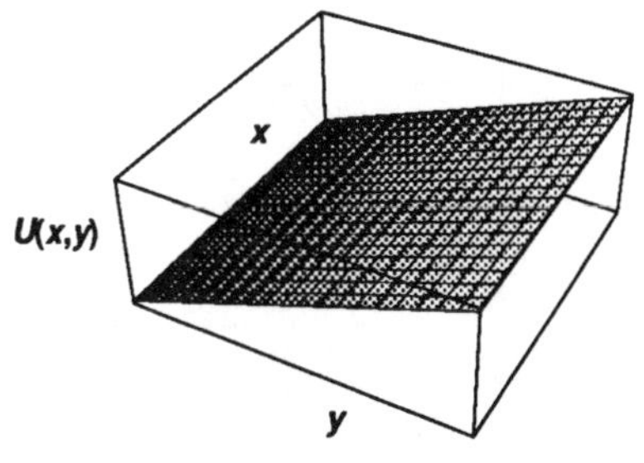

An object that moves on a trajectory determined only by the force of gravity near the surface of the Earth has a potential energy function that varies only with changes in height y.

Figure 4.12

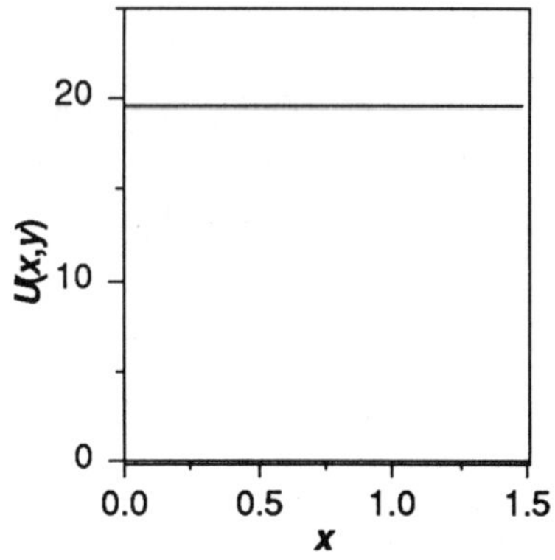

XY-PLOT

Data
Horizontal (*x*) axis: cells E13 - S13
Vertical (*y*) axis: cells E30 - S30

Figure 4.13

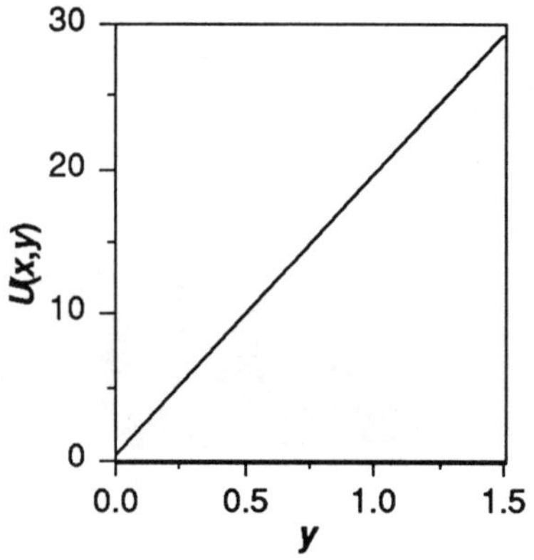

XY-PLOT

Data
Horizontal (*x*) axis: cells D14 - D28
Vertical (*y*) axis: cells B14 - B28

$$U(x,y) = mgy \tag{4.39}$$

where y is the height measured above some convenient reference plane, such as the surface of the Earth.[3]

As we write this book, few electronic spreadsheet programs are capable of plotting three-dimensional surfaces such as Equation 4.39 (Figure 4.11). Models 4.3 and 4.5, however, offer the opportunity to explore such surfaces with programs such as Lotus 1-2-3, Version 2.0, and Excel, Version 1.5. By the time you read this book, this situation may have changed and you should use whatever capabilities are available.

a. Build and verify Model 4.3.

The array of cells defined by Formula Set IV generates the value of $U(x,y)$ in Equation 4.39 for a grid of x-, y-values. You may expand this region to include a larger number of cells if you wish. The range of values included in this region are determined by the values of x_0, y_0, Δx, and Δy in cells B6 - B7, B9 - B10. The acceleration of gravity g and the mass of the object m are entered into cells B3 and B4, respectively.

Cells B14 - B28 (Formula Set II) reproduce a single column of data from Formula Set IV in a form that can be plotted with current spreadsheet programs. You can determine the level of the cut along the y-axis (value of x) through the surface of $U(x,y)$ by changing the value of x entered into cell B13.

Cells E30 - S30 (Formula Set V) reproduce a single row of data from Formula Set IV so that you determine the level of cut along the x-axis (value of y) through the surface of $U(x,y)$ by changing the value of y entered into cell D30.

b. Build two on-screen graphic images to plot $U(x,y)$ vs x and $U(x,y)$ vs y for different values of y and x (Figures 4.12 and 4.13) and explore these plots for different values of y and x until you can visualize the three-dimensional surface (Figure 4.11) of $U(x,y)$.

c. What is the shape of the kinetic energy K surface that forces $K + U$ to remain constant? What is the shape of the $K + U$ surface?

Build models similar to Model 4.3 that confirm your answer.

d. In conservative systems, Newton's second law of motion implies that the total mechanical energy $K + U$ remains constant. This result has a very interesting qualitative feature. If a particle's motion brings it back to some particular place it has visited before, *its speed will be exactly the same, although its velocity may be in a different direction.* Nonconservative forces do not produce motions that display this interesting property. It's one thing to understand in an abstract sense that this is true, and quite another to see it with your own eyes!

Model 4.4 takes Equations 4.38a and 4.38b and uses a Runge-Kutta method to solve this equation for position and velocity. The model is very similar to the two-dimensional Runge-Kutta models you built in Chapter 3. Model 4.4, however, includes three new columns at the extreme left that uses the output of your Runge-Kutta solution to calculate kinetic energy K (column B), potential energy U (column C), and total mechanical energy $K + U$ (column A).

Plot x vs y (Figure 4.14) for a variety of initial starting positions and velocities and correlate these plots with plots of K, U, and $K + U$ vs time (Figure 4.15).

As the object reaches its maximum height, what are the values of K, U, and $K + U$, respectively?

What are the values of K, U, and $K + U$ just before the object strikes the ground?

At these points and at all other points along the trajectory is total mechanical energy conserved?

e. Recall the shape of the $U(x,y)$ surface (Figure 4.11). Pick points at random along this surface and identify the state of the object with this amount of potential energy. How fast is it traveling? What is its distance from the ground? Can you determine whether the object is moving up or down?

In Exercise 4.4, you explore in greater detail what the potential energy function can tell you about the position and velocity of an object.

2. Consider an object of mass m orbiting a much more massive spherical object of mass M at distances where Newton's law of gravitation must be applied. If coordinate axes are chosen such that

Figure 4.14

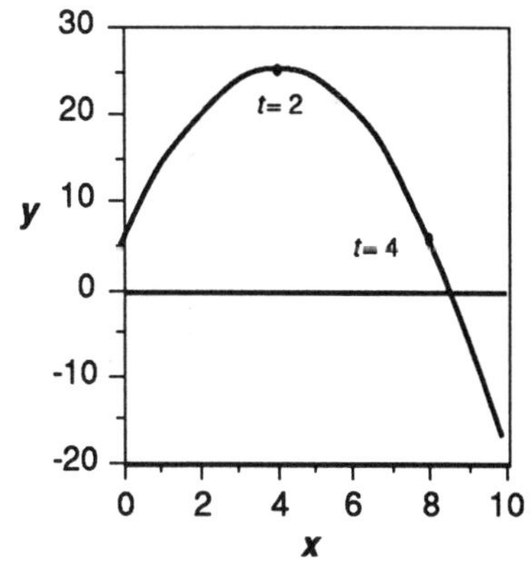

XY-PLOT

Data
Horizontal (x) axis: cells E10 - E60
Vertical (y) axis: cells G10 - G60

Figure 4.15

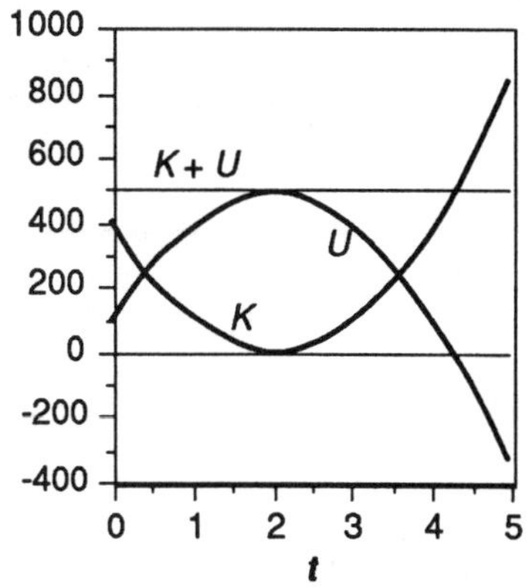

XY-PLOT

Data
Horizontal (x) axis: cells D10 - D60
Vertical (y) axis: cells A10 - A60 (K+U), B10 - B60 (K), C10 - C60 (U)

Figure 4.16

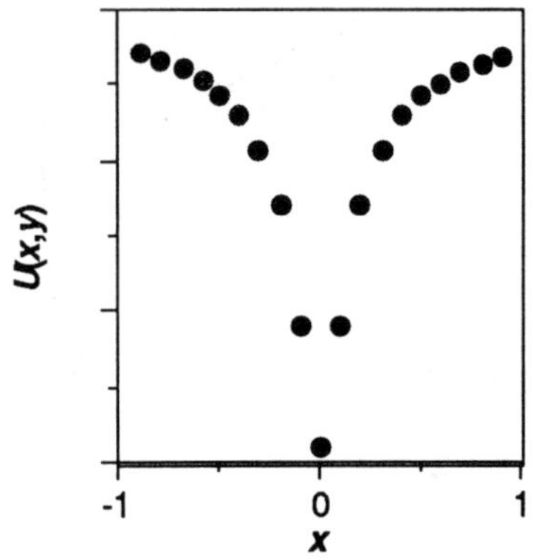

XY-PLOT

Data
Horizontal (*x*) axis: cells E14 - S14
Vertical (*y*) axis: cells E31 - S31

Notice that only the data points are presented. The relatively large step size of the model is such that continuous lines seriously misrepresent the shape of the function.

Figure 4.17

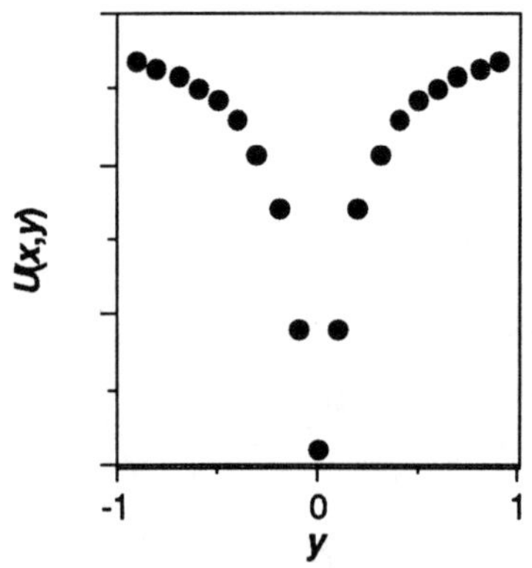

XY-PLOT

Data
Horizontal (*x*) axis: cells D15 - D29
Vertical (*y*) axis: cells B15 - B29

Notice that only the data points are presented. The relatively large step size of the model is such that continuous lines seriously misrepresent the shape of the function.

the orbit is confined to the *x*- ,*y*-plane and if the only force acting on this object is the force of gravity, then

$$F = -G\frac{Mm}{(x^2+y^2)^{3/2}}(xi + yj) \tag{4.40}$$

where *x*, *y* are the coordinates of the object of mass *m* in the *x*-, *y*-plane ($x = 0$ and $y = 0$ are taken to be the center of the object of mass *M*).

The potential energy function for the object of mass *m* is

$$U(x,y) = -G\frac{Mm}{\sqrt{x^2+y^2}} \tag{4.41}$$

The shape of $U(x,y)$ is shown in Figure 4.10.

a. Build and verify Model 4.5. Model 4.5 calculates the location of points along the surface $U(x,y)$ in Equation 4.41 in a fashion identical to that described previously for Model 4.3. Build two on-screen graphic images to plot $U(x,y)$ vs *x* and $U(x,y)$ vs *y* for different values of *y* and *x* (Figures 4.16 and 4.17) and explore these plots for different values of *y* and *x* until you can visualize the three-dimensional surface (Figure 4.10) of $U(x,y)$.

   Alternatively, you can plot several different cross sections simultaneously and get some idea of the three-dimensional representation.

b. What is the shape of the kinetic energy *K* surface that forces $K + U$ to remain constant? What is the shape of the $K + U$ surface?

   Build models similar to Model 4.5 that confirm your answer.

c. Build and verify Model 4.6

   Model 4.6 takes Equation 4.40 and uses a Runge-Kutta method to solve this equation for position and velocity. The model uses the output of your Runge-Kutta solution to calculate kinetic energy *K* (column B), potential energy *U* (column C), and total mechanical energy $K + U$ (column A).

   Plot *x* vs *y* (Figure 4.18) for a variety of initial starting positions and velocities and compare these plots with plots of *K*, *U*, and $K + U$ vs time (Figure 4.19).

Explore the values of K, U, and $K + U$ at various points along the object's orbit. Include opposite points of the ellipse in your explorations.

At these points and at all other points along the orbit, is total mechanical energy conserved?

d. Recall the shape of the $U(x,y)$ surface (Figure 4.10). Pick points at random along this surface and identify the state of the object with this amount of potential energy.

Notice that when the object returns to a place it has been before, it has not only the same *speed* (a consequence of the conservation of energy) but also the same *direction*. This latter characteristic is a special and distinctive feature of the inverse square force law when the total energy is negative (assuming that the potential energy has been defined to be zero at infinity). Even the slightest deviation from the inverse square form eliminates this special feature (as you will see in the problems).

Figure 4.18

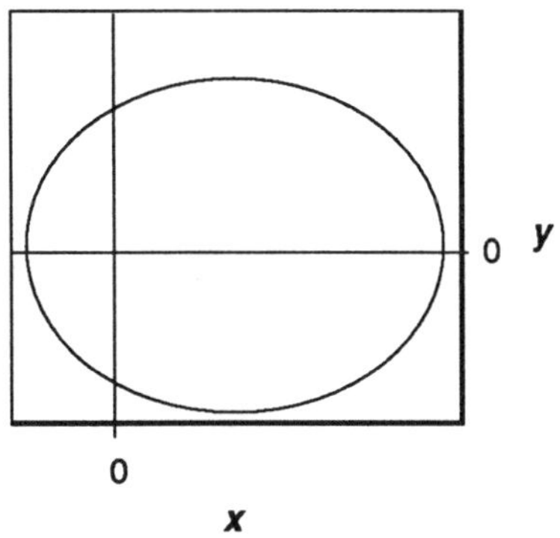

XY-PLOT

Data
Horizontal (x) axis: cells E11 - E110
Vertical (y) axis: cells G11 - G110

PROBLEMS

1. The motion of an object with a mass of 1 kg is driven by the force described in Equation 4.42:

$$\boldsymbol{F} = Ax^3\boldsymbol{i} + By\boldsymbol{j} \tag{4.42}$$

where $A = -2.3\ \text{J m}^{-4}$ and $B = -6.8\ \text{J m}^{-2}$.

The potential energy function for this force is

$$U(x,y) = -\frac{Ax^4}{4} - \frac{By^2}{2} + C \tag{4.43}$$

where $C = 8.4$ J.

a. Build, verify, and explore models that plot the shape (or cross sections of the shape) of F and U.

b. Build, verify, and explore a model that solves Equation 4.42 for position and velocity using a Runge-Kutta method. Demonstrate that total mechanical energy is conserved in this model.

c. Plot y vs x (the trajectory of the motion). Also plot v_x, v_y, and K vs x. Find two times at which the particle passes through the same location. Notice that the two velocity components are not

Figure 4.19

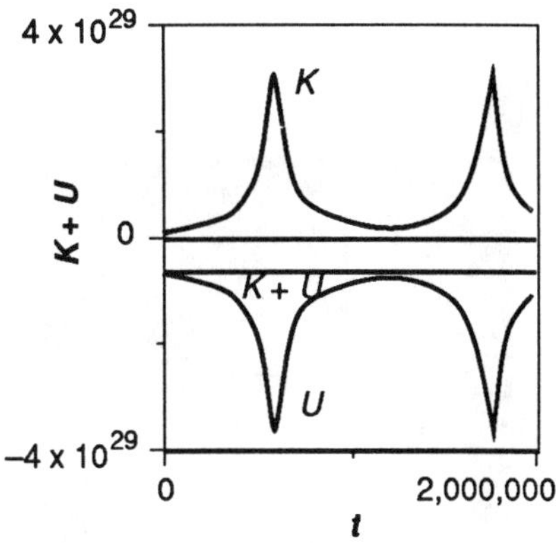

XY-PLOT

Data
Horizontal (x) axis: cells D11 - D110
Vertical (y) axis: cells A11 - A110 (K+U), B11 - B110 (K), C11 - C110 (U)

necessarily the same but that the kinetic energy (and so the speed of the particle) is unchanged.

2. The motion of an object is driven by the force described in Equation 4.44:

$$F = xy\boldsymbol{i} + y^3\mathbf{j} \tag{4.44}$$

Notice that this force is nonconservative (i.e., Equation 4.44 depends only on position but does not satisfy the conditions established by Equation 4.32).

   a. Build and verify a model that solves Equation 4.44 for position and velocity using a Runge-Kutta method.

   b. Find initial conditions such that the particle's trajectory passes through the same point in space more than once. For example, if the particle is initially at rest at $x_0 = 2$ and $y_0 = -2$ the particle occupies the point (0.502, 0.400) at both $t = 1.068$ and $t = 4.775$. Notice that the kinetic energy is very different at these two times. Thus, you can conclude that there is no function $U(x,y)$ such that $K + U$ is a constant.

   c. Add one or more columns to your model to calculate the work $W(t)$ done by the force between time 0 and any time t (Equation 4.45):

$$W(t) = \int_0^t (F_x v_x + F_y v_y)dt \tag{4.45}$$

   Notice that you have all the ingredients in your model to calculate this integral by Simpson's rule. How does the work done by this nonconservative force relate to kinetic energy (i.e., does Equation 4.18 of Exercise 4.1 apply)?

Figure 4.20

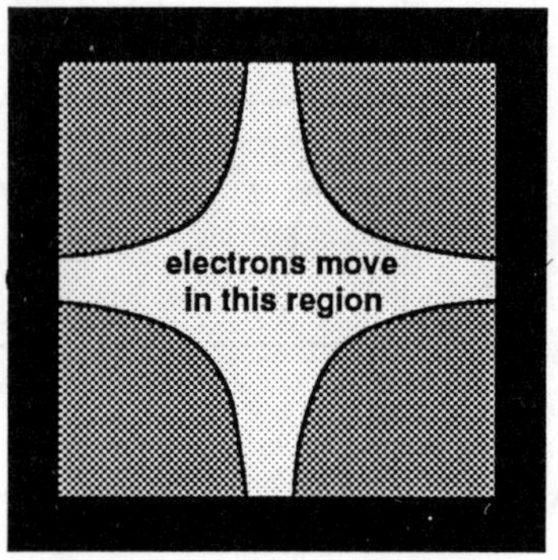

Problem 3 discusses an electronic device that consists of the space between two metallic hyperbolas.

3. Suppose an electronic device consists of the space inside two sets of hyperbolas made out of metal (Equation 4.46, Figure 4.20):

$$xy = a^2 \quad \text{and} \quad xy = -a^2 \tag{4.46}$$

where $a = 2$ cm. The force of an electron inside this device is

$$F_x = \frac{eVy}{a^2} \tag{4.47a}$$

$$F_y = \frac{eVx}{a^2} \tag{4.47b}$$

where $e = -1.6 \times 10^{-19}$ J V^{-1} (the amount of electric charge on the electron) and $V = 100$ V. The potential energy function from which this force law can be derived is

$$U(x,y) = -\frac{eVxy}{a^2} \qquad (4.48)$$

Electrons are released from the surface of these two hyperbolas by shining light on the metal. The kinetic energy of the electrons as they leave the surface is $2 \times 1.6 \times 10^{-19}$ J (i.e., 2 electron-volts), and their velocity is directed toward the inside of the device.

a. For various places on the electrode and for various initial directions, find the trajectory of the electron.

b. Demonstrate that the total energy is conserved in any such motion.

c. Find the time it takes for the electrons to go from one plate to another. (A convenient measure of time is 10^{-9} s = 1 ns = the time it takes light to travel about 1 ft.)

   What place and what initial direction produces the longest transit time through the device?

4. When the flattening of the Earth at the poles is considered, the gravitational potential energy function for a mass m at distance r from the Earth's center is

$$U(r,\theta) = \frac{-GMm}{r}\left[1 - 5.4 \times 10^{-4}\left(\frac{R_E}{r}\right)^2(3\cos^2(\theta) - 1)\right] \qquad (4.49)$$

where θ is the angle measured from the North Pole and $r > r_E$ is the Earth's radius.

$U(r,\theta)$ can be rewritten in terms of an x-, y-coordinate frame by assuming that the x-, y-plane passes through the poles of the Earth and that the y-axis passes through the North Pole. Under these conditons the relations

$$r = \sqrt{x^2 + y^2} \qquad (4.50)$$

$$\cos\theta = \frac{y}{r}, \quad \sin\theta = \frac{x}{r} \qquad (4.51)$$

can be used to derive the x-, and y-components of the force

Figure 4.21

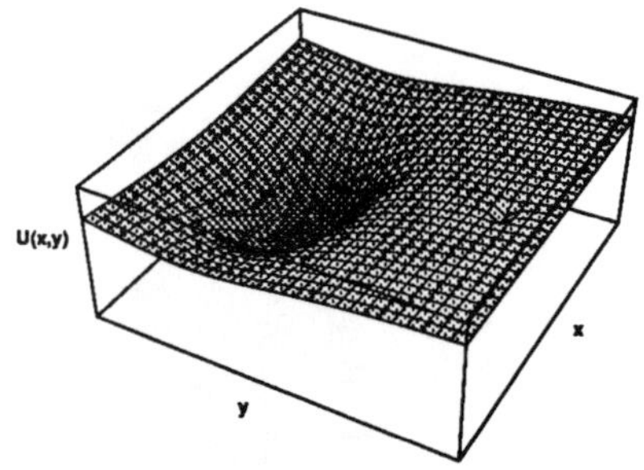

A spacecraft influenced by the gravitational fields of both the Earth and the Moon has an interesting potential energy function. You'll have to look carefully to find the Moon's influence on this combined field.

$$F_x = -\left(\frac{GMmx}{r^3}\right)\left(1 + \frac{3A}{r^2} - \frac{15Ay^2}{r^4}\right) \tag{4.52}$$

$$F_y = -\left(\frac{GMmy}{r^3}\right)\left(1 + \frac{9A}{r^2} - \frac{15Ay^2}{r^4}\right) \tag{4.53}$$

where $A = 5.4 \times 10^{-4} R_E^2$, $R_E = 6.38 \times 10^6$ m, $M = 5.98 \times 10^{24}$ kg, and $G = 6.673 \times 10^{-11}$ N m^2 kg^{-2}

Build, verify, and explore a model that solves for position, velocity, kinetic energy, potential energy, and total mechanical energy of a satellite of mass *m* starting at a distance of 200 km above the North Pole with a velocity such that the orbit would be circular if the Earth were a perfect sphere. You may wish to increase the value of *A* to see more clearly the qualitative effect of the Earth being out-of-round. Is the system conservative?

5. A spacecraft under the influence of both the Moon and the Earth experiences the gravitational attractions of both bodies. If the coordinate system origin is the center of the Earth, then the potential energy function for the spacecraft can be approximated by

$$U(x,y) = \frac{-GMm}{\sqrt{x^2+y^2}} + \frac{-GM'm}{\sqrt{(X-x)^2 + (Y-y)^2}} \tag{4.54}$$

where $x = y = 0$ is the center of the Earth, *X* and *Y* are the coordinates of the center of the Moon, *M* is the mass of the Earth (5.98×10^{24} kg), *M'* is the mass of the Moon (7.36×10^{22} kg), and the gravitational constant $G = 6.673 \times 10^{-11}$ N m^2 kg^{-2}. For simplicity, assume that the Moon is stationary at coordinates $X = 0.38 \times 10^9$ and $Y = 0$ m.

a. Analytically determine *F* and then build, verify, and explore a Runge-Kutta model of this system. Is the force conservative?

b. Build a model that permits you to explore the shape of *U(x,y)*. Compare your shape with Figure 4.21.

c. Explore the orbit of the spacecraft under a variety of initial conditions.

d. Allow the Moon to move in a circular orbit as a first approximation to its actual motion. Assume a constant tangential velocity v_{Moon} of 1023 m s^{-1}. Explore the orbital motion of the spacecraft leaving the Earth and going around the Moon when the Moon is moving.

MODELS

FORMULAS

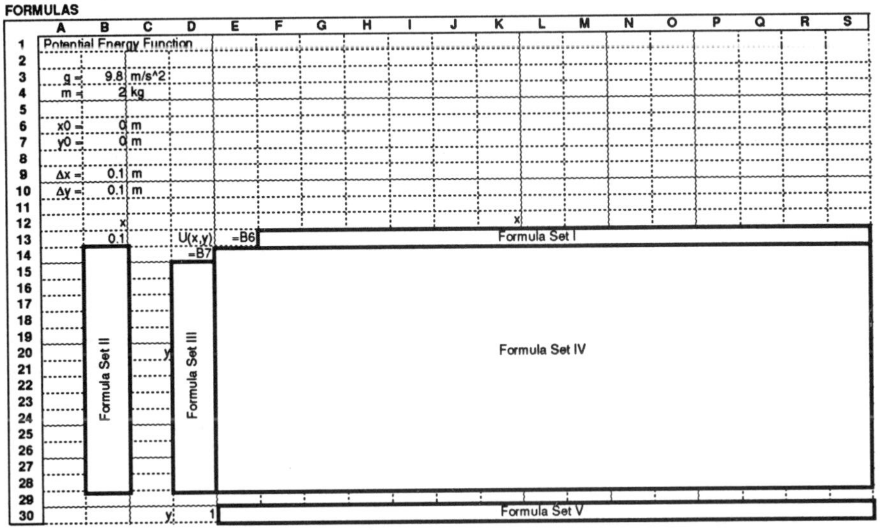

VALUES

	A	B	C	D	E	F	G	H	I	J	K	L	M	N	O	P	Q	R	S
1	Potential Energy Function																		
2																			
3	g =	9.8	m/s^2																
4	m =	2	kg																
5																			
6	x0 =	0	m																
7	y0 =	0	m																
8																			
9	Δx =	0.1	m																
10	Δy =	0.1	m																
11																			
12		x									x								
13		0.1		U(x,y)	0	0.1	0.2	0.3	0.4	0.5	0.6	0.7	0.8	0.9	1	1.1	1.2	1.3	1.4
14		0		0	0	0	0	0	0	0	0	0	0	0	0	0	0	0	0
15		-1.96		0.1	-1.96	-1.96	-1.96	-1.96	-1.96	-1.96	-1.96	-1.96	-1.96	-1.96	-1.96	-1.96	-1.96	-1.96	-1.96
16		-3.92		0.2	-3.92	-3.92	-3.92	-3.92	-3.92	-3.92	-3.92	-3.92	-3.92	-3.92	-3.92	-3.92	-3.92	-3.92	-3.92
17		-5.88		0.3	-5.88	-5.88	-5.88	-5.88	-5.88	-5.88	-5.88	-5.88	-5.88	-5.88	-5.88	-5.88	-5.88	-5.88	-5.88
18		-7.84		0.4	-7.84	-7.84	-7.84	-7.84	-7.84	-7.84	-7.84	-7.84	-7.84	-7.84	-7.84	-7.84	-7.84	-7.84	-7.84
19		-9.8		0.5	-9.8	-9.8	-9.8	-9.8	-9.8	-9.8	-9.8	-9.8	-9.8	-9.8	-9.8	-9.8	-9.8	-9.8	-9.8
20		-11.8	y	0.6	-11.8	-11.8	-11.8	-11.8	-11.8	-11.8	-11.8	-11.8	-11.8	-11.8	-11.8	-11.8	-11.8	-11.8	-11.8
21		-13.7		0.7	-13.7	-13.7	-13.7	-13.7	-13.7	-13.7	-13.7	-13.7	-13.7	-13.7	-13.7	-13.7	-13.7	-13.7	-13.7
22		-15.7		0.8	-15.7	-15.7	-15.7	-15.7	-15.7	-15.7	-15.7	-15.7	-15.7	-15.7	-15.7	-15.7	-15.7	-15.7	-15.7
23		-17.6		0.9	-17.6	-17.6	-17.6	-17.6	-17.6	-17.6	-17.6	-17.6	-17.6	-17.6	-17.6	-17.6	-17.6	-17.6	-17.6
24		-19.6		1	-19.6	-19.6	-19.6	-19.6	-19.6	-19.6	-19.6	-19.6	-19.6	-19.6	-19.6	-19.6	-19.6	-19.6	-19.6
25		-21.6		1.1	-21.6	-21.6	-21.6	-21.6	-21.6	-21.6	-21.6	-21.6	-21.6	-21.6	-21.6	-21.6	-21.6	-21.6	-21.6
26		-23.5		1.2	-23.5	-23.5	-23.5	-23.5	-23.5	-23.5	-23.5	-23.5	-23.5	-23.5	-23.5	-23.5	-23.5	-23.5	-23.5
27		-25.5		1.3	-25.5	-25.5	-25.5	-25.5	-25.5	-25.5	-25.5	-25.5	-25.5	-25.5	-25.5	-25.5	-25.5	-25.5	-25.5
28		-27.4		1.4	-27.4	-27.4	-27.4	-27.4	-27.4	-27.4	-27.4	-27.4	-27.4	-27.4	-27.4	-27.4	-27.4	-27.4	-27.4
29																			
30			y:	1	-19.6	-19.6	-19.6	-19.6	-19.6	-19.6	-19.6	-19.6	-19.6	-19.6	-19.6	-19.6	-19.6	-19.6	-19.6

Formula Set I Prototype Cell Is F13	Formula Set II Prototype Cell Is B14	Formula Set III Prototype Cell Is D15
=E13+B9	=B4*B3*D14	=D14+B10

Formula Set IV Prototype Cell Is E14	Formula Set V Prototype Cell Is E30
=B4*B3*D14	=B4*B3*D30

Model 4.3

BUILD and verify Model 4.3.

ENTER data using cells B3 and B4 (acceleration of gravity, mass of object), B6 and B7 (starting values for *x* and *y*), B9 and B10 (increment values for *x* and *y*), B13 (*x*-value for cut through surface parallel to the *y*-axis), D30 (*y*-value for cut through surface parallel to *x*-axis).

READ output from the model at cells E14 - S28 to obtain an overview of the values on the surface $U(x,y)$. Perhaps by the time you read this book, your spreadsheet program will be able to plot this surface.

READ output from the model at cells B14 - B28 to obtain a plot of the values on the surface $U(x,y)$ on the cut surface parallel to the *y*-axis and at the *x*-depth specified at cell B13.

READ output from the model at cells E30 - S30 to obtain a plot of the values on the surface $U(x,y)$ on the cut surface parallel to the *x*-axis and at the *y*-depth specified at cell D30.

Model 4.4

BUILD and verify Model 4.4.

ENTER data using cells E3 (Δt), H2 and H3 (acceleration of gravity, mass of object), E5 and E6 (initial *x*- and *y*-positions of object), H5 and H6 (initial *x*- and *y*-velocities of object).

READ output from the model from columns A - E and G to obtain total mechanical energy, kinetic energy, potential energy, time, *x*-position, and *y*-position, respectively.

NOTICE that the model is quite similar to the two-dimensional Runge-Kutta models that you explored in Chapter 3.

FORMULAS

	A	B	C	D	E	F	G	H	I	J	K	L
1				2D Conservation								
2							g =	9.8	m/s^2			
3				Δt =	0.1	s	m =	2	kg			
4												
5				s0x =	0	m	v0x =	2	m/s			
6				s0y =	5	m	v0y =	20	m/s			
7												
8	K+U	K	U	t	sx	vx	sy	vy	ksx0	kvx0	ksy0	kvy0
9	J	J	J	sec	m	m/s	m	m/s				
10				0	=E5	=H5	=E6	=H6				
11–14	Form Set I	Form Set II	Form Set III	Form Set IV	Form Set V	Form Set VI	Form Set VII	Form Set VIII	Form Set IX	Form Set X	Form Set XI	Form Set XII

FORMULAS (Contd.)

	M	N	O	P	Q	R	S	T	U	V	W	X
8	ksx1	kvx1	ksy1	kvy1	ksx2	kvx2	ksy2	kvy2	ksx3	kvx3	ksy3	kvy3
9												
10												
11–14	Form Set XIII	Form Set XIV	Form Set XV	Form Set XVI	Form Set XVII	Form Set XVIII	Form Set XIX	Form Set XX	Form Set XXI	Form Set XXII	Form Set XXIII	Form Set XXIV

VALUES

	A	B	C	D	E	F	G	H	I	J	K	L
1				2D Conservation								
2							g =	9.8	m/s^2			
3				Δt =	0.1	s	m =	2	kg			
4												
5				s0x =	0	m	v0x =	2	m/s			
6				s0y =	5	m	v0y =	20	m/s			
7												
8	K+U	K	U	t	sx	vx	sy	vy	ksx0	kvx0	ksy0	kvy0
9	J	J	J	sec	m	m/s	m	m/s				
10	502	404	98	0	0	2	5	20				
11	502	365.76	136.24	0.1	0.2	2	6.951	19.02	0.2	0	2	-0.98
12	502	329.44	172.56	0.2	0.4	2	8.804	18.04	0.2	0	1.902	-0.98
13	502	295.04	206.96	0.3	0.6	2	10.559	17.06	0.2	0	1.804	-0.98
14	502	262.57	239.43	0.4	0.8	2	12.216	16.08	0.2	0	1.706	-0.98

VALUES (Contd.)

	M	N	O	P	Q	R	S	T	U	V	W	X
8	ksx1	kvx1	ksy1	kvy1	ksx2	kvx2	ksy2	kvy2	ksx3	kvx3	ksy3	kvy3
9												
10												
11	0.2	0	1.951	-0.98	0.2	0	1.951	-0.98	0.2	0	1.902	-0.98
12	0.2	0	1.853	-0.98	0.2	0	1.853	-0.98	0.2	0	1.804	-0.98
13	0.2	0	1.755	-0.98	0.2	0	1.755	-0.98	0.2	0	1.706	-0.98
14	0.2	0	1.657	-0.98	0.2	0	1.657	-0.98	0.2	0	1.608	-0.98

Model 4.4 (continued)

Formula Set I
Prototype Cell is A10

=B10+C10

Formula Set II
Prototype Cell is B10

=**H3** *(F10^2+H10^2)/2

Formula Set III
Prototype Cell is C10

= **H3*H2** *G10

Formula Set IV
Prototype Cell is D11

=D10+**E3**

Formula Set V
Prototype Cell is E11

=E10+(I11+2*M11+2*Q11+U11)/6

Formula Set VI
Prototype Cell is F11

=F10+(J11+2*N11+2*R11+V11)/6

Formula Set VII
Prototype Cell is G11

=G10+(K11+2*O11+2*S11+W11)/6

Formula Set VIII
Prototype Cell is H11

=H10+(L11+2*P11+2*T11+X11)/6

Formula Set IX
Prototype Cell is I11

=**E3** *F10

Formula Set X
Prototype Cell is J11

0

Formula Set XI
Prototype Cell is K11

=**E3** *H10

Formula Set XII
Prototype Cell is L11

=**-E3*H2**

Formula Set XIII
Prototype Cell is M11

=**E3** *(F10+J11/2)

Formula Set XIV
Prototype Cell is N11

0

Formula Set XV
Prototype Cell is O11

=**E3** *(H10+L11/2)

Formula Set XVI
Prototype Cell is P11

=**-E3*H2**

Formula Set XVII
Prototype Cell is Q11

=**E3** *(F10+N11/2)

Formula Set XVIII
Prototype Cell is R11

0

Formula Set XIX
Prototype Cell is S11

=**E3** *(H10+P11/2)

Formula Set XX
Prototype Cell is T11

=**-E3*H2**

Formula Set XXI
Prototype Cell is U11

=**E3** *(F10+R11)

Formula Set XXII
Prototype Cell is V11

0

Formula Set XXIII
Prototype Cell is W11

=**E3** *(H10+T11)

Formula Set XXIV
Prototype Cell is X11

=**-E3*H2**

Model 4.5

BUILD and verify Model 4.5.

ENTER data using cells B3 - B5 (acceleration of gravity, mass of Earth, mass of orbiting object), B7 and B8 (starting values for x and y), B10 and B11 (increment values for x and y), B14 (x-value for cut through surface parallel to the y-axis), D31 (y-value for cut through surface parallel to x-axis).

READ output from the model at cells E15 - S29 to obtain an overview of the values on the surface $U(x,y)$. Perhaps by the time you read this book, your spreadsheet program will be able to plot this surface.

READ output from the model at cells B15 - B29 to obtain a plot of the values on the surface $U(x,y)$ on the cut surface parallel to the y-axis and at the x-depth specified at cell B14.

READ output from the model at cells E31 - S31 to obtain a plot of the values on the surface $U(x,y)$ on the cut surface parallel to the x-axis and at the y-depth specified at cell D31.

FORMULAS

	A	B	C	D	E	F	G	H	I	J	K	L	M	N	O	P	Q	R	S
1	Potential Energy Function																		
2																			
3	G =	6.67E-11	Nm^2/kg^2																
4	M =	5.98E+24	kg																
5	m =	7.34E+22	kg																
6																			
7	x0 =	-0.5	m																
8	y0 =	-0.5	m																
9																			
10	Δx =	0.1	m																
11	Δy =	0.1	m																
12																			
13		x:									x:								
14		0.1		U(x,y):	=B7	Formula Set I													
15		Formula Set II		=B8	Formula Set IV														
16				Formula Set III															
17																			
18																			
19																			
20																			
21			y:																
22																			
23																			
24																			
25																			
26																			
27																			
28																			
29																			
30																			
31			y:	0.1	Formula Set V														

VALUES

	A	B	C	D	E	F	G	H	I	J	K	L	M	N	O	P	Q	R	S
1	Potential Energy Function																		
2																			
3	G =	6.67E-11	Nm^2/kg^2																
4	M =	5.98E+24	kg																
5	m =	7.34E+22	kg																
6																			
7	x0 =	-0.5	m																
8	y0 =	-0.5	m																
9																			
10	Δx =	0.1	m																
11	Δy =	0.1	m																
12																			
13		x:									x:								
14		0.1		U(x,y):	-0.5	-0.4	-0.3	-0.2	-0.1	-3E-17	0.1	0.2	0.3	0.4	0.5	0.6	0.7	0.8	0.9
15		-5.7E+37		-0.5	-4E+37	-5E+37	-5E+37	-5E+37	-6E+37	-6E+37	-6E+37	-5E+37	-5E+37	-5E+37	-4E+37	-4E+37	-3E+37	-3E+37	-3E+37
16		-7.1E+37		-0.4	-5E+37	-5E+37	-6E+37	-7E+37	-7E+37	-7E+37	-7E+37	-7E+37	-6E+37	-5E+37	-5E+37	-4E+37	-4E+37	-3E+37	-3E+37
17		-9.3E+37		-0.3	-5E+37	-6E+37	-7E+37	-8E+37	-9E+37	-1E+38	-9E+37	-8E+37	-7E+37	-6E+37	-5E+37	-4E+37	-4E+37	-3E+37	-3E+37
18		-1.3E+38		-0.2	-5E+37	-7E+37	-8E+37	-1E+38	-1E+38	-1E+38	-1E+38	-1E+38	-8E+37	-7E+37	-5E+37	-5E+37	-4E+37	-4E+37	-3E+37
19		-2.1E+38		-0.1	-6E+37	-7E+37	-9E+37	-1E+38	-2E+38	-3E+38	-2E+38	-1E+38	-9E+37	-7E+37	-6E+37	-5E+37	-4E+37	-4E+37	-3E+37
20		-2.9E+38		0	-6E+37	-7E+37	-1E+38	-1E+38	-3E+38	-7E+53	-3E+38	-1E+38	-1E+38	-7E+37	-6E+37	-5E+37	-4E+37	-4E+37	-3E+37
21		-2.1E+38	y:	0.1	-6E+37	-7E+37	-9E+37	-1E+38	-2E+38	-3E+38	-2E+38	-1E+38	-9E+37	-7E+37	-6E+37	-5E+37	-4E+37	-4E+37	-3E+37
22		-1.3E+38		0.2	-5E+37	-7E+37	-8E+37	-1E+38	-1E+38	-1E+38	-1E+38	-1E+38	-8E+37	-7E+37	-5E+37	-5E+37	-4E+37	-4E+37	-3E+37
23		-9.3E+37		0.3	-5E+37	-6E+37	-7E+37	-8E+37	-9E+37	-1E+38	-9E+37	-8E+37	-7E+37	-6E+37	-5E+37	-4E+37	-4E+37	-3E+37	-3E+37
24		-7.1E+37		0.4	-5E+37	-5E+37	-6E+37	-7E+37	-7E+37	-7E+37	-7E+37	-7E+37	-6E+37	-5E+37	-5E+37	-4E+37	-4E+37	-3E+37	-3E+37
25		-5.7E+37		0.5	-4E+37	-5E+37	-5E+37	-5E+37	-6E+37	-6E+37	-6E+37	-5E+37	-5E+37	-5E+37	-4E+37	-4E+37	-3E+37	-3E+37	-3E+37
26		-4.8E+37		0.6	-4E+37	-4E+37	-4E+37	-5E+37	-5E+37	-5E+37	-5E+37	-5E+37	-4E+37	-4E+37	-4E+37	-3E+37	-3E+37	-3E+37	-3E+37
27		-4.1E+37		0.7	-3E+37	-4E+37	-4E+37	-4E+37	-4E+37	-4E+37	-4E+37	-4E+37	-4E+37	-4E+37	-3E+37	-3E+37	-3E+37	-3E+37	-3E+37
28		-3.6E+37		0.8	-3E+37	-3E+37	-3E+37	-4E+37	-4E+37	-4E+37	-4E+37	-4E+37	-3E+37	-3E+37	-3E+37	-3E+37	-3E+37	-3E+37	-2E+37
29		-3.2E+37		0.9	-3E+37	-3E+37	-3E+37	-3E+37	-3E+37	-3E+37	-3E+37	-3E+37	-3E+37	-3E+37	-3E+37	-3E+37	-3E+37	-2E+37	-2E+37
30																			
31			y:	0.1	-6E+37	-7E+37	-9E+37	-1E+38	-2E+38	-3E+38	-2E+38	-1E+38	-9E+37	-7E+37	-6E+37	-5E+37	-4E+37	-4E+37	-3E+37

Formula Set I Prototype Cell is F14
=E14+ **B10**

Formula Set II Prototype Cell is B15
=-(**B3*****B4*****B5**)/SQRT(B14 ^2+**D**15^2)

Formula Set III Prototype Cell is D16
=D15+ **B11**

Formula Set IV Prototype Cell is E15
=-(**B3*****B4*****B5**)/SQRT(E14 ^2+**D**15^2)

Formula Set V Prototype Cell is E31
=-(**B3*****B4*****B5**)/SQRT(E14 ^2+**D** 31^2)

FORMULAS

	A	B	C	D	E	F	G	H	I	J	K	L
1				Orbits (2D)			G =	6.67E-11	Nm^2/kg^2			
2							m1 =	5.98E+24	kg	M =	=H2+H3	kg
3				Δt =	20000		m2 =	7.34E+22	kg	–GM =	=-H1*K2	
4												
5				s0x =	4E+08	m	v0x =	0	m/s			
6				s0y =	0	m	v0y =	600	m/s			
7				s0z =	0	m	v0z =	0	m/s			
8												
9	K+U	K	U	t	sx	vx	sy	vy	ksx0	kvx0	ksy0	kvy0
10				sec	m	m/s	m	m/s				
11	Form Set I	Form Set II	Form Set III	0	=E5	=H5	=E6	=H6				
12–15				Form Set IV	Form Set V	Form Set VI	Form Set VII	Form Set VIII	Form Set IX	Form Set X	Form Set XI	Form Set XII

FORMULAS (Contd.)

	M	N	O	P	Q	R	S	T	U	V	W	X
9	ksx1	kvx1	ksy1	kvy1	ksx2	kvx2	ksy2	kvy2	ksx3	kvx3	ksy3	kvy3
10												
11												
12–15	Form Set XIII	Form Set XIV	Form Set XV	Form Set XVI	Form Set XVII	Form Set XVIII	Form Set XIX	Form Set XX	Form Set XXI	Form Set XXII	Form Set XXIII	Form Set XXIV

VALUES

	A	B	C	D	E	F	G	H	I	J	K	L
1				Orbits (2D)			G =	6.67E-11	Nm^2/kg^2			
2							m1 =	5.98E+24	kg	M =	6E+24	kg
3				Δt =	20000		m2 =	7.34E+22	kg	–GM =	-4E+14	
4												
5				s0x =	4E+08	m	v0x =	0	m/s			
6				s0y =	0	m	v0y =	600	m/s			
7				s0z =	0	m	v0z =	0	m/s			
8												
9	K+U	K	U	t	sx	vx	sy	vy	ksx0	kvx0	ksy0	kvy0
10				sec	m	m/s	m	m/s				
11	-6E+28	1E+28	-7E+28	0	4E+08	0	0	600				
12	-6E+28	1E+28	-7E+28	20000	3.99E+08	-50.49	11994950	599.2422	0	-50.47	1E+07	0
13	-6E+28	1E+28	-7E+28	40000	3.98E+08	-101.099	23959526	596.9595	-1E+06	-50.53	1E+07	-1.517
14	-6E+28	1E+28	-7E+28	60000	3.95E+08	-151.946	35862984	593.1242	-2E+06	-50.71	1E+07	-3.053
15	-6E+28	1E+28	-7E+28	80000	3.92E+08	-203.156	47673829	587.6886	-3E+06	-51.01	1E+07	-4.626

VALUES (Contd.)

	M	N	O	P	Q	R	S	T	U	V	W	X
9	ksx1	kvx1	ksy1	kvy1	ksx2	kvx2	ksy2	kvy2	ksx3	kvx3	ksy3	kvy3
10												
11												
12	-5E+05	-50.45	1E+07	-0.757	-5E+05	-50.52	1E+07	-0.758	-1E+06	-50.53	1E+07	-1.517
13	-2E+06	-50.57	1E+07	-2.28	-2E+06	-50.64	1E+07	-2.283	-2E+06	-50.71	1E+07	-3.052
14	-3E+06	-50.81	1E+07	-3.831	-3E+06	-50.87	1E+07	-3.836	-3E+06	-51.01	1E+07	-4.626
15	-4E+06	-51.17	1E+07	-5.429	-4E+06	-51.24	1E+07	-5.437	-4E+06	-51.43	1E+07	-6.256

Model 4.6

BUILD and verify Model 4.6.

ENTER data using cells E3 (Δt), H1 - H3 (Newton's gravitational constant, mass of Earth, mass of orbiting object), E5 and E6 (initial x and y positions of object), H5 and H6 (initial x- and y-velocities of object).

READ output from the model from columns A - E and G to obtain total mechanical energy, kinetic energy, potential energy, time, x-position, and y-position, respectively.

NOTICE that the model is quite similar to the two-dimensional Runge-Kutta models that you explored in Chapter 3.

Model 4.6 (continued)

Formula Set I
Prototype Cell is A11

=B11+C11

Formula Set II
Prototype Cell is B11

=**H3** *(F11^2+H11^2)/2

Formula Set III
Prototype Cell is C11

=-**H1** ***H2** ***H3** /SQRT(E11^2+G11^2)

Formula Set IV
Prototype Cell is D12

=D11+**E3**

Formula Set V
Prototype Cell is E12

=E11+(I12+2*M12+2*Q12+U12)/6

Formula Set VI
Prototype Cell is F12

=F11+(J12+2*N12+2*R12+V12)/6

Formula Set VII
Prototype Cell is G12

=G11+(K12+2*O12+2*S12+W12)/6

Formula Set VIII
Prototype Cell is H12

=H11+(L12+2*P12+2*T12+X12)/6

Formula Set IX
Prototype Cell is I12

=**E3** *F11

Formula Set X
Prototype Cell is J12

=**E3** ***K3** *E11/(E11^2+G11^2)^1.5

Formula Set XI
Prototype Cell is K12

=**E3** *H11

Formula Set XII
Prototype Cell is L12

=**E3** ***K3** *G11/(E11^2+G11^2)^1.5

Formula Set XIII
Prototype Cell is M12

=**E3** *(F11+J12/2)

Formula Set XIV
Prototype Cell is N12

=**E3** ***K3** *(E11+I12/2)/((E11+I12/2)^2+(G11+K12/2)^2)^1.5

Formula Set XV
Prototype Cell is O12

=**E3** *(H11+L12/2)

Formula Set XVI
Prototype Cell is P12

=**E3** ***K3** *(G11+K12/2)/((E11+I12/2)^2+(G11+K12/2)^2)^1.5

Formula Set XVII
Prototype Cell is Q12

=**E3** *(F11+N12/2)

Formula Set XVIII
Prototype Cell is R12

=**E3** ***K3** *(E11+M12/2)/((E11+M12/2)^2+(G11+O12/2)^2)^1.5

Formula Set XIX
Prototype Cell is S12

=**E3** *(H11+P12/2)

Formula Set XX
Prototype Cell is T12

=**E3** ***K3** *(G11+O12/2)/((E11+M12/2)^2+(G11+O12/2)^2)^1.5

Formula Set XXI
Prototype Cell is U12

=**E3** *(F11+R12)

Formula Set XXII
Prototype Cell is V12

=**E3** ***K3** *(E11+Q12)/((E11+Q12)^2+(G11+S12)^2)^1.5

Formula Set XXIII
Prototype Cell is W12

=**E3** *(H11+T12)

Formula Set XIV
Prototype Cell is X12

=**E3** ***K3** *(G11+S12)/((E11+Q12)^2+(G11+S12)^2)^1.5

EXERCISE **4.3**

THE CONSERVATION OF ENERGY: NONCONSERVATIVE SYSTEMS

TO GET THE MOST OUT OF THIS EXERCISE, YOU SHOULD ALREADY KNOW

The definitions of the terms:
potential energy
kinetic energy
total mechanical energy

How to solve second-order differential equations using Runge-Kutta methods.

The previous two exercises explored mechanical systems free of dissipative forces such as friction (and satisfying other special conditions). Such systems are called conservative because the total mechanical energy E_0 is constant (i.e., "conserved"). Furthermore E_0 is partitioned between two forms of energy: kinetic energy K and potential energy U. For a single object subject to conservative forces

$$E_0 = K + U = \frac{mv^2}{2} + U(x) \tag{4.55}$$

where m is the mass of the object, v is the velocity of the object, and $U(x)$ is the potential energy function.

In the real world, large-scale (macroscopic) mechanical systems are nonconservative. Friction and other dissipative forces are always present and often very useful.[1] The total macroscopic mechanical energy of such systems decreases with time because forces such as friction transmute a portion of the energy into forms other than macroscopic kinetic energy and potential energy (e.g., heat and light).[2] Total energy is still conserved, but a new term must be added to the right side of Equation 4.55 (Equation 4.56):

$$E_{tot} = K + U + E_{nc} \tag{4.56}$$

Figure 4.22

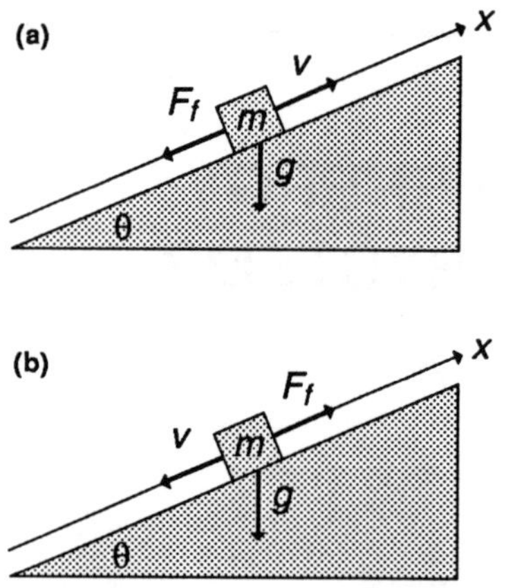

A block of mass *m* moves along an incline of angle θ. The *x*-axis is chosen to coincide with the incline with *x*-values increasing to the right. The origin of the *x*-axis is taken as the base of the incline.

(a) When the velocity of the block ***v*** is directed up the incline, the force of friction ***F**f* is directed down the incline.

(b) When the velocity of the block ***v*** is directed down the incline, the force of friction ***F**f* is directed up the incline.

The friction force therefore works against the motion of the block both when it slides up the incline and when it slides down the incline. The friction force is nonconservative.

where E_{tot} is the total energy and E_{nc} is the energy associated with nonconservative forces. It was shown by James Prescott Joule in the 1840s that the mechanical energy lost due to friction shows up as heat. Other forms of energy were subsequently discovered, and it is these physical facts that elevate Equation 4.56 from a mere mathematical definition of the symbol E_{nc} to a physical law.

Consider the example of a block of mass *m* sliding down a surface inclined from the horizontal by angle θ and subject to friction determined by μ_k, the coefficient of kinetic friction (Figure 4.22).

The net force acting on the block is

$$\Sigma F = F_{grav} + F_{frict} = -mg \sin \theta - \varepsilon(v)\, \mu_k\, mg \cos \theta \tag{4.57}$$

where F_{grav} is the gravitational force and F_{frict} is the frictional force and where

$$\varepsilon(v) = \begin{cases} +1 \text{ when } v = dx/dt > 0 \\ 0 \text{ when } v = 0 \\ -1 \text{ when } v < 0 \end{cases} \tag{4.58}$$

Notice that this discontinuous function is necessary to take account of the fact that the friction force always acts to *resist* the motion. The formula given here is an approximate representation of sliding friction that is useful for many purposes. It does not cover such effects as static friction or those, for example, that lead to the awful screech that sometimes happens when chalk moves across a blackboard. You will consider static friction in Problem 1.

This is the first time you have encountered a force law that is not a continuous function. In the real world, the change in the friction force as the velocity changes sign is actually smooth but occurs over a very small range of velocities. In physics and other sciences, discontinuous functions are often used to represent very rapid change when extreme precision is unnecessary. However, as you will see, you must be careful when using numeric techniques such as the Runge-Kutta method because they implicitly assume that the functions fed into them are smooth. When used with care, they are still quite serviceable despite this violation of one of the assumptions of their derivation. It is sometimes necessary to take special precautions to avoid the difficulties shown in the left side of Figure 2.23 of Exercise 2.4. You will see techniques that automate this process of step alignment later.

You can determine the position and velocity of the block at any time by solving Equation 4.57 using either analytic or numeric (e.g., Runge-Kutta) methods.

As the block moves along the incline, work is performed both by the conservative force F_{grav} (W_{grav}) and the nonconservative force F_{frict} (W_{frict}). W_{grav} changes the potential energy ΔU of the block (Equation 4.59) and W_{frict} directly increases the "lost" mechanical energy associated with nonconservative forces ΔE_{nc} (Equation 4.60).[3]

$$\Delta U(x) = -W_{grav} = -\int_{x_0}^{x} F_{grav}\, dx$$

$$= \int_{x_0}^{x} mg \sin\theta \; dx = mg(x - x_0)\sin\theta \qquad (4.59)$$

$$\Delta E_{nc} = -W_{frict} = -\int_{x_0}^{x} F_{frict}\, dx$$

$$= \int_{x_0}^{x} \varepsilon(v)\mu_k mg \; \cos\theta \; dx \qquad (4.60)$$

where x_0, x are the starting and ending positions along the incline.

Notice that you cannot calculate the integral in Equation 4.60 until *the initial conditions have been specified, the equation of motion solved, and the sense of the velocity determined.* This is in sharp contrast to the case of the gravity force where the work integral can always be evaluated even in the absence of any knowledge about the motion.

If the nonconservative energy of the block is assigned value of zero at $t = 0$ and if the potential energy U of the block is assigned a value of zero when the block is resting on the ground (and if $x = 0$ when this is true), then these two forms of energy may be calculated using Equations 4.61 and 4.62:

$$U = mgx \sin\theta \qquad (4.61)$$

$$E_{frict} = \int_{x_0}^{x} \varepsilon(v)\mu_k \mathrm{mg} \cos\theta \; \mathrm{dx} \qquad (4.62)$$

In this exercise you explore the movements of objects subject to friction forces and discover that total macroscopic mechanical energy decreases when friction is taken into account. In this exercise, you can only give this reduction in the mechanical energy a name, E_{frict}; it requires experiment to verify that this amount of mechanical energy turns into

Figure 4.23

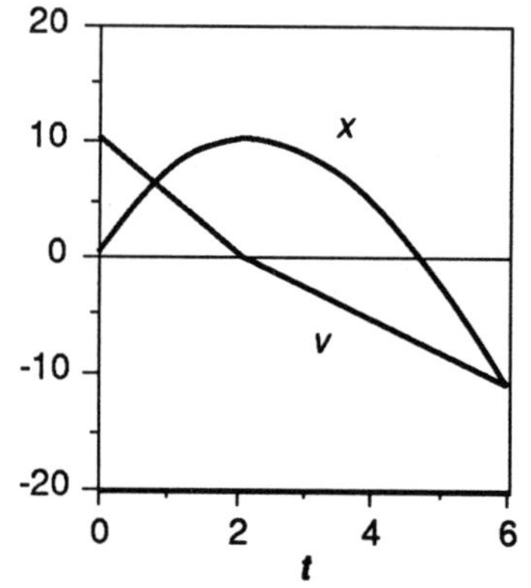

XY-PLOT

Data
Horizontal (*x*) axis: cells J13 - J75 (*t*)
Vertical (*y*) axis: cells K13 - K75 (*x*), L13 - L75 (*v*)

Figure 4.24

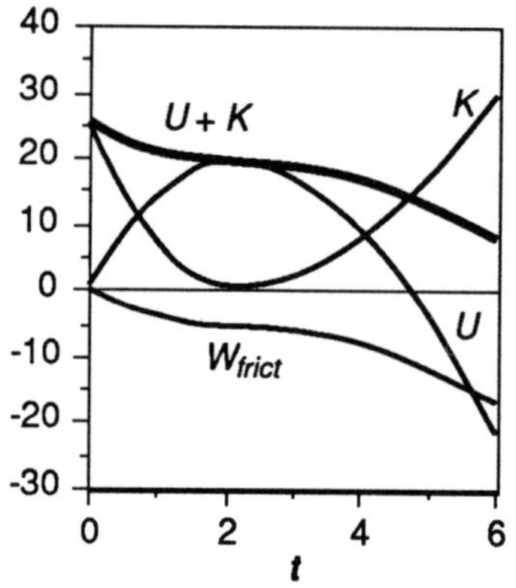

XY-PLOT

Data
Horizontal (*x*) axis: cells J13 - J75 (*t*)
Vertical (*y*) axis: cells A13 - A75 (*K*), B13 - B75 (*U*), C13 - C75 (*K* + *U*), F13 - F75 (W_f)

heat. With more elaborate experiments and more profound models of Nature, it can be shown that most of this heat actually ends up as microscopic kinetic energy, microscopic potential energy, and radiant energy. The study of heat and its effects is called *thermodynamics*, and the careful investigation of atoms and molecules in intimate contact is called *condensed matter physics*.

LESSONS

Build and verify Model 4.7. Model 4.7 uses Equations 4.57 and a Runge-Kutta method to calculate the position and velocity of a block sliding on an incline of angle θ. The model also monitors the kinetic energy of the block K, the work done by the gravitational force W_g, the frictional force W_f, and their sum, the potential energy due to gravity U, the total mechanical energy $(K + U)$, and the total mechanical energy plus heat $(K + U - W_f)$.

Build on-screen graphics that plot the position and velocity of the block vs time (Figure 4.23), the energy partitions, and the work done by the friction force vs time (Figure 4.24).

1. Begin your exploration of the behavior of the system by starting a block of mass 0.5 kg at the bottom of the ramp with a velocity of 10 m s^{-1} going up the ramp ($x_0 = 0$, $v_0 = 10$ m s^{-1}). The ramp has an incline θ of 23° and a coefficient of kinetic friction μ_k of 0.

   a. What is the total mechanical energy of the block as it starts up the ramp? What is the total mechanical energy of the block when it completes its journey and returns to the bottom of the ramp? Is the system conservative?

   b. Monitor the velocity of the block as it starts up the ramp, comes to a halt, and then slides back down to its starting position. Is the rate of change of velocity a constant throughout?

2. Introduce a friction force into your system by setting the coefficient of kinetic friction μ_k to values of 0.06, 0.12, and 0.18. In each case,

   a. What is the total mechanical energy of the block as it starts up the ramp? What is the total mechanical energy of the block when it completes its journey and returns to the bottom of the ramp? Is the system conservative?

   Your observations should verify that, when friction is present, there cannot be any function of position only (a potential energy function) that can be added to kinetic energy to produce an energy conservation law.

How does total mechanical energy compare to the work done by the frictional force? Explain in simple words why this relationship exists.

b. Monitor the velocity of the block as it starts up the ramp, comes to a halt, and then slides back down to its starting position. Notice that the rate of change of velocity changes when velocity changes direction. Explain in simple words why this happens.

3. While μ_k has a value of 0.12, gradually increase θ until it has a value of 90°. What happens to $K + U$ vs time as θ increases? Is total mechanical energy conserved when θ equals 90°? Explain why.

4. Investigate your model when θ is so small that the force of gravity is less than the friction force. Examine the model carefully under conditions where velocity is near zero.

   a. Do you believe that your model is giving an accurate representation of Nature in this case?

   b. Does making Δt very, very small give realistic motions when the velocity is near (but not at) zero?

   In Problem 1, you will build a model that more accurately represents the actual behavior of sliding friction.

5. Input a variety of different initial conditions into your model and explore the motion of the block and the partition of its various forms of energy. Explain in simple language what happens to the block and its energy partitions under each set of conditions that you test.

   Include in your exploration large values of Δt (e.g., 5 s). Notice what happens to the MaxErr/K_0 cell (what does this cell calculate?), and study the four *kvf* columns of your worksheet. With large values Δt, the Runge-Kutta calculation can become unstable and produce nonsense because of the discontinuity in the force function. You will see this phenomenon again in later examples.

PROBLEMS

1. If you push a brick up a ramp with a small slope, it is unlikely to slide back down. To make it return to the bottom of the ramp, you might

either increase the slope of the ramp until it starts slipping or you might give it a shove to "get it going." This is the phenomenon of *static* friction.

When two solids press against one another (e.g., the block and the ramp) and their relative velocity is zero, they tend to stick together and a certain, minimum applied force $F_{critical}$ is necessary to get them moving relative to each other. For forces less than $F_{critical}$ the friction force adjusts itself to oppose and exactly match the applied force (yielding a net force of zero). The value of $F_{critical}$ may be calculated in terms of a coefficient of static friction μ_s:

$$F_{critical} = \mu_s F_{norm} \tag{4.63}$$

where F_{norm} is the component of the force normal to the surface between the two solids. It is important to realize that μ_s is always larger than μ_k.

A more accurate model of the friction force acting in the x-direction to resist the motion of a block of mass m along a ramp of angle θ is then

$$F_{friction} = \begin{cases} mg \sin\theta & \text{if } mg\sin\theta \le F_{critical} = \mu_s\, mg\cos\theta \text{ and } v = 0 \\ -\mu_k\, mg\sin\theta & \text{if } mg\sin\theta > F_{critical} \text{ and } v > 0 \\ +\mu_k\, mg\sin\theta & \text{if } mg\sin\theta < F_{critical} \text{ and } v < 0 \end{cases} \tag{4.64}$$

Model 4.8 demonstrates one method for including the phenomenon of static friction into the Runge-Kutta calculations for the motion of a block on a ramp. Equation 4.64 is implemented by using the built-in spreadsheet functions AND and NOT—functions that you have not yet encountered in this book. Their use in the model should be apparent by inspection, but you may also wish to read what your software reference manual has to say about logical functions. You should realize that in a numeric calculation the exact condition $v = 0$ rarely happens because v can "step over" 0 within the interval Δt. In the model therefore, note is taken of the times in which the various velocities in the Runge-Kutta calculation *change sign*, and it uses this change as the indicator for the $v = 0$ condition.[4]

Build and verify Model 4.8. While Model 4.8 offers a better description of sliding friction than that shown in the lessons, it provides nothing new toward your understanding of energy and its conservation. Model 4.8 does, however, show how to incorporate more complicated circumstances about Nature into your computer simulations.

a. Consider the behavior of your system when it consists of a block of mass 0.5 kg at the bottom of the ramp with a velocity of 10 m s^{-1} going up the ramp. The ramp has an incline $\theta = 23°$, the coefficient of static friction $\mu_s = 0.2$ and the coefficient of kinetic friction $\mu_k = 0$.

   Explore the behavior of the various works and energies of the system and verify that the system is conservative.

b. Introduce a friction force to your system by setting the coefficient of kinetic friction μ_k to values of 0.06, 0.12, and 0.18 and repeat your explorations.

c. Holding μ_k constant at 0.12, gradually increase μ_s until the block stops at the top of its motion. What value of μ_k just brings the block to a stop (to two significant figures)? What happens if you gradually increase the angle θ of the ramp?

e. Under various initial conditions, determine the fraction of initial kinetic energy that is converted to heat in those cases where the block stops.

f. What happens to your model when $\mu_s < \mu_k$ (a nonphysical condition)?

2. A block of mass 1 kg moves on a table with sliding friction. It is attached to a horizontal spring that is anchored to the far wall. The spring constant $k = 5$ N m^{-1} and, as usual, the acceleration of gravity = 9.8 m s^{-2}.

   Modify either Model 4.7 or 4.8 to study the motion of the block as you increase the coefficient of kinetic friction from 0 to 0.2. Assume that the block is initially at rest and is released a distance 1 m from its resting position. If you include static friction in your model, set its coefficient 20% higher than kinetic friction.

Plot position, velocity, and energies vs time on the same graph for about 6 s of motion. You may also wish to plot velocity vs time separately in order to observe its structure when the block's velocity passes through zero and when it finally stops. When there is friction, are the time intervals between moments of zero velocity equal? Discuss.

3. Modify your model in Problem 2 to include a velocity-dependent retarding force due to an oil-filled damping pot (a shock absorber) attached to the block.

$$F_{shock} = -bv \tag{4.65}$$

where b is constant related to the properties of the damping pot.

   a. Begin your investigations with no kinetic sliding friction and observe the effects of this new friction force varying b from 0 to 5 N m^{-1} s.

   b. Set the coefficient of sliding friction to a value of 0.05 and find the smallest value of b (to two significant figures) such that the block just stops the first time the velocity approaches zero after being released.

4. An automobile experiences a retarding force (Equation 4.66) as it ploughs through the air.

$$F_{retard} = -\frac{A\rho v^2 \varepsilon(v)}{2} \tag{4.66}$$

where A is the effective frontal area of the car (2 m^2), ρ is the density of the air (1.293 kg m^{-3} at sea level and 0°C), v is the velocity of the car, and $\varepsilon(v)$ is the discontinuous function shown in Equation 4.58.

If the automobile has a mass of 1100 kg and is acted upon by both this retarding force and a constant forward force of friction of 1750 N (from the interface between the car's tires and the road when the engine is running flat out), build and verify a model that can determine the position, velocity, and work done by both the driving force and the retarding force. (This exercise occurs on the Bonneville flats, and there are no potential energies in this problem).

   a. Assuming that the automobile starts from rest, plot its position, velocity, kinetic energy, and the work done by the two forces as a function of time.[5]

b. Add other columns to your model to calculate and plot the power associated with the retarding force and the drive force (power is force times velocity) in horsepower. One horsepower = 745 J s^{-1}.

c. Calculate the velocity of the car in miles per hour (you can convert from meters per second by knowing that 1 mile = 5280 ft and 1 in = 2.54 cm).

d. How fast can this car travel, and how long does it take to attain 90% of its fastest speed.[6]

e. Take your automobile off the flats and drive it along a slope of your choosing (adding a gravitational component of force that either retards or augments its motion). What are the maximum uphill and downhill speeds of your car on different terrains?

f. What parameters do you think might reasonably describe an automobile such as a TestaRossa or Lamberghini?

Model 4.7

BUILD and verify Model 4.7.

ENTER data using cells B5 - B7 (Δt, initial position, initial velocity) and E5 - E8 (angle of incline, mass of object, acceleration of gravity, and kinetic friction coefficient).

READ output from the model from columns J - L (time, position, velocity) and A - G (kinetic energy, potential energy, total mechanical energy, total energy, work of gravity, work of friction, total work).

MODELS

FORMULAS

	A	B	C	D	E	F	G	H	I	J	K	L	M	N	O
1	Up and down motion of a mass on a slide with sliding friction														
2	NOTE: Effect of static friction is not included														
3	(Sliding Friction force = mu*(normal force) directed opposite to velocity)														
4															
5	Δt =	0.1	s	theta =	23	°	=E5*PI()/180		Fg/m =	=E7*SIN(G5)					
6	x0 =	0	m	m =	0.5	kg			Ff/m =	=E7*COS(G5)*E8					
7	v0 =	10	m/s	g =	9.8	m/s^2			MaxError/K0 =	=MAX(MAX(I13:I75),-MIN(I13:I75))					
8				μk =	0.12										
9															
10				K+U-			Tot.	MaxError Queries					0		
11	K	U	K+U	W(f)	W(g)	W(f)	Work	?K	?K-W	t	x	v	kx	kvg	kvf
12	(J)	(J)	(J)	(J)	(J)	(J)	(J)	(J)	(J)	(s)	(m)	(m/s)			
13					0	0		0		0	=B6	=B7			
14–16	Form Set I	Form Set II	Form Set III	Form Set IV	Form Set V	Form Set VI	Form Set VII	Form Set VIII	Form Set IX	Form Set X	Form Set XI	Form Set XII	Form Set XIII	Form Set XIV	Form Set XV

FORMULAS (Contd.)

	P	Q	R	S	T	U	V	W	X	Y	Z	AA	AB	AC	AD
10			1					2					3		
11	x	v	kx	kvg	kvf	x	v	kx	kvg	kvf	x	v	kx	kvg	kvf
12															
13–16	Form Set XVI	Form Set XVII	Form Set XVIII	Form Set XIX	Form Set XX	Form Set XXI	Form Set XXII	Form Set XXIII	Form Set XXIV	Form Set XXV	Form Set XXVI	Form Set XXVII	Form Set XXVIII	Form Set XXIX	Form Set XXX

VALUES

	A	B	C	D	E	F	G	H	I	J	K	L	M	N	O
1	Up and down motion of a mass on a slide with sding friction														
2	NOTE: Effect of static friction is not included														
3	(Sliding Friction force = mu*(normal force) directed opposite to velocity)														
4															
5	Δt =	0.1	s	theta =	23	°	0.4014	rad	Fg/m =	3.8292	m/s^2				
6	x0 =	0	m	m =	0.5	kg			Ff/m =	1.0825	m/s^2				
7	v0 =	10	m/s	g =	9.8	m/s^2			MaxError/K0 =	65.749	<- gets big when vel changes sign				
8				μk =	0.12										
9															
10				K+U-			Tot.	MaxError Queries					0		
11	K	U	K+U	W(f)	W(g)	W(f)	Work	?K	?K-W	t	x	v	kx	kvg	kvf
12	(J)	(J)	(J)	(J)	(J)	(J)	(J)	(J)	(J)	(s)	(m)	(m/s)			
13	25	0	25	25	0	0	0	0	0	0	0	10	1	-0.4	-0.1
14	22.604	1.8676	24.472	25	-1.868	-0.528	-2.396	-47.6	-45.21	0.1	0.9754	9.5088	0.95	-0.4	-0.1
15	20.33	3.6411	23.971	25	-3.641	-1.029	-4.67	-45.33	-40.66	0.2	1.9018	9.0177	0.9	-0.4	-0.1
16	18.175	5.3206	23.496	25	-5.321	-1.504	-6.825	-43.18	-36.35	0.3	2.779	8.5265	0.85	-0.4	-0.1

VALUES (Contd.)

	P	Q	R	S	T	U	V	W	X	Y	Z	AA	AB	AC	AD
10			1					2					3		
11	x	v	kx	kvg	kvf	x	v	kx	kvg	kvf	x	v	kx	kvg	kvf
12															
13	0.5	9.75	0.98	-0.4	-0.1	0.49	9.75	0.98	-0.4	-0.1	0.98	9.51	0.95	-0.4	-0.1
14	1.45	9.26	0.93	-0.4	-0.1	1.44	9.26	0.93	-0.4	-0.1	1.9	9.02	0.9	-0.4	-0.1
15	2.35	8.77	0.88	-0.4	-0.1	2.34	8.77	0.88	-0.4	-0.1	2.78	8.53	0.85	-0.4	-0.1
16	3.21	8.28	0.83	-0.4	-0.1	3.19	8.28	0.83	-0.4	-0.1	3.61	8.04	0.8	-0.4	-0.1

Model 4.7 (continued)

Formula Set I	Prototype Cell is A13	=0.5*E6*L13^2
Formula Set II	Prototype Cell is B13	=J5*E6*(K13)
Formula Set III	Prototype Cell is C13	=A13+B13
Formula Set IV	Prototype Cell is D13	=C13-F13
Formula Set V	Prototype Cell is E14	=E13+E6*(N13*L13+2*(S13*Q13+X13*V13)+AC13*AA13)/6
Formula Set VI	Prototype Cell is F14	=F13+E6*(O13*L13+2*(T13*Q13+Y13*V13)+AD13*AA13)/6
Formula Set VII	Prototype Cell is G13	=E13+F13
Formula Set VIII	Prototype Cell is H14	=-A14-A13
Formula Set IX	Prototype Cell is I13	=H13-G13
Formula Set X	Prototype Cell is J14	=J13+B5
Formula Set XI	Prototype Cell is K14	=K13+(M13+2*(R13+W13)+AB13)/6
Formula Set XII	Prototype Cell is L14	=L13+(N13+O13+2*(S13+T13+X13+Y13)+AC13+AD13)/6
Formula Set XIII	Prototype Cell is M13	=L13*B5
Formula Set XIV	Prototype Cell is N13	=-B5*(J5)
Formula Set XV	Prototype Cell is O13	=-B5*(IF(L13=0,0,IF(L13>0,1,-1)*J6))
Formula Set XVI	Prototype Cell is P13	=K13+M13/2
Formula Set XVII	Prototype Cell is Q13	=L13+(N13+O13)/2
Formula Set XVIII	Prototype Cell is R13	=Q13*B5
Formula Set XIX	Prototype Cell is S13	=-B5*(J5)
Formula Set XX	Prototype Cell is T13	=-B5*(IF(Q13=0,0,IF(Q13>0,1,-1)*J6))
Formula Set XXI	Prototype Cell is U13	=K13+R13/2
Formula Set XXII	Prototype Cell is V13	=L13+(S13+T13)/2
Formula Set XXIII	Prototype Cell is W13	=V13*B5
Formula Set XXIV	Prototype Cell is X13	=-B5*(J5)
Formula Set XXV	Prototype Cell is Y13	=-B5*(IF(V13=0,0,IF(V13>0,1,-1)*J6))
Formula Set XXVI	Prototype Cell is Z13	=K13+W13
Formula Set XXVII	Prototype Cell is AA13	=L13+(X13+Y13)
Formula Set XXVIII	Prototype Cell is AB13	=AA13*B5
Formula Set XXIX	Prototype Cell is AC13	=-B5*(J5)
Formula Set XXX	Prototype Cell is AD13	=-B5*(IF(AA13=0,0,IF(AA13>0,1,-1)*J6))

Model 4.8

BUILD and verify Model 4.8.

ENTER data using cells B5 - B7 (Δt, initial position, initial velocity) and E5 - E9 (angle of incline, mass of object, acceleration of gravity, static friction coefficient, and kinetic friction coefficient).

READ output from the model from columns J - L (time, position, velocity) and A - G (kinetic energy, potential energy, total mechanical energy, total energy, work of gravity, work of friction, total work).

NOTICE that the built-in logical functions of this model must be written differently depending on your spreadsheet program.

FORMULAS

	A	B	C	D	E	F	G	H	I	J	K	L	M	N	O
1	Up and down motion of a mass on a slide with sliding friction														
2	NOTE: Effect of static friction is not included														
3	(Sliding Friction force = mu*(normal force) directed opposite to velocity)														
4															
5	Δt =	0.1	s	theta	23	°	=E5*PI()/180		Fg/m =	=E7*SIN(G5)					
6	x0 =	0	m	m =	0.5	kg			Ffs/m =	=J7*E8/E9					
7	v0 =	10	m/s	g =	9.8	m/s^2			Ffk/m =	=E7*COS(G5)*E9					
8				µs =	0.2				kvgcrit =	=J7*(E8/E9)*B5					
9				µk =	0.12				MxErr/K0 =	=MAX(MAX(I14:I76),-MIN(I14:I76))					
10															
11				K+U-			Tot.	MaxError Queries					0		
12	K	U	K+U	W(f)	W(g)	W(f)	Work	?K	?K-W	t	x	v	kx	kvg	kvf
13	(J)	(J)	(J)	(J)	(J)	(J)	(J)	(J)	(J)	(s)	(m)	(m/s)			
14					0	0		0		0	=B6	=B7			
15–17	Form Set I	Form Set II	Form Set III	Form Set IV	Form Set V	Form Set VI	Form Set VII	Form Set VIII	Form Set IX	Form Set X	Form Set XI	Form Set XII	Form Set XIII	Form Set XIV	Form Set XV

FORMULAS (Contd.)

	P	Q	R	S	T	U	V	W	X	Y	Z	AA	AB	AC	AD	AE
12	x	v	kx	kvg	kvf	x	v	kx	kvg	kvf	x	v	kx	kvg	kvf	delv
13																
14–17	Form Set XVI	Form Set XVII	Form Set XVIII	Form Set XIX	Form Set XX	Form Set XXI	Form Set XXII	Form Set XXIII	Form Set XXIV	Form Set XXV	Form Set XXVI	Form Set XXVII	Form Set XXVIII	Form Set XXIX	Form Set XXX	Form Set XXXI

VALUES

	A	B	C	D	E	F	G	H	I	J	K	L	M	N	O
1	Up and down motion of a mass on a slide with sdling friction														
2	NOTE: Effect of static friction is not included														
3	(Sliding Friction force = mu*(normal force) directed opposite to velocity)														
4															
5	Δt =	0.1	s	theta	23	°	0.4014	rad	Fg/m =	3.8292	m/s^2				
6	x0 =	0	m	m =	0.5	kg			Ffs/m =	1.8042	m/s^2				
7	v0 =	10	m/s	g =	9.8	m/s^2			Ffk/m =	1.0825	m/s^2				
8				µs =	0.2				kvgcrit =	0.1804	m/s				
9				µk =	0.12				MxErr/K0 =	0.0029	<- gets big when vel changes sign				
10															
11				K+U-			Tot.	MaxError Queries					0		
12	K	U	K+U	W(f)	W(g)	W(f)	Work	?K	?K-W	t	x	v	kx	kvg	kvf
13	(J)	(J)	(J)	(J)	(J)	(J)	(J)	(J)	(J)	(s)	(m)	(m/s)			
14	25	0	25	25	0	0	0	0	0	0	0	10	1	-0.4	-0.1
15	22.604	1.8676	24.472	25	-1.868	-0.528	-2.396	-2.396	4E-15	0.1	0.9754	9.5088	0.95	-0.4	-0.1
16	20.33	3.6411	23.971	25	-3.641	-1.029	-4.67	-4.67	6E-15	0.2	1.9018	9.0177	0.9	-0.4	-0.1
17	18.175	5.3206	23.496	25	-5.321	-1.504	-6.825	-6.825	9E-15	0.3	2.779	8.5265	0.85	-0.4	-0.1

VALUES (Contd.)

	P	Q	R	S	T	U	V	W	X	Y	Z	AA	AB	AC	AD	AE
12	x	v	kx	kvg	kvf	x	v	kx	kvg	kvf	x	v	kx	kvg	kvf	delv
13																
14	0.5	9.75	0.98	-0.4	-0.1	0.49	9.75	0.98	-0.4	-0.1	0.98	9.51	0.95	-0.4	-0.1	-0.5
15	1.45	9.26	0.93	-0.4	-0.1	1.44	9.26	0.93	-0.4	-0.1	1.9	9.02	0.9	-0.4	-0.1	-0.5
16	2.35	8.77	0.88	-0.4	-0.1	2.34	8.77	0.88	-0.4	-0.1	2.78	8.53	0.85	-0.4	-0.1	-0.5
17	3.21	8.28	0.83	-0.4	-0.1	3.19	8.28	0.83	-0.4	-0.1	3.61	8.04	0.8	-0.4	-0.1	-0.5

Model 4.8 (continued)

This model contains logic functions that cannot be tested with one set of input data. Portions of Model 4.8 are shown here using different inputs to test different portions of the logic.

	D	E
8	μs =	0.2
9	μk =	0.12

	K		K
12	x	45	8.5938
13	(m)	46	8.2847
14	0	47	7.9482
15	0.9754	48	7.5843
16	1.9018	49	7.1929
17	2.779	50	6.774
18	3.6071	51	6.3276
19	4.386	52	5.8538
20	5.1159	53	5.3525
21	5.7966	54	4.8237
22	6.4283	55	4.2675
23	7.0108	56	3.6838
24	7.5442	57	3.0727
25	8.0284	58	2.434
26	8.4636	59	1.7679
27	8.8496	60	1.0744
28	9.1866	61	0.3534
29	9.4744	62	-0.395
30	9.7131	63	-1.171
31	9.9026	64	-1.975
32	10.043	65	-2.805
33	10.134	66	-3.664
34	10.177	67	-4.55
35	10.173	68	-5.463
36	10.139	69	-6.404
37	10.077	70	-7.372
38	9.9879	71	-8.368
39	9.8711	72	-9.391
40	9.7269	73	-10.44
41	9.5552	74	-11.52
42	9.356	75	-12.62
43	9.1294	76	-13.76
44	8.8753		

	D	E
8	μs =	0.45
9	μk =	0.12

	K		K
12	x	45	10.185
13	(m)	46	10.185
14	0	47	10.185
15	0.9754	48	10.185
16	1.9018	49	10.185
17	2.779	50	10.185
18	3.6071	51	10.185
19	4.386	52	10.185
20	5.1159	53	10.185
21	5.7966	54	10.185
22	6.4283	55	10.185
23	7.0108	56	10.185
24	7.5442	57	10.185
25	8.0284	58	10.185
26	8.4636	59	10.185
27	8.8496	60	10.185
28	9.1866	61	10.185
29	9.4744	62	10.185
30	9.7131	63	10.185
31	9.9026	64	10.185
32	10.043	65	10.185
33	10.134	66	10.185
34	10.177	67	10.185
35	10.185	68	10.185
36	10.185	69	10.185
37	10.185	70	10.185
38	10.185	71	10.185
39	10.185	72	10.185
40	10.185	73	10.185
41	10.185	74	10.185
42	10.185	75	10.185
43	10.185	76	10.185
44	10.185		

Model 4.8 (continued)

The built-in logical functions AND and NOT have two different representations depending upon your spreadsheet program. Programs such as Excel represent these operations as functions where the arguments are included within parentheses (e.g., AND(predicate 1, predicate 2, ...)). The formula sets shown on this page use this notation.

Programs such as Lotus 1-2-3 represent the operations AND and NOT using a form of *infix* notation (e.g., predicate 1 #AND# predicate 2 #AND# ...). If your program uses infix notation, replace formula sets 12, 15, 17, 20, 22, 25, 27, and 30 with the formula sets on the next page.

Formula Set I
Prototype Cell is A14
```
=0.5*E6*L14^2
```

Formula Set II
Prototype Cell is B14
```
=J5*E6*(K14)
```

Formula Set III
Prototype Cell is C14
```
=A14+B14
```

Formula Set IV
Prototype Cell is D14
```
=C14-F14
```

Formula Set V
Prototype Cell is E15
```
=E14+E6*(N14*L14+2*(S14*Q14+X14*V14)+AC14*AA14)/6
```

Formula Set VI
Prototype Cell is F15
```
=F14+E6*(O14*L14+2*(T14*Q14+Y14*V14)+AD14*AA14)/6
```

Formula Set VII
Prototype Cell is G14
```
=E14+F14
```

Formula Set VIII
Prototype Cell is H15
```
=A15-A14
```

Formula Set IX
Prototype Cell is I14
```
=H14-G14
```

Formula Set X
Prototype Cell is J15
```
=J14+B5
```

Formula Set XI
Prototype Cell is K15
```
=K14+(M14+2*(R14+W14)+AB14)/6
```

Formula Set XII
Prototype Cell is L15
```
=IF(AND(ABS(N14)<J8,NOT((L14+AE14)*L14>0)),0,L14+AE14)
```

Formula Set XIII
Prototype Cell is M14
```
=L14*B5
```

Formula Set XIV
Prototype Cell is N14
```
=-B5*(J5)
```

Formula Set XV
Prototype Cell is O14
```
=IF(AND(L14=0,ABS(N14)<J8),-N14,-B5*(IF(L14>0,1,-1)*J7))
```

Formula Set XVI
Prototype Cell is P14
```
=K14+M14/2
```

Formula Set XVII
Prototype Cell is Q14
```
=IF(AND(ABS(N14)<J8,NOT((L14+(N14+O14)/2)*L14>0)),0,L14+(N14+O14)/2)
```

Formula Set XVIII
Prototype Cell is R14
```
=Q14*B5
```

Formula Set XIX
Prototype Cell is S14
```
=-B5*(J5)
```

Formula Set XX
Prototype Cell is T14
```
=IF(AND(Q14=0,ABS(S14)<J8),-S14,-B5*(IF(Q14>0,1,-1)*J7))
```

Formula Set XXXI
Prototype Cell is AE14
```
=(N14+O14+2*(S14+T14+X14+Y14)+AC14+AD14)/6
```

Formula Set XXI
Prototype Cell is U14
```
=K14+R14/2
```

Formula Set XXII
Prototype Cell is V14
```
=IF(AND(ABS(S14)<J8,NOT((L14+(S14+T14)/2)*L14>0)),0,L14+(S14+T14)/2)
```

Formula Set XXIII
Prototype Cell is W14
```
=V14*B5
```

Formula Set XXIV
Prototype Cell is X14
```
=-B5*(J5)
```

Formula Set XXV
Prototype Cell is Y14
```
=IF(AND(V14=0,ABS(X14)<J8),-X14,-B5*(IF(V14>0,1,-1)*J7))
```

Formula Set XXVI
Prototype Cell is Z14
```
=K14+W14
```

Formula Set XXVII
Prototype Cell is AA14
```
=IF(AND(ABS(X14)<J8,NOT((L14+(X14+Y14))*L14>0)),0,L14+(X14+Y14))
```

Formula Set XXVIII
Prototype Cell is AB14
```
=AA14*B5
```

Formula Set XXIX
Prototype Cell is AC14
```
=-B5*(J5)
```

Formula Set XXX
Prototype Cell is AD14
```
=IF(AND(AA14=0,ABS(AC14)<J8),-AC14,-B5*(IF(AA14>0,1,-1)*J7))
```

Formula Set XII
Prototype Cell is L15

```
=IF(ABS(N14)<J8 #AND##NOT#(L14+AE14)*L14>
0,0,L14+AE14)
```

Formula Set XV
Prototype Cell is O14

```
=IF(L14=0#AND#ABS(N14)< J8 ,-N14,-
B5 *(@IF(L14>0,1,-1)* J7 ))
```

Formula Set XVII
Prototype Cell is Q14

```
=IF(ABS(N14)<J8 #AND##NOT#(L14+(N14+O14)/
2)*L14>0,0,L14+(N14+O14)/2)
```

Formula Set XX
Prototype Cell is T14

```
=IF(Q14=0#AND#ABS(S14)< J8 ,-S14,-
B5 *(IF(Q18>0,1,-1)* J7 ))
```

Formula Set XXII
Prototype Cell is V14

```
=IF(ABS(S14)<J8 #AND##NOT#(L14+(S14+T14)/
2)*L14>0,0,L14+(S14+T14)/2)
```

Formula Set XXV
Prototype Cell is Y14

```
=IF(V14=0#AND#ABS(X14)< J8 ,-X14,-
B5 *(IF(V14>0,1,-1)* J7 ))
```

Formula Set XXVII
Prototype Cell is AA14

```
=IF(ABS(X14)<J8 #AND##NOT#(L14+(X14+Y14))*
L14>0,0,L14+(X14+Y14))
```

Formula Set XXX
Prototype Cell is AD14

```
=IF(AA14=0#AND#ABS(AC14)< J8 ,-AC14,-
B5 *(IF(AA14>0,1,-1)* J7 ))
```

Model 4.8 (continued)

The built-in logical functions AND and NOT have two different representations depending on your spreadsheet program. Programs such as Excel represent these operations as functions where the arguments are included within parentheses (e.g., AND(predicate 1, predicate 2, ...)). The formula sets shown on the previous page use this notation.

Programs such as Lotus 1-2-3 represent the operations AND and NOT using a form of *infix* notation (e.g., predicate 1 #AND# predicate 2 #AND# ...). If your program uses infix notation, all formula sets on the previous page are correct except for formula sets 12, 15, 17, 20, 22, 25, 27 and 30. Replace those formula sets with the formula sets on this page.

EXERCISE **4.4**

THE ENERGY INTEGRAL

TO GET THE MOST OUT OF THIS EXERCISE, YOU SHOULD ALREADY KNOW

The definitions of the terms:
potential energy
kinetic energy
total mechanical energy

How to numerically integrate functions using Simpson's rule.

Newton's second law of motion permits you to determine the position, velocity, and acceleration of an object from a knowledge of the forces acting on it by directly solving the second-order differential equations of motion. In one-dimensional conservative systems, the law of conservation of mechanical energy makes it possible to obtain the motion of an object by a computationally simpler method if you have a simple functional form for the potential energy. Of course, this method is exactly equivalent to directly solving Newton's differential equations of motion, but part of the job has already been done when the potential energy function is derived from the force law. In one dimension, you can easily use this extra knowledge to reduce the amount of computation necessary to obtain the motion. In two and three dimensions, this also can be done sometimes, but not always.

Consider a one-dimensional mechanical system that has all of its energy partitioned between kinetic energy K and potential energy U. The total energy of the system E is

$$E = K + U = \frac{mv^2}{2} + U(x) \tag{4.67}$$

Equation 4.67 can be solved for v:

$$v = \frac{dx}{dt} = \pm \frac{\sqrt{E - U(x)}}{\sqrt{m/2}} \tag{4.68}$$

Two signs are shown because the velocity can be either positive or negative. Energy conservation is never sufficient to get the direction of a velocity. Further information must be added, perhaps in the form of an initial condition, to choose one of the two possible signs. You can integrate Equation 4.68 to get

$$\int_{t_0}^{t} dt = t - t_0 = \pm \sqrt{m/2} \int_{x_0}^{x} \frac{dx}{\sqrt{E - U(x)}} \tag{4.69}$$

If you choose the plus sign, then choosing $x > x_0$ yields $t > t_0$ — exactly what you would expect if $v > 0$. Similarly, choosing the minus sign implies that $x < x_0$, $t > t_0$, and $v < 0$.

Equation 4.69 is often called the *energy integral*, and it rarely has an analytic solution. Elementary physics textbooks discuss it only briefly to say that solutions depend on "sophisticated computer techniques." You, of course, have access to these techniques through your spreadsheet program and your knowledge of numeric integration methods such as Simpson's rule.

A word of caution is necessary, however. When all energy of the system is held as potential energy, the denominator of the energy integral equals zero and the value of the integrand is indeterminate. This happens any time the velocity is zero, a common initial condition. Even if the initial velocity of a motion is not zero, it can become zero later. For example, if a mass is attached to a spring so that it oscillates, its velocity becomes zero at regular intervals — when the spring is at maximum extension and the mass changes its direction of motion. This is called a *turning point* of the motion. In other situations, an object's velocity can asymptotically approach zero, stopping only after an infinite time duration. This latter case is an idealization, of course, and you are unlikely to live long enough to ever see it.

In the lessons that follow you will use your computer and numeric integration techniques to solve for the one-dimensional motion of an object from a knowledge of its potential and kinetic energies.

LESSON

A ball is thrown into the air on a windless day. The potential energy of the system can be expressed by Equation 4.70.

Figure 4.25

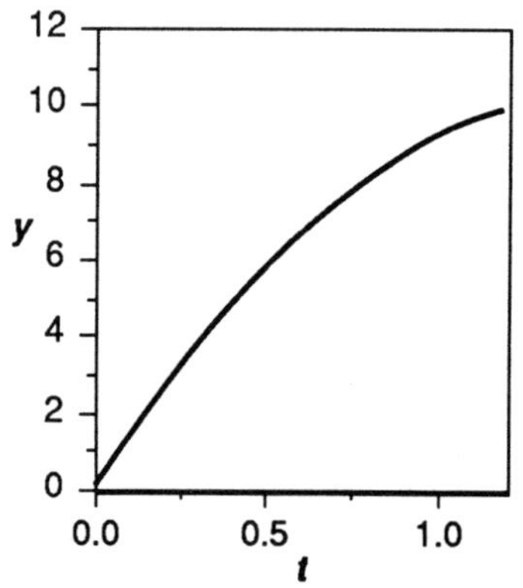

XY-PLOT

Data
Horizontal (*x*) axis: cells G13 - G33 (time)
Vertical (*y*) axis: cells A13 - A33 (position)

Figure 4.26

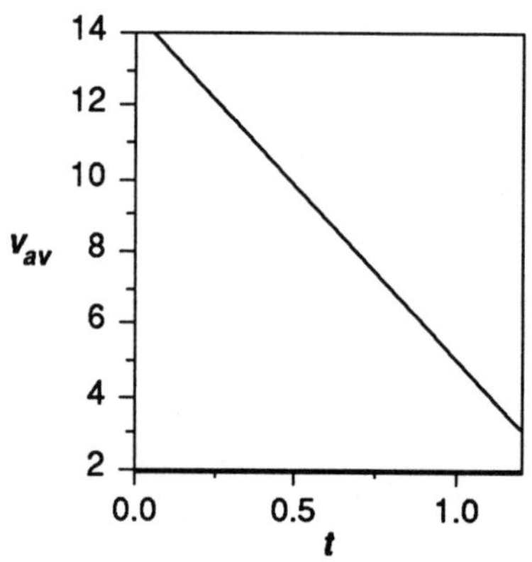

XY-PLOT

Data
Horizontal (*x*) axis: cells G14 - G33 (time)
Vertical (*y*) axis: cells H14 - H33 (average velocity)

$$U(y) = mgy \tag{4.70}$$

where y is the distance above the ground.

1. Build and verify Model 4.9. This model employs Simpson's rule to calculate values of $t - t_0$ for the positive values of Equation 4.69 where $U(y) = mgy$. It is organized in a format similar to previous Runge-Kutta calculations.

   a. Explore the model and its structure. From Equation 4.68 it should be clear that Equation 4.69 is integrating the function $1/v$. The labels in columns B, D, and F take explicit note of this fact. The model is read by noting the time (column G) that it takes to reach height y (column A) — backward, of course, to your experience until now.

   b. The average value for velocity is computed in column G.

2. Plot t vs y (Figure 4.25) and, if you wish, average velocity vs y (Figure 4.26).

   a. Examine your plots and compare them with those obtained in previous exercises in which you directly solved the second-order differential equations arising from the second law of motion.

   b. For what values of y does the model fail to evaluate t? Why?

   c. Increase the total energy of the system E fourfold to 120 J. What happens to the maximum attainable height?

3. Early in this exercise, you were cautioned that when all the energy of a one-dimensional mechanical system is held as potential energy, the denominator of the energy integral equals zero and the value of the integrand is indeterminate. This happens any time the velocity is zero — a frequent occurrence, for example, in periodic motions such as a mass oscillating on the end of a spring.

   Model 4.9 does not adequately deal with such conditions, and the flood of error messages can be quite frustrating. You may wish to proceed directly to Problem 8 (or Model 4.10) in this exercise to discover one way around this problem.

SPREADSHEET TECHNIQUE

Finding the Period of Oscillation: Lookup Tables

Several of the problems in this exercise ask you to find the period or frequency of an oscillatory motion. You can, of course, locate by eye a maximum and minimum displacement in the position column and then identify the time interval between them. The period is twice this time interval, and frequency is the reciprocal of the period. Perhaps, however, you believe that computers should be assigned such menial tasks and that your spreadsheet should display the period or frequency directly. If so, this spreadsheet technique is for you.

The trick to directly calculating the period of a motion involves the use of lookup tables and their corresponding built-in functions — VLOOKUP, HLOOKUP, or LOOKUP. Your program may not have all these functions, and your program may not use these functions in precisely the same way. Thus, the procedure described here may not work without modification. Once you understand the principle of the technique, however, you should have no trouble implementing it in your environment.

- First, create two columns somewhere to the left of your position column *x* and use them to locate the maxima and minima in the displacements of the motion. To find a maximum, use an IF statement to create the expression "IF [x(t) > x(t–1) AND x(t) > x(t+1)] THEN mark this cell with a 1." Minimum displacements are found and marked similarly.
- Next, construct a single formula to calculate the period using the built-in function VLOOKUP. The goal is to "lookup" in one column the first instance of the number 1 and to retrieve the time value associated with this 1 from another column. Unfortunately, the behavior of VLOOKUP is so different that we'll merely confuse you by trying to explain all alternatives for all spreadsheets here. For this reason, we almost chose not to include this spreadsheet technique at all! Still, if you enjoy puzzle solving, explore the example given below while reading from your software reference manual about lookup functions. The example describes the structure of a period calculator in Excel and in Lotus 1-2-3.

```
=2*ABS(VLOOKUP(0.9,A5:D18,3)-VLOOKUP(0.9,B5:D18,2)) [Excel-like Syntax]
=2*ABS(VLOOKUP(0.9,D5..D18,-3)-VLOOKUP(0.9,D5..D18,-2)) [1-2-3-like Syntax]
```

	A	B	C	D
1	period =	1.6		
2			t	x(t)
3	max	min	sec	m
4			0	0.9093
5	0	0	0.2	0.335
6	0	0	0.4	-0.443
7	0	1	0.6	-0.952
8	0	0	0.8	-0.883
9	0	0	1	-0.279
10	0	0	1.2	0.4941
11	1	0	1.4	0.9679
12	0	0	1.6	0.8546
13	0	0	1.8	0.2229
14	0	0	2	-0.544
15	0	1	2.2	-0.981
16	0	0	2.4	-0.823
17	0	0	2.6	-0.166
18			2.8	0.5921

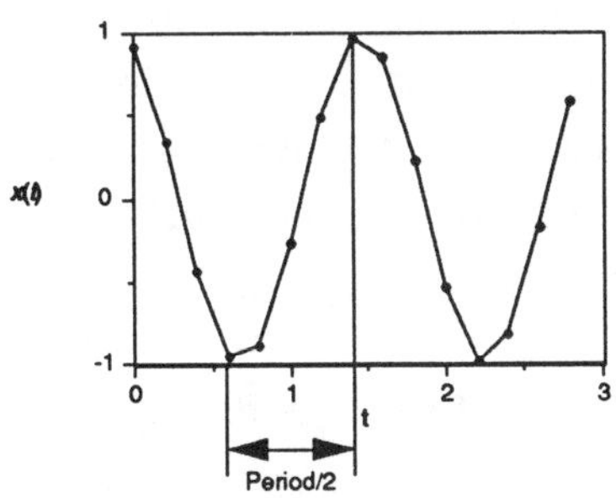


```
=IF(AND(D17<D16,D17<D18),1,0)    [Excel-like Syntax]
=IF(AND(D17>D16,D17>D18),1,0)

=IF(D17<D16 #AND# D17<D18,1,0)   [1-2-3-like Syntax]
=IF(D17>D16 #AND# D17>D18,1,0)
```

Figure 4.27

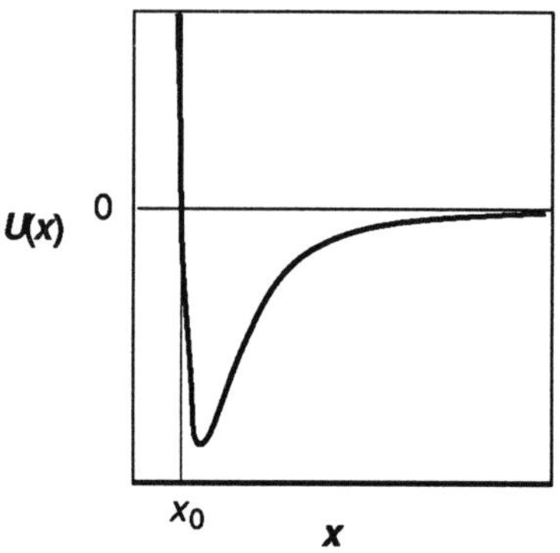

The Lennard-Jones potential energy function is often useful in predicting the behavior of two interacting atoms.

PROBLEMS

1. A body of mass 106 kg moves radially toward the center of the Earth from a height of 1000 km and with an initial downward velocity of 1000 m s^{-1}. Ignoring air resistance, determine the motion using the energy integral (Equation 3.95) and Simpson's rule and calculate the moment when the body strikes the Earth for each of the following assumptions:

   a. $U(y) = mgy$ where g equals GM/R^2 (G is Newton's gravitational constant, 6.673×10^{-11} N m^2 kg^{-2}; M is the mass of the Earth, 5.98×10^{24} kg; and R is the radius of the Earth, $4/2\pi \times 10^7$ m) and y is the distance above the Earth's surface.[1] Assume first that the initial velocity of the body is 2500 m s^{-1} downward; then that it is 2500 m s^{-1} upward.

   b. $U(y) = mgR\,[1 - R/(R + y)]$ (this is the exact Newtonian potential with its zero point set at the Earth's surface) and the same initial conditions as above. How do the times compare to those in Problem 1a? Why?

   c. Now take the initial velocity as 10500 m s^{-1} upward using the potentials in both Problem 1a and 1b. Why is the difference so dramatic?

2. The interaction between two atoms is often approximated using the Lennard-Jones potential (Equation 4.71) (Figure 4.27):

$$U(x) = U_0\left[\left(\frac{x_0}{x}\right)^{12} - \left(\frac{x_0}{x}\right)^{6}\right] \qquad (4.71)$$

   where x is the relative distance between two atoms, U_0 = 106 eV (1eV = 1.6×10^{-19} J/eV), and $x_0 = 0.63 \times 10^{-10}$ m. These are reasonable values for atoms near the middle of the periodic table of elements. When two particles of masses m_1 and m_2 interact, their relative motion can be calculated using the interaction potential energy and the reduced mass $m = m_1m_2/(m_1 + m_2)$.

   Suppose each of the two atoms has a mass of 56×10^9 eV/c^2, where $c = 3 \times 10^{10}$ cm s^{-1} (3000 nm/fs).[2]

   a. Find and plot the distance between the two atoms when their total energy of relative motion is –5eV, their initial separation is 1 nm, and they are initially moving toward one another. Use nanometers (nm) and femtoseconds (fs) as your units. What is their distances of closest and farthest approach? How long do they take to go between these two extremes of motion and what is the frequency of the oscillatory motion? Do they spend more time close to each other or relatively far apart? Discuss.

b. Suppose the total energy is –25eV and the initial separation is 0.7517 nm. Find and plot the motion. Can you find a reasonable approximation to the motion? What is its frequency? Does the frequency depend on the energy?

c. Estimate the greatest negative energy of relative motion that the system can have. What happens to your numeric calculation if you choose a greater negative energy?

d. If the energy is +5eV, the initial separation is 1nm, and the initial velocity acts to decrease the separation, calculate and plot the motion. What is the distance of closest approach? Then what happens?

3. A ball rolls over an undulating surface such that $U(x)$ can be described by Figure 4.28.

Plot x in meters vs time in seconds when the total energy of the ball is 20 J and it starts first at $x = 0$ and then at $x = 20$, moving toward the highest point in both cases. Does the ball get over the bump, and if so, how long does it take it to move between $x = 0$ and $x = 20$ or from $x = 20$ to $x = 0$? (In this problem, you are ignoring the energy of rotation associated with a real ball.)

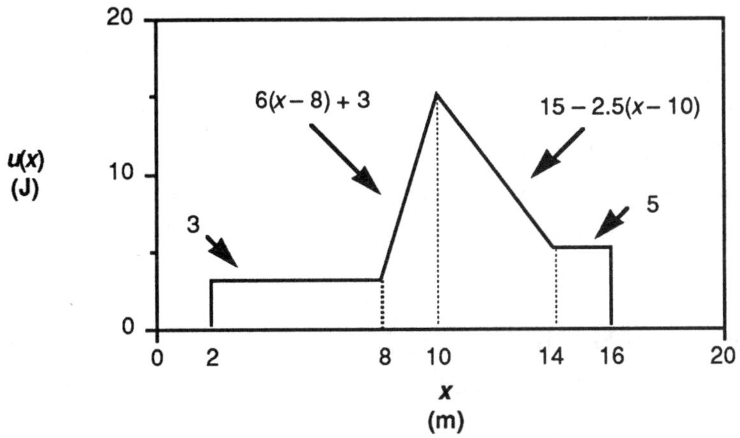

Figure 4.28

The potential energy function for a ball moving over an undulating surface has the same shape as the surface itself.

4. The cushion of a pool table acts like a spring with a spring constant of $k = 10^3$ N m^{-1}. Thus, if the edge of the cushion is at $x = 0$, the potential of a billiard ball can be modeled as

$$U(x) = \begin{cases} \frac{kx^2}{2} & \text{when} \quad x < 0 \\ 0 & \text{when} \quad x \geq 0 \end{cases} \tag{4.72}$$

If a billiard ball of mass 0.1 kg moving at 1.3 m s^{-1} strikes the cushion, what is the maximum compression of the cushion, and how long in milliseconds is the ball in contact with it? How do these results change if the ball's velocity is doubled? Tripled?

5. Suppose that the pool table cushion in the previous problem has a restoring spring constant that increases as it is compressed. Measurements indicate that a good approximation to the magnitude of the restoring force is

$$F_{restore} = k_0 e^{\alpha x} x \tag{4.73}$$

where $k_0 = 105$ N m^{-1} and $\alpha = 200$ m^{-1}.

With this force law, the potential function is

$$U(x) = \begin{cases} \left(\frac{k_0}{\alpha^2}\right)\left[1 - (1 - \alpha x)\varepsilon^{\alpha x}\right] & \text{when} \quad x < 0 \\ 0 & \text{when} \quad x \geq 0 \end{cases} \tag{4.74}$$

For a ball of mass 0.1 kg and an initial velocity of 1.3 m s^{-1}, what is the maximum compression of the cushion and how long is the ball in contact with it? How do these results change if the ball's velocity is doubled? Tripled? Comment on your observations and compare them to your results in Problem 4. Explain the differences qualitatively.

6. An electron is held in the field of two atoms which are 10 nm apart. A crude model of the potential that the electron finds itself in is

$$U(x) = U_0\left(\frac{x^2}{a^2} - 1\right)^2 \tag{4.75}$$

where $U_0 = 20$ eV and $a = 5$ nm.

The mass of an electron is 0.511 MeV/c^2.

MeV means million electron volts (1eV = 1.6 x 10^{-19} J), nm means nanometers (1 nm = 10^{-10} m), and c = 3000 nm/fs is the velocity of light [fs means femtoseconds (10^{-15} s)].

a. Calculate and plot the potential well $U(x)$ between $x = -10$ nm and +10 nm.

b. If the electron is initially at $x = 5$ nm and has a total energy of 30 eV, calculate and plot its motion using units of nm and fs. What are the maximum and minimum positions of the electron? What is the period of its motion?

c. Do the same if its energy is 20.01 eV, just a little larger than the height of the potential barrier between the two atoms. Where does the electron spend most of its time in this case? Comment.

Is the motion qualitatively different if the electron starts out at $x = -5$ nm?

d. Repeat when the energy is 5 eV. How is the motion qualitatively different when the starting point is –5 nm rather than +5 nm?

7. A hydrogen atom trapped between two molecules has the potential energy

$$U(x) = U_0\left(\frac{x}{a} + \frac{a}{x}\right) \tag{4.76}$$

where $U_0 = 5.6$ eV, $a = 2.4$ nm, and x is in nanometers.

The mass of the trapped hydrogen atom is 938 MeV/c^2 where c = 3000 nm/fs is the velocity of light.

a. Plot $U(x)$ between $x = 0.5$ nm and 12 nm. What is the minimum potential energy? Can the atom have less energy than this?

b. If the atom has an energy of 14 eV and starts its motion at the minimum of the potential well, calculate and plot its motion. What are its minimum and maximum positions (in nanometers)? What is its period of oscillation (in femtoseconds)?

c. What happens if the energy is raised to 50 eV? Where does the hydrogen atom spend most of its time?

8. Model 4.10 calculates the one-dimensional motion of a particle in a uniform gravitational field by integrating the law of conservation of mechanical energy (Equation 4.69). Unlike Model 4.9, however, this model handles three details that are necessary to translate the formal mathematical relation into a fully practical numerical procedure. All these details are related to the *turning points of a motion* — points where the velocity goes to zero and the energy integral becomes indeterminate.[3]

    - First, the model properly chooses the sign of the distance steps in the numeric integration (Simpson's rule is used) — column A.

    - Next, the model detects when a turning point is in one of the normal integration steps and then calculates a shorter step to just reach it rather than overstepping it — column B.

    - Finally, the model uses a special formula to calculate the time interval when a distance step either begins or ends on a turning point — column I.

    The model also includes a calculation of the exact vertical displacement as a function of time (column J) to compare with the y-column (column C).

    Model 4.10 can be adapted to many but not all one-dimensional motions where a potential energy function exists. In those very special cases in which the particle approaches a place where both the velocity and the force are zero, the model will fail. Such motions are just mathematical curiosities, however, and never happen in real life because it would take a particle an infinite time to reach that special place!

    Use Model 4.10 to explore the previous problems in this exercise.

MODELS

Model 4.9

BUILD and verify Model 4.9.

ENTER data into cells B3 - B5 (initial time, initial position, position increment) and E3 - E5 (total mechanical energy, mass of object, acceleration of gravity).

READ output from columns A, G, and H (position, time, average velocity).

FORMULAS

	A	B	C	D	E	F	G	H
1	U(y) = mgy							
2								
3	t0 =	0	s	E =	30	J		
4	y0 =	0	m	m =	0.3	kg		
5	Δy	0.5	m	g =	9.8	m/s^2		
6								
7	ABS(Δy) =	=ABS(B5)	m					
8	mg =	=E4*E5						
9	SQRT(m/2) =	=SQRT(E4/2)						
10								
11	y	1/√(y)	y+Δy/2	1/√(y+Δy/2)	y+Δy	1/√(y+Δy)	t	avg. v
12	(m)	(s/m)	(m)	(s/m)	(m)	(s/m)	(s)	(m/s)
13	=B4						0	
14–17	Formula Set I	Formula Set II	Formula Set III	Formula Set IV	Formula Set V	Formula Set VI	Formula Set VII	Formula Set VIII

=B9/SQRT(E3-B8*A13) (formula in cell B13)

VALUES

	A	B	C	D	E	F	G	H
1	U(y) = mgy							
2								
3	t0 =	0	s	E =	30	J		
4	y0 =	0	m	m =	0.3	kg		
5	Δy	0.5	m	g =	9.8	m/s^2		
6								
7	ABS(Δy) =	0.5	m					
8	mg =	2.94						
9	SQRT(m/2) =	0.3872983						
10								
11	y	1/√(y)	y+Δy/2	1/√(y+Δy/2)	y+Δy	1/√(y+Δy)	t	avg. v
12	(m)	(s/m)	(m)	(s/m)	(m)	(s/m)	(s)	(m/s)
13	0	0.0707107	0.25	0.0715931	0.5	0.0725095	0	
14	0.5	0.0725095	0.75	0.0734619	1	0.0744529	0.0357994	13.966718
15	1	0.0744529	1.25	0.0754851	1.5	0.0765615	0.0725336	13.611304
16	1.5	0.0765615	1.75	0.0776853	2	0.07886	0.1102798	13.24635
17	2	0.07886	2.25	0.0800898	2.5	0.0813788	0.1491267	12.871043

Formula Set I Prototype Cell is A14
=A13+**B5**

Formula Set II Prototype Cell is B14
=F13

Formula Set III Prototype Cell is C13
=A13+**B5**/2

Formula Set IV Prototype Cell is D13
=**B9**/SQRT(**E3-B8** *C13)

Formula Set V Prototype Cell is E13
=A13+**B5**

Formula Set VI Prototype Cell is F13
=**B9**/SQRT(**E3-B8** *E13)

Formula Set VII Prototype Cell is G14
=G13+ABS(**B5**)*(B13+4*D13+F13)/6

Formula Set VIII Prototype Cell is H14
=(A14-A13)/(G14-G13)

Model 4.10

BUILD and verify Model 4.10.

ENTER data using cells B3 - B8 (normal *y*-increment [always positive]; initial time, position, and velocity; particle mass; acceleration of gravity). A copy of the contents of cell B5 is placed in cell G5 to facilitate copying the contents of cells H5 into cell H6 and into columns D, F, and H.

READ output from columns I (time) and J (position).

FORMULAS

(Cell A12: =IF(B6=0,IF(H6<H5,-1,1),B6/ABS(B6)))

	A	B	C	D	E	F	G	H	I	J	K
1	Motion in a Uniform Gravitational field - U(y) =mgy										
2											
3	Δymax =	0.05	m			Escl =	=0.001*ABS(H6)				
4	t0 =	0	s			E =	=0.5*B7*B6^2+B7*B8*B5				
5	y0 =	0	m			y0 =	=B5	FS I	<= E-U @ y0		
6	v0 =	5	m/s			y0+Δymax =	=B5+B3	FS I	<= E-U @ y0+Δymax		
7	m =	0.3	kg			SQRT(m/2) =	=SQRT(B7/2)				
8	g =	9.8	m/s^2								
9											
10	SIGN	ABS(Δy)	y	E-U	y+Δy/2	E-U	y+Δy	E-U	t	yexact	error
11			(m)	(J)	(m)	(J)	(m)	(J)	(s)		
12		=B3	=B5	Formula Set V	Formula Set VI	Formula Set VII	Formula Set VIII	Formula Set IX	0	Formula Set XI	Formula Set XII
13–16	Formula Set II	Formula Set III	Formula Set IV	Formula Set V	Formula Set VI	Formula Set VII	Formula Set VIII	Formula Set IX	Formula Set X	Formula Set XI	Formula Set XII

VALUES

	A	B	C	D	E	F	G	H	I	J	K
1	Motion in a Uniform Gravitational field - U(y) =mgy										
2											
3	Δymax =	0.05	m			Escl =	0.0036	J			
4	t0 =	0	s			E =	3.75	J			
5	y0 =	0	m			y0 =	0	3.75	<= E-U @ y0		
6	v0 =	5	m/s			y0+Δymax =	0.05	3.603	<= E-U @ y0+Δymax		
7	m =	0.3	kg			SQRT(m/2) =	0.3873				
8	g =	9.8	m/s^2								
9											
10	SIGN	ABS(Δy)	y	E-U	y+Δy/2	E-U	y+Δy	E-U	t	yexact	error
11			(m)	(J)	(m)	(J)	(m)	(J)	(s)		
12	1	0.05	0	3.75	0.025	3.6765	0.05	3.603	0	0	0
13	1	0.05	0.05	3.603	0.075	3.5295	0.1	3.456	0.0101	0.05	2.9E-10
14	1	0.05	0.1	3.456	0.125	3.3825	0.15	3.309	0.02041	0.1	6.2E-10
15	1	0.05	0.15	3.309	0.175	3.2355	0.2	3.162	0.03094	0.15	1E-09
16	1	0.05	0.2	3.162	0.225	3.0885	0.25	3.015	0.0417	0.2	1.5E-09

Formula Set I
Prototype Cell is H5
=**G4-B7** *__B8__*G5

Formula Set II
Prototype Cell is A13
=IF(B12=**B3**,A12,-A12)

Formula Set III
Prototype Cell is B13
=IF(B12<>**B3**,1,IF(H12<**G3**,D12/(D12-H12),1))* **B3**

Formula Set IV
Prototype Cell is C13
=C12+A13*B13

Formula Set V
Prototype Cell is D12
=**G4-B7** *__B8__*C12

Formula Set VI
Prototype Cell is E12
=C12+A12*B12/2

Formula Set VII
Prototype Cell is F12
=**G4-B7** *__B8__*E12

Formula Set VIII
Prototype Cell is G12
=C12+A12*B12

Formula Set IX
Prototype Cell is H12
=**G4-B7** *__B8__*G12

Formula Set X
Prototype Cell is I13
=I12+IF(MIN(D12,H12)<**G3**,2/SQRT(ABS(D13-H13)),(1/SQRT(D12)+4/SQRT(F12)+1/SQRT(H12))/6)*B13* **G7**

Formula Set XI
Prototype Cell is J12
=-0.5* **B8***I12^2+**B6***I12+**B5**

Formula Set XII
Prototype Cell is K12
=J12-C12

NOTICE the following characteristics of the model. *You must understand them to adapt the model to a different potential energy function.*

Cell G6 (y_0+Δy_{max}) calculates the position of the first step if the sign of this step turns out to be positive. It is used to help decide the initial direction of the steps.

Cells H5 and H6 calculate the values of the "kinetic energy" at the initial position and one normal step away. "Kinetic energy" is in quotes because these values can be negative! If the initial velocity is zero, then the direction of the motion is chosen to increase the particle's kinetic energy. The initial direction of the position steps is determined by the relative values of these two cells when the initial velocity is zero. *If this model is adapted to another potential energy function, cell H5 should be modified and then copied into H6 and into columns D, F, and H of the model.*

Cell G3 (Escl) contains the energy scale of the problem. It is set to 0.1% of the value of the kinetic energy at $y_0+\Delta y_{max}$. This number is used to decide when the kinetic energy in the main calculation is so small that a turning point has been found.

Cell G4 (E) contains the total mechanical energy of the system. It is calculated as the sum of the initial kinetic energy and the initial potential energy. *If you change the potential energy function in the model, you must change the formula in this cell.*

The first row of the main table of calculations is special. Cell A12 selects the initial direction of the change of position. If the initial velocity is not zero, it is just the sign of the initial velocity; if the initial velocity is zero, it gives the direction that increases kinetic energy. Cell B12 copies Δy_{max}; cell C12 copies y_0; and cell D12 contains the initial kinetic energy. Cell E12 calculates the center of the first distance interval and is calculated from columns A and B. Cell F12 contains the kinetic energy at $y_0+\Delta y_{max}/2$. Cell G12 contains the end of the first position interval, and cell H12 contains the kinetic energy at this position. Cell I15 is simply the initial time.

The second and subsequent rows of the main table of calculations work as follows. Cell A13 assures that *y* moves in the same direction as the previous row when the previous step size is normal. If the step size is not normal, the model is at a turning point and the direction of the integration must be changed. Cell B13 adjusts the step size, if necessary. (If the previous step size is abnormal, then this step size must be normal; but if the previous step size is normal, there may be a turning point in it. This is checked by comparing the calculated kinetic energy with Escl. If a turning point is found, the present step is shortened so that it will just reach the turning point.) Cell I13 contains the heart of the calculation and contains two cases. If the previous step did not contain a turning point, the formula simply selects the usual Simpson's rule calculation for the integral. If the previous step did contain a turning point, a special formula is used that gives a very good approximation to the correct value of the integral when one of the ends of the integral is a turning point. This formula is special to energy integrals, but the same idea can be used anytime an integrand becomes infinite at one of the end points of an integral.[4]

EXERCISE **4.5**

THE CONSERVATION OF MOMENTUM: COLLISIONS IN ONE DIMENSION

TO GET THE MOST OUT OF THIS EXERCISE, YOU SHOULD ALREADY KNOW

The definitions of the terms:
momentum
conservation of energy

How to solve simultaneous equations.

How to solve differential equations using the Runge-Kutta method.

Until now, we have always written Newton's second law for a particle of mass m as

$$\boldsymbol{F} = m\boldsymbol{a} = m\frac{d\boldsymbol{v}}{dt} \tag{4.77}$$

where $\boldsymbol{F}$ is the force acting on the particle and $\boldsymbol{a}$ and $\boldsymbol{v}$ are the acceleration and velocity of the particle, respectively.

When m is a constant, the usual case in classical physics, Equation 4.77 can also be written

$$\boldsymbol{F} = \frac{d(m\boldsymbol{v})}{dt} = \frac{d\boldsymbol{p}}{dt} \tag{4.78}$$

where

$$\boldsymbol{p} = m\boldsymbol{v} \tag{4.79}$$

is the linear momentum of the particle. Notice that $\boldsymbol{p}$ is a vector because it is the product of m, a scalar, and $\boldsymbol{v}$, a vector.

Equations 4.77 and 4.78 are equivalent only when the mass of the particle is constant with time. When m changes with time (a rocket losing propellant, for example), experiment shows that only Equation

4.78 describes the motion correctly. Newton himself recognized this important physical fact and stated his second law in its more general form (Equation 4.78).[1]

When a system contains several particles, each particle has its own momentum

$$\boldsymbol{p}_j = m_j \boldsymbol{v}_j \qquad j = 1, n \tag{4.80}$$

and each is acted upon by its own force $\boldsymbol{F}_j$. The use of the index, and the notation $j = 1, n$ means simply that the equation as written really stands for n equations with j taking on each of the values 1 to n.

Adding together the equations of motion for each particle (Equation 4.78) yields

$$\boldsymbol{F} = \sum_{j=1}^{n} \boldsymbol{F}_j = \sum_{j=1}^{n} \frac{d\boldsymbol{p}_j}{dt} = \frac{d\boldsymbol{P}}{dt} \tag{4.81}$$

where $\boldsymbol{F}$ is the total, or net, force on the system and

$$\boldsymbol{P} = \sum_{j=1}^{n} \boldsymbol{p}_j \tag{4.82}$$

is the total momentum.

When no force acts on a single particle, its momentum is constant. Similarly, when no total, or net, force acts on a system of particles, the total momentum of the system is constant, Furthermore, if Newton's third law is applicable to the forces of interaction between the particles (as is the case for a wide class of actual interactions in Nature) then the total force will be zero if no *external* forces act on the system. This is so because the internal forces cancel in pairs to leave only external forces. In other words, *for a system subject to no net external force its total momentum is conserved.* The distribution of the total momentum among the several particles of the system will in general vary with time due to the action of the internal forces, however.

This property is exploited with particular elegance in the study of collisions. Indeed, studying the collisions of atoms, nuclei, and elementary particles (and application of the appropriate conservation laws) has yielded a great deal of our knowledge of the subatomic world. You will discover that the constraints imposed on the motions of objects by the laws of conservation of energy and momentum allow you to analyze some features of collisions without knowing the intimate details of the interactions that occur during the collision. The only requirements on the interaction are that it satisfy Newton's laws and that the potential

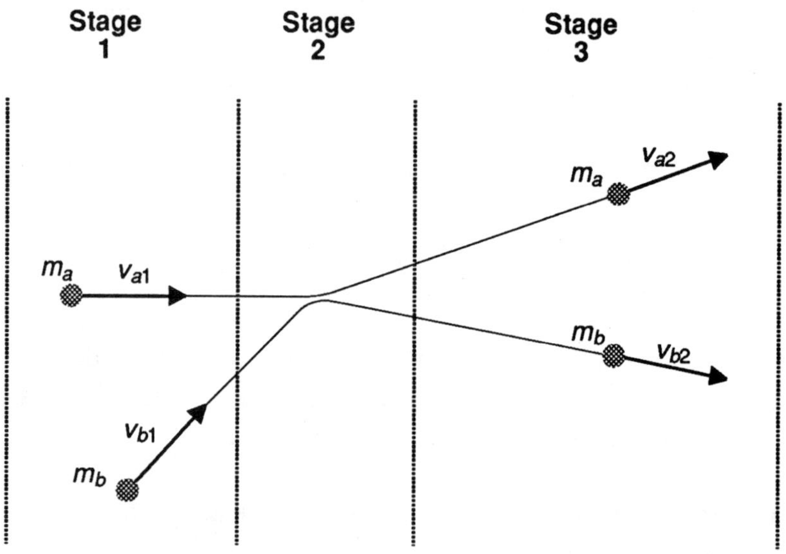

Figure 4.29

Collisions occur in three stages. Suppose two particles of masses m_a and m_b move toward each other with velocities v_{a1} and v_{b1}. First (stage 1), the particles are far apart, and any interaction is so small that it can be ignored. Second (stage 2), the particles move closely together, and forces of interaction change the trajectory of both. Third (stage 3), the particles leave the interaction zone with velocities v_{a2} and v_{b2}.

energy of the interaction is a constant when the objects are far apart. This constant is usually taken as zero.

In the next two exercises you will consider only collisions between two objects where no mass is exchanged during the collision. You should think of such collisions as occurring in three stages (Figure 4.29).

- Stage 1 consists of two independent objects of known masses (m_a, m_b) and velocities (v_{a1}, v_{b1}). Because the mass and velocity of each object are known, the kinetic energy (K_a, K_b) and momentum ($p_a = m_a v_{a1}$, $p_b = m_b v_{b1}$) are also known.

- In stage 2, the two objects move closely together, and interaction forces change the velocity of the objects from v_{a1} and v_{b1} to v_{a2} and v_{b2}. During this transition, the total energy and momentum of the system remain constant although the energy and momentum of each individual object change.

- In stage 3, the two objects move away from each other (and from the possibility for interaction), and the two objects return to straight-line trajectories. Of course, the final kinetic energy and momentum of each object will, in general, differ from their values before the collision.

You can analyze the collision of these two objects without any detailed knowledge of the forces of interaction in stage 2 by noting that momentum and energy are conserved during the collision. For elastic collisions, kinetic energy is conserved (because we assume that the potential energy is the same long before the collision and long after it) and Equations 4.83 - 4.84 apply:

$$m_a v_{a1} + m_b v_{b1} = m_a v_{a2} + m_b v_{b2} \tag{4.83}$$

$$\frac{m_a v_{a1}^2}{2} + \frac{m_b v_{b1}^2}{2} = \frac{m_a v_{a2}^2}{2} + \frac{m_b v_{b2}^2}{2} \tag{4.84}$$

where v_{a1}, v_{b1}, v_{a2}, and v_{b2} are the magnitudes of the velocities $\mathbf{v}_{a1}$, $\mathbf{v}_{b1}$, $\mathbf{v}_{a2}$, and $\mathbf{v}_{b2}$.

In Equations 4.83 and 4.84, variables m_a, $\mathbf{v}_{a1}$, m_b, and $\mathbf{v}_{b1}$ are known quantities while $\mathbf{v}_{a2}$ and $\mathbf{v}_{b2}$ are unknown. It might appear that you have two equations and two unknowns and that the system is completely specified, but the vector nature of velocity renders this thought illusory. Indeed, for two or three dimensions, there are always more unknowns than equations. Only in one dimension do you really have two equations and two unknowns. In this exercise, you consider only this simplest case and defer collisions in two or three dimensions to Exercise 4.6.

Consider the case of two objects with movements confined to a single coordinate axis. Then the velocities in Equation 4.83 can be treated as scalar quantities, and as noted above, you have two equations with two unknowns. There are two solutions to these equations, either

$$v_{a2} = v_{a1} \tag{4.85}$$

$$v_{b2} = v_{b1} \tag{4.86}$$

or

$$v_{a2} = v_{a1}\left(\frac{m_a - m_b}{m_a + m_b}\right) + v_{b1}\left(\frac{2m_b}{m_a + m_b}\right) \tag{4.87}$$

$$v_{b2} = v_{a1}\left(\frac{2m_a}{m_a + m_b}\right) + v_{b1}\left(\frac{m_b - m_a}{m_a + m_b}\right) \tag{4.88}$$

Which of these two solutions actually occurs in a particular collision depends on the detailed interaction between the two objects. By carefully noting the initial conditions at which the final velocities suddenly jump from one solution to the other, you can learn a bit about the interaction.

Before starting your explorations one additional line of thought is important. For inelastic collisions kinetic energy is not conserved since part of the energy is converted into other forms (e.g., heat and sound in the case of colliding billiard balls).[2] Usually this converted energy is represented by the symbol Q and Equation 4.84 becomes

Figure 4.30

Two pucks slide freely (and without friction) along a pair of parallel air troughs. When the pucks are in close proximity to one another they interact via a massless spring attached to one of them. The troughs are separated by distance d, and the resting length of the spring is r_0. If the distance between the pucks is r, then the spring gets compressed when $r < r_0$. Of course, this can happen only if $d < r_0$. Because the troughs compel the pucks to move in just one dimension, only the component of the spring's force parallel to the troughs is effective in affecting the motion (Equations 4.89 - 4.91).

If the two pucks move toward each other, then they collide in an environment in which all of the initial conditions and interaction forces are known.

(a) Stage 1 of the collision is the time during which the pucks move toward each other, but $r > r_0$ so the spring is not compressed.

(b) Stage 2 of the collision occurs when $r < r_0$ and the spring is compressed.

(c) Stage 3 of the collision occurs when the pucks have once again separated to a distance greater than r_0.

Notice that the figure shows only one of the possible outcomes of the collision. If the force of the spring is great enough, the two pucks will be unable to pass and will bounce off each other and return along their approach path.

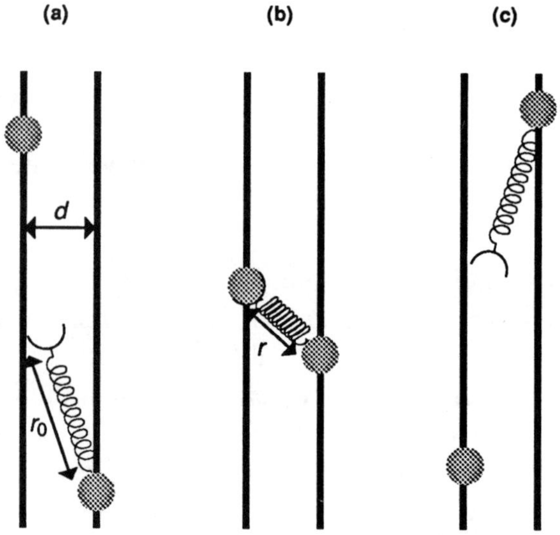

$$\frac{m_a v_{a1}^2}{2} + \frac{m_b v_{b1}^2}{2} = \frac{m_a v_{a2}^2}{2} + \frac{m_b v_{b2}^2}{2} + Q \tag{4.89}$$

If the value of Q is known nothing changes but the complexity of the algebra and arithmetic. In the one-dimensional case, for example, Equations 4.83 and 4.89 can be solved simultaneously to yield

$$v_{a2} = \frac{m_a^{3/2} v_{a1} + m_a^{1/2} m_b v_{b1} - m_b^{1/2} F}{m_a^{1/2}(m_a + m_b)} \tag{4.90}$$

$$v_{b2} = \frac{m_b^{3/2} v_{b1} + m_a m_b^{1/2} v_{a1} + m_a^{1/2} F}{m_b^{1/2}(m_a + m_b)} \tag{4.91}$$

where $F = (-2Qm_a - 2Qm_b + m_a m_b v_{a1}^2 - 2m_a m_b v_{a1} v_{b1} + m_a m_b v_{b1}^2)^{1/2}$. This is only one of two solutions, of course. Notice that when $Q = 0$ Equations 4.90 - 4.91 revert to their elastic counterparts (Equations 4.85 - 4.88).

LESSONS

The laws of conservation of energy and momentum let you analyze some of the features of a collision without knowing the details of the interaction between the particles. Nevertheless, it should be clear that these conservation laws do not constitute new rules of behavior but are simply a consequence of Newton's laws of motion. Consider two pucks sliding toward each other along parallel air troughs and interacting via springs, shock absorbers, and other mechanical devices during stage 2 of their collision (Figure 4.30). If the details of this device are fully specified, you need only Newton's laws of motion to determine the motion of the pucks before, during, and after their interaction with each other. If the details of the device are not fully known, you can still use conservation laws to glean some information about the object's motion.

In several of the exercises that follow, you will use your personal computer and the Runge-Kutta method to solve Newton's second law of motion for the behavior of two colliding objects when the nature of the forces of interaction are fully known. You will also use the laws of conservation of energy and momentum to "predict" behavior that the Runge-Kutta method has already calculated for you.

1. Build and verify Model 4.11.This model simulates the behavior of two pucks sliding along a pair of parallel air troughs and interacting via a massless spring attached to one of them (Figure 4.30). The geometry is described in the caption of Figure 4.30. Because the troughs compel the pucks to move in just one dimension, only the component of the spring's force parallel to the troughs is effective in affecting the motion during stage 2 of the collision (Equation 4.92 - 4.93):

$$F_a = k(r - r_0)\cos\theta = \frac{k(r - r_0)(x_b - x)_a}{r} \quad (4.92)$$

$$F_b = -F_a \quad (4.93)$$

$$r = \sqrt{(x_a - x_b)^2 + d^2} \quad (4.94)$$

where F_a is the force on puck a, F_b is the force on puck b, x_a and x_b are positions of pucks a and b, r is the distance between puck a and puck b, θ is the the angle between the line joining the pucks and the troughs, and d is the distance between the troughs.

The model is constructed so that, if there were no spring, the pucks would pass each other at $x_a = x_b = 0$ at time = t_{coll}, whose value is chosen so that the pucks are initially separated by a distance of 1.5 r_0. The time step is chosen to be 1/10 of t_{coll}, and so the table should consist of at least thirty to forty time steps.

Prepare plots of the positions of the two masses x_a and x_b vs time (Figure 4.31) and of total kinetic energy and momentum vs time (Figure 4.32).

a. Explore the model until you understand its structure.

b. Using the initial model conditions described for Model 4.11, describe in simple language the motion of the two pucks as they traverse their course. Why are the pucks unable to pass each other? What happens to the kinetic energy of the pucks during the collision? The momentum?

Figure 4.31

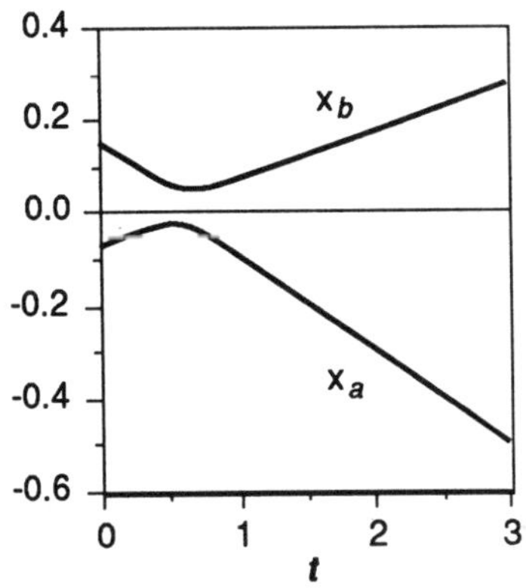

XY-PLOT

Data
Horizontal (*x*) axis: cells C16 - C57 (*t*)
Vertical (*y*) axis: cells D16 - D57 (x_a), E16 - E57 (x_b)

Figure 4.32

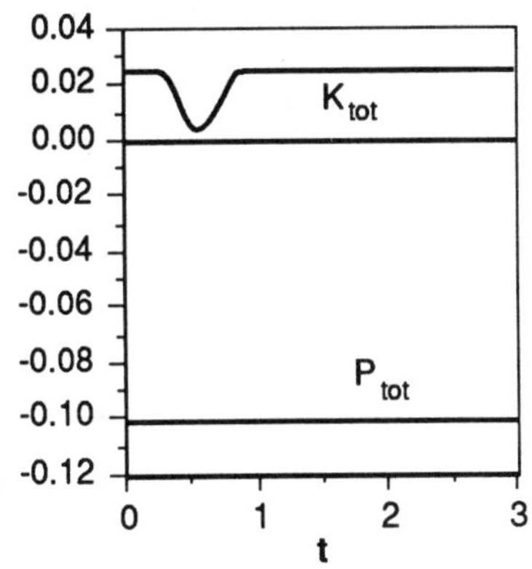

XY-PLOT

Data
Horizontal (*x*) axis: cells C16 - C57 (*t*)
Vertical (*y*) axis: cells A16 - A57 (K_{tot}), B16 - B57 (P_{tot})

c. Increase the mass of puck a to 20 kg and describe the motion of the two pucks as they traverse their course. Now describe the motion of the pucks. Comment on the kinetic energy and momentum of the system during each stage of the collision.

d. Initialize your system such that $m_a = 20$ kg, $m_b = 1$ kg, and the initial velocities of pucks a and b are 0.10 and –0.20 m s^{-1}, respectively. Now increase the initial velocity of puck a by steps of 0.02 m s^{-1}, up to a maximum velocity of 0.22 m s^{-1}.

Observe carefully what happens in each case, being sure to notice that the observed final velocities always agree (to within numeric calculational errors) to one of the two possibilities predicted by conservation laws alone.

At what velocity does the character of the final motion change dramatically? Find this special velocity to a precision of three significant figures. Can you explain why there is a sudden change in the character of the motion?

Notice that by determining initial velocity at which the character of the motion changes, you learn just a little bit about the potential energy of interaction. What is the value of the kinetic energy of relative motion corresponding to this velocity?

e. When $m_a = 20$ kg and $m_b = 1$ kg (and using the initial velocities of 0.1m s^{-1} and -0.2 m s^{-1}), increase m_b in steps of 2 kg up to 20 kg. Observe carefully the motions of the two pucks.

How does the character of the motion change as m_b increases?

Explain why it is qualitatively very different from the previous case. Does the behavior agree with your intuition?

f. If you were to perform a one-dimensional "scattering" experiment like this with one mass much heavier than the other, what would you have to do to learn something about the interaction potential?

For $m_a = 20$ kg, $m_b = 1$ kg, and $v_a = 0.1$m s^{-1}, explore what happens as v_b becomes more negative. At what velocity does the character of the motion change? Calculate the kinetic energy of relative motion in this case and compare it to the corresponding value found Lesson 1b above.

g. Add cells to your model to compute the relative velocity $(v_b - v_a)$ before and after the collision. What happens to the magnitude of this quantity under a large number of different initial conditions?

h. Experiment with many different collisions of different initial conditions and write a brief report on your observations.

2. Build and verify Model 4.12. This model is the same as Model 4.11 except that a shock absorber has been installed that acts in line with the spring. The shock absorber is much like that used on automobiles and consists of a plunger moving in a viscous fluid. The force-resisting motion in such a device is proportional to the velocity at which the plunger is moved. During a collision, the interaction forces on pucks a and b are

$$F_a = k(r - r_0)\cos\theta - \gamma(v_a - v_b)\cos^2\theta$$

$$= \frac{k(r - r_0)(x_b - x_a)}{r} - \gamma\frac{(v_a - v_b)(x_b - x_a)^2}{r^2} \tag{4.95}$$

$$F_b = -F_a \tag{4.96}$$

where r and θ are as in Model 4.11 and γ is the damping constant of the "shock."

The model is constructed along the same lines as Model 4.11, and should consist of at least 30 or 40 rows.

Prepare plots of the positions of the two masses x_a and x_b vs time (Figure 4.33) and of total kinetic energy and momentum vs time (Figure 4.34).

a. Explore the model until you understand its structure.

b. Set γ to 0 and verify that, under these conditions, Model 4.12 behaves identically to Model 4.11.

c. Leave γ at zero but adjust all other initial conditions to those presented in the description of the model. Gradually increase the value of γ from 0 to 5 N s m^{-1}.

Verify that Newton's second law of motion and the conservation laws for energy and momentum give comparable results under all conditions. Describe the changes that occur in puck

Figure 4.33

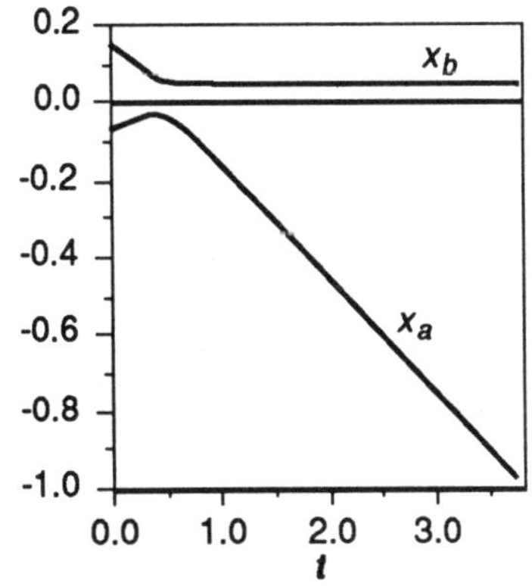

XY-PLOT

Data
Horizontal (*x*) axis: cells C21 - C62 (time)
Vertical (*y*) axis: cells D21 - D62 (x_a), E21 - E62 (x_b)

Figure 4.34

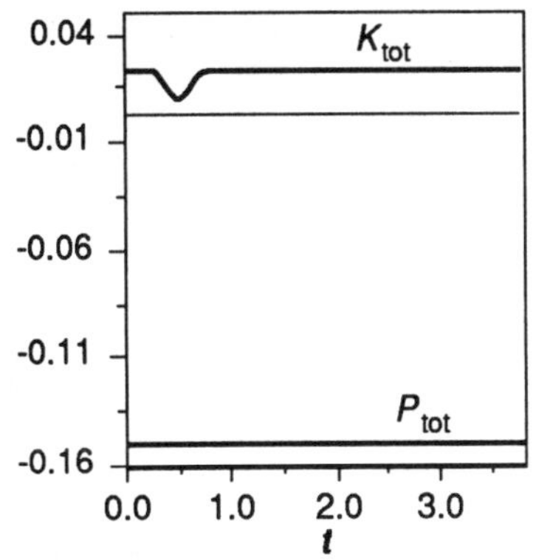

XY-PLOT

Data
Horizontal (*x*) axis: cells C21 - C62 (time)
Vertical (*y*) axis: cells A21 - A62 (K_{tot}), B21 - B62 (P_{tot})

motion and total kinetic energy as damping increases. Why do the two pucks "stick together" when $\gamma = 5$ N s m^{-1}.

d. Set γ equal to 0 and find a velocity v_a (very near 0.472 m s^{-1}) where the two pucks don't bounce off each other but almost stop for an extended period of time. Choose a value such that the "fin 1" conservation solution gives the correct result.

Now increase γ to 0.1. What happens? Why?

Continue increasing γ up to about 5 N s m^{-1}. You may find it interesting to plot the force acting on puck a (available in column K).

e. Set γ to 1 N s m^{-1} and increase the initial velocity of puck a from 0.1 m s^{-1} to 0.8 m s^{-1} in steps of 0.1 m s^{-1}. Observe carefully what happens in each case, being sure to notice that the observed final velocities always agree with one of the two calculations obtained from conservation laws. At what velocity does the character of the final motion change from one solution set to the other?

f. Using initial velocities of 0.1 m s^{-1} and –0.2 m s^{-1}, increase the mass of puck a to 20 kg in steps of 2 kg while carefully observing the motions of the two pucks. How does the character of the motion change? Explain why these changes are qualitatively very different from those in Lesson 2e. Does the behavior agree with your intuition?

g. If you perform this one-dimensional "scattering" experiment using one mass much heavier than the other, what must you do to learn something about the interaction potential? For $m_a = 20$ kg, $m_b = 1$ kg, and $v_a = 0.1$ m s^{-1}, explore what happens as v_b becomes more negative.

At what velocity does the character of the motion change? Calculate the kinetic energy of relative motion in this case and compare it to the corresponding value found in Lesson 2b.

h. Add cells to your model to compute the relative velocity $(v_b - v_a)$ before and after the collision. What happens to the magnitude of this quantity under a large number of different initial conditions?

i. Experiment with many different collisions of different initial conditions and write a brief report on your observations.

PROBLEMS

1. It was claimed in the discussion of one-dimensional scattering that the set of possible values of the final velocities does not depend on the scattering force so long as its associated potential is a constant when the two objects are very far apart and Newton's third law is satisfied. However, the choice between the two possible outcomes *does* depend on the force law and the initial conditions. Verify these claims by modifying your copy of Model 4.11 to have a special spring with a force law of

$$F = F_0\left(e^{\frac{k(r-r_0)}{F_0}} - 1\right) \tag{4.97}$$

where F_0 gives (intuitively) the approximate force at which the exponential "takes over." This spring behaves like a normal one so long as its displacement is small but rapidly becomes stiffer as the displacement becomes larger. For the parameters given earlier, make a graph of this force law for F_0 = 1 N, 2 N, 5 N and compare them to that of an ordinary spring of Lesson 1. Study the motions with this special spring for these several values of F_0.

2. In nuclear physics, energy stored in a nucleus can be released if it is hit hard enough by another particle. This release of stored energy constitutes what is called an exothermic reaction and can be modeled by changing Model 4.11 so that it includes a second, "cocked" spring whose energy is released in sufficiently hard collisions. Specifically, the second spring (also of length r_0 but with a different spring constant k_2) is initially compressed; it releases its potential energy the first time that the distance between the pucks equals some triggering value r_{trig}. This situation can be modeled by two pucks interacting via a single spring whose spring constant k is a function of the history of puck motions (Equation 4.98):

$$k = \begin{cases} k_1 & \text{if } r \text{ never equals } r_{trig} \\ k_1 + k_2 & \text{if } r \text{ ever equals } r_{trig} \end{cases} \tag{4.98}$$

Because this force law depends on history, it is clearly not a function of position, and so the total kinetic energy of the pucks is not conserved. Rather total mechanical energy increases during a *superelastic collision.*

Model 4.13 is our model of this situation. Notice that the sharp discontinuity in the spring constant is aligned with the time steps of the model by introducing a special column to the model in which Δt is reduced, when necessary, to assure that the discontinuity always falls at the boundary of one time interval and the next.

Also, notice the method used to make the value of the spring constant "stick."

Build, verify, and explore the structure of Model 4.13. Also construct appropriate on-screen graphics to answer the questions below.

a. Using the initial conditions described for Model 4.13, verify that the cocked spring is not triggered when puck a is initially at rest and that your results are the same as those generated by Model 4.11 (Lesson 1).

   Gradually increase the initial velocity of puck a towards puck b until it just triggers the release of the cocked spring. What velocity is required? What is its final velocity if the initial velocity of puck a is just slightly below this critical value? What is its final velocity if the initial velocity of puck a is just slightly above this critical value?

   What is the critical value of the kinetic energy of relative motion that triggers the exothermic reaction?

b. If puck b is initially at rest and has a mass of 938 kg, what critical initial velocity of puck a triggers the reaction? What are the final velocities of puck a when its initial velocity is just below and just above its "critical" velocity? To what critical kinetic energy of relative motion does this correspond? Why is this energy the same as in the previous problem?

   Can you see why it is more useful to speak in terms of energies rather than velocities in discussing such problems?

   Is the motion of the heavy puck much influenced by each of the previous collisions? (The mass ratio here is the same as that between an electron and an a hydrogen atom.)

c. Using the same conditions as those of Problem 2b, determine the initial velocity of puck a required for it to move past puck b?

How much energy does puck a acquire from puck b in its journey? Does this result seem reasonable you?

3. In some atomic and nuclear reactions, kinetic energy is absorbed in a collision and stored in one or both of the reactants. Such collisions are examples of *endothermic reactions*. An endothermic reaction, in a manner entirely analogous to the exothermic reaction of Problem 2, also has an energy threshold that must be surmounted before it is triggered. This situation can be modeled using Model 4.13 by setting k_2 to a negative number with an absolute value that is less than k_1. In the "pucks in parallel air troughs" metaphor, such values correspond to two springs initially relaxed and arranged such that if a collision brings the two pucks closer together than r_{trig} one of the springs "sticks" in a compressed state.

   Explore your model for initial conditions where $k_1 = 15$ N m^{-1} and $k_2 = -10$ N m^{-1}, $m_a = 0.5$ kg, and $m_b = 1$ kg.

MODELS

Model 4.11

BUILD and verify Model 4.11.

ENTER data using B4 and B5 (mass of pucks a and b), B9 - C9 (initial velocities of pucks a and b), and F4 - F6 (separation between air troughs, spring constant, relaxed length of spring).

READ output from the model at cells B10 - C12 (row 10 displays the final velocities of the two pucks obtained by solving the equations of motion, rows 11 and 12 display the two, alternate final velocities obtained from the conservation laws for energy and momentum). Cell F8 calculates the kinetic energy of *relative* motion between the two pucks (i.e., total kinetic energy less the kinetic energy of the center of mass). Columns A and B show the total kinetic energy and total momentum of the system while columns C - H yield the time, positions, and velocities of the two pucks and the distance between them.

NOTICE that the formulas in cells B10 and C10 assume that the model extends to 57 rows.

FORMULAS

	A	B	C	D	E	F	G	H	I	J	K	L	M
1	Elastic Trough												
2													
3		PUCKS				SPRING							
4	ma =	1	kg		d =	0.05	m						
5	mb =	1	kg		k =	15	N/m						
6					r0 =	0.15	m						
7	vel. =>	va	vb										
8		(m/sec)	(m/sec)		Krel =	=0.5*B4*B5/(B4+B5)*(B9-C9)^2							
9	initial =>	0.1	-0.2										
10	final =>	=F57	=G57		tcol =	=SQRT(2.25*F6^2-F4^2)/ABS(B9-C9)							
11	fin1 =>	=B9	=C9		Δt =	=F10/F12							
12	fin2 =>	=(B9*(B4-B5)+C9*2*B5)/(B4+B5)	=(2*B9*B4+C9*(B5-B4))/(B4+B5)		cnt =	10							
13									0				
14	K(tot)	P(tot)	t	xa	xb	va	vb	r	kxa	kxb	kva	kvb	xa
15	(J)		(sec)	(m)	(m)	(m/sec)	(m/sec)	(m)					
16			0	=-F10*B9	=-F10*C9	=B9	=C9						
17–19	Form Set I	Form Set II	Form Set III	Form Set IV	Form Set V	Form Set VI	Form Set VII	Form Set VIII	Form Set IX	Form Set X	Form Set XI	Form Set XII	Form Set XIII

FORMULAS (Contd.)

	N	O	P	Q	R	S	T	U	V	W	X	Y	Z
13					1								
14	xb	va	vb	r	kxa	kxb	kva	kvb	xa	xb	va	vb	r
15													
16–19	Form Set XIV	Form Set XV	Form Set XVI	Form Set XVII	Form Set XVIII	Form Set XIX	Form Set XX	Form Set XXI	Form Set XXII	Form Set XXIII	Form Set XXIV	Form Set XXV	Form Set XXVI

FORMULAS (Contd.)

	AA	AB	AC	AD	AE	AF	AG	AH	AI	AJ	AK	AL	AM
13	2									3			
14	kxa	kxb	kva	kvb	xa	xb	va	vb	r	kxa	kxb	kva	kvb
15													
16–19	Form Set XXVII	Form Set XXVIII	Form Set XXIX	Form Set XXX	Form Set XXXI	Form Set XXXII	Form Set XXXIII	Form Set XXXIV	Form Set XXXV	Form Set XXXVI	Form Set XXXVII	Form Set XXXVIII	Form Set XXXIX

Model 4.11 (continued)

VALUES

	A	B	C	D	E	F	G	H	I	J	K	L	M
1	Elastic Trough												
2													
3		PUCKS				SPRING							
4	ma =	1	kg		d =	0.05	m						
5	mb =	1	kg		k =	15	N/m						
6					r0 =	0.15	m						
7	vel. =>	va	vb										
8		(m/sec)	(m/sec)		Krel =	0.0225	J						
9	initial =>	0.1	-0.2										
10	final =>	-0.2	0.1005		tcol =	0.7312	sec						
11	fin1 =>	0.1	-0.2		Δt =	0.0731	sec						
12	fin2 =>	-0.2	0.1		cnt =	10							
13									0				
14	K(tot)	P(tot)	t	xa	xb	va	vb	r	kxa	kxb	kva	kvb	xa
15	(J)		(sec)	(m)	(m)	(m/sec)	(m/sec)	(m)					
16	0.025	-0.1	0	-0.07	0.146	0.1	-0.2	0.225	0.007	-0.01	0	0	-0.07
17	0.025	-0.1	0.0731	-0.07	0.132	0.1	-0.2	0.204	0.007	-0.01	0	0	-0.06
18	0.025	-0.1	0.1462	-0.06	0.117	0.1	-0.2	0.182	0.007	-0.01	0	0	-0.05
19	0.025	-0.1	0.2194	-0.05	0.102	0.1	-0.2	0.161	0.007	-0.01	0	0	-0.05

VALUES (Contd.)

	N	O	P	Q	R	S	T	U	V	W	X	Y	Z
13					1								
14	xb	va	vb	r	kxa	kxb	kva	kvb	xa	xb	va	vb	r
15													
16	0.139	0.1	-0.2	0.214	0.007	-0.01	0	0	-0.07	0.139	0.1	-0.2	0.214
17	0.124	0.1	-0.2	0.193	0.007	-0.01	0	0	-0.06	0.124	0.1	-0.2	0.193
18	0.11	0.1	-0.2	0.172	0.007	-0.01	0	0	-0.05	0.11	0.1	-0.2	0.172
19	0.095	0.1	-0.2	0.151	0.007	-0.01	0	0	-0.05	0.095	0.1	-0.2	0.151

VALUES (Contd.)

	AA	AB	AC	AD	AE	AF	AG	AH	AI	AJ	AK	AL	AM
13	2									3			
14	kxa	kxb	kva	kvb	xa	xb	va	vb	r	kxa	kxb	kva	kvb
15													
16	0.007	-0.01	0	0	-0.07	0.132	0.1	-0.2	0.204	0.007	-0.01	0	0
17	0.007	-0.01	0	0	-0.06	0.117	0.1	-0.2	0.182	0.007	-0.01	0	0
18	0.007	-0.01	0	0	-0.05	0.102	0.1	-0.2	0.161	0.007	-0.01	0	0
19	0.007	-0.01	0	0	-0.04	0.088	0.1	-0.2	0.141	0.007	-0.01	-0.01	0.009

Model 4.11 (continued)

Formula Set I	Prototype Cell is A16	=0.5*(B4*F16^2+B5*G16^2)
Formula Set II	Prototype Cell is B16	=B4*F16+B5*G16
Formula Set III	Prototype Cell is C17	=C16+F11
Formula Set IV	Prototype Cell is D17	=D16+(I16+2*R16+2*AA16+AJ16)/6
Formula Set V	Prototype Cell is E17	=E16+(J16+2*S16+2*AB16+AK16)/6
Formula Set VI	Prototype Cell is F17	=F16+(K16+2*T16+2*AC16+AL16)/6
Formula Set VII	Prototype Cell is G17	=G16+(L16+2*U16+2*AD16+AM16)/6
Formula Set VIII	Prototype Cell is H16	=SQRT((E16-D16)^2+F4^2)
Formula Set IX	Prototype Cell is I16	=F11*F16
Formula Set X	Prototype Cell is J16	=F11*G16
Formula Set XI	Prototype Cell is K16	=IF(H16>F6,0,F5*(H16-F6)*(E16-D16)/H16)*F11/B4
Formula Set XII	Prototype Cell is L16	=-K16*B4/B5
Formula Set XIII	Prototype Cell is M16	=D16+I16/2
Formula Set XIV	Prototype Cell is N16	=E16+J16/2
Formula Set XV	Prototype Cell is O16	=F16+K16/2
Formula Set XVI	Prototype Cell is P16	=G16+L16/2
Formula Set XVII	Prototype Cell is Q16	=SQRT((N16-M16)^2+F4^2)
Formula Set XVIII	Prototype Cell is R16	=F11*O16
Formula Set XIX	Prototype Cell is S16	=F11*P16
Formula Set XX	Prototype Cell is T16	=IF(Q16>F6,0,F5*(Q16-F6)*(N16-M16)/Q16)*F11/B4
Formula Set XXI	Prototype Cell is U16	=-T16*B4/B5
Formula Set XXII	Prototype Cell is V16	=D16+R16/2
Formula Set XXIII	Prototype Cell is W16	=E16+S16/2
Formula Set XXIV	Prototype Cell is X16	=F16+T16/2
Formula Set XXV	Prototype Cell is Y16	=G16+U16/2
Formula Set XXVI	Prototype Cell is Z16	=SQRT((W16-V16)^2+F4^2)
Formula Set XXVII	Prototype Cell is AA16	=F11*X16
Formula Set XXVIII	Prototype Cell is AB16	=F11*Y16
Formula Set XXIX	Prototype Cell is AC16	=IF(Z16>F6,0,F5*(Z16-F6)*(W16-V16)/Z16)*F11/B4
Formula Set XXX	Prototype Cell is AD16	=-AC16*B4/B5
Formula Set XXXI	Prototype Cell is AE16	=D16+AA16
Formula Set XXXII	Prototype Cell is AF16	=E16+AB16
Formula Set XXXIII	Prototype Cell is AG16	=F16+AC16
Formula Set XXXIV	Prototype Cell is AH16	=G16+AD16
Formula Set XXXV	Prototype Cell is AI16	=SQRT((AF16-AE16)^2+F4^2)
Formula Set XXXVI	Prototype Cell is AJ16	=F11*AG16
Formula Set XXXVII	Prototype Cell is AK16	=F11*AH16
Formula Set XXXVIII	Prototype Cell is AL16	=IF(AI16>F6,0,F5*(AI16-F6)*(AF16-AE16)/AI16)*F11/B4
Formula Set XXXIX	Prototype Cell is AM16	=-AL16*B4/B5

Model 4.12

BUILD and verify Model 4.12.

ENTER data using B4 and B5 (mass of pucks a and b), B11 - C11 (initial velocities of pucks a and b), and F4 - F7 (separation between air troughs, spring constant, relaxed length of spring, damping coefficient of the shock absorber).

READ output from the model at cells B12 - C14. (Row 12 displays the final velocities of the two pucks obtained by solving the equations of motion, rows 13 and 14 display the two alternate final velocities obtained from the conservation law for momentum and that of energy, taking into account the energy loss due to the shock absorber. Notice that the *Q*-value for this collision can only be obtained from the actual motion.) Cell F11 calculates the kinetic energy of *relative* motion between the two pucks (i.e., total kinetic energy less the kinetic energy of the center of mass). Columns A and B show the total kinetic energy and total momentum of the system while columns C - H yield the time, positions, and velocities of the two pucks and the distance between them.

NOTICE that the formulas in cells B12, C12, and F12 assume that the model extends to row 62.

=(F14+B5*F13)/B7
=(F14-B5*F13)/B7

FORMULAS

	A	B	C	D	E	F	G	H	I	J	K	L	M
1	Gliders in an air trough interacting via a spring and shock absorber												
2													
3		PUCKS				SPRING							
4	ma =	0.5	kg	<initially	d =	0.05	m						
5	mb =	1	kg	0.5kg	k =	15	N/m						
6	m =	=B4*B5/B7			r0 =	0.15	m						
7	M =	=B4+B5			gam =	0	N/(m/sec)						
8													
9	vel. =>	va	vb										
10		(m/sec)	(m/sec)										
11	initial =>	0.1	-0.2		Krel =	=0.5*B6*(B11-C11)^2							
12	final =>	=F62	=G62		Q =	=A21-A62							
13	fin 1 =>		=(F14-B4*F13)/B7		Δ =	=SQRT(2/B6*(F11-F12))							
14	fin 2 =>		=(F14+B4*F13)/B7		P =	=B21							
15													
16	Δt =	=F16/B17			tcol =	=SQRT(2.25*F6^2-F4^2)/ABS(B11-C11)							
17	cnt =	8					=-F16*B11						
18							=-F16*C11		0				
19	K(tot)	P(tot)	t	xa	xb	va	vb	r	kxa	kxb	kva	kvb	xa
20	(J)		(sec)	(m)	(m)	(m/sec)	(m/sec)						
21						=B11	=C11						
22–24	Form Set I	Form Set II	Form Set III	Form Set IV	Form Set V	Form Set VI	Form Set VII	Form Set VIII	Form Set IX	Form Set X	Form Set XI	Form Set XII	Form Set XIII

FORMULAS (Contd.)

	N	O	P	Q	R	S	T	U	V	W	X	Y	Z
18					1								
19	xb	va	vb	r	kxa	kxb	kva	kvb	xa	xb	va	vb	r
20													
21–24	Form Set XIV	Form Set XV	Form Set XVI	Form Set XVII	Form Set XVIII	Form Set XIX	Form Set XX	Form Set XXI	Form Set XXII	Form Set XXIII	Form Set XXIV	Form Set XXV	Form Set XXVI

FORMULAS (Contd.)

	AA	AB	AC	AD	AE	AF	AG	AH	AI	AJ	AK	AL	AM
18	2									3			
19	kxa	kxb	kva	kvb	xa	xb	va	vb	r	kxa	kxb	kva	kvb
20													
21–24	Form Set XXVI I	Form Set XXVI II	Form Set XXIX	Form Set XXX	Form Set XXXI	Form Set XXXI I	Form Set XXXI II	Form Set XXXI V	Form Set XXX V	Form Set XXX VI	Form Set XXX VII	Form Set XXX VIII	Form Set XXXI X

Model 4.12 (continued)

VALUES

	A	B	C	D	E	F	G	H	I	J	K	L	M
1	Gliders in an air trough interacting via a spring and shock absorber												
2													
3		PUCKS				SPRING							
4	ma =	0.5	kg	<-initially	d =	0.05	m						
5	mb =	1	kg	0.5kg	k =	15	N/m						
6	m =	0.3333	kg		r0 =	0.15	m						
7	M =	1.5	kg		gam =	0	N/(m/sec)						
8													
9	vel. =>	va	vb										
10		(m/sec)	(m/sec)										
11	initial =>	0.1	-0.2		Krel =	0.015	J						
12	final =>	-0.296	-0.002		Q =	0.0005							
13	fin 1 =>	0.0964	-0.198		Δ =	0.2946							
14	fin 2 =>	-0.296	-0.002		P =	-0.15							
15													
16	Δt =	0.0914	sec		tcol =	0.7312	sec						
17	cnt =	8											
18									0				
19	K(tot)	P(tot)	t	xa	xb	va	vb	r	kxa	kxb	kva	kvb	xa
20	(J)		(sec)	(m)	(m)	(m/sec)	(m/sec)						
21	0.0225	-0.15	0	-0.07	0.146	0.1	-0.2	0.225	0.009	-0.02	0	0	-0.07
22	0.0225	-0.15	0.0914	-0.06	0.128	0.1	-0.2	0.198	0.009	-0.02	0	0	-0.06
23	0.0225	-0.15	0.1828	-0.05	0.11	0.1	-0.2	0.172	0.009	-0.02	0	0	-0.05
24	0.02224	-0.15	0.2742	-0.05	0.091	0.0983	-0.199	0.146	0.009	-0.02	-0.01	0.005	-0.04

VALUES (Contd.)

	N	O	P	Q	R	S	T	U	V	W	X	Y	Z
18					1								
19	xb	va	vb	r	kxa	kxb	kva	kvb	xa	xb	va	vb	r
20													
21	0.137	0.1	-0.2	0.212	0.009	-0.02	0	0	-0.07	0.137	0.1	-0.2	0.212
22	0.119	0.1	-0.2	0.185	0.009	-0.02	0	0	-0.06	0.119	0.1	-0.2	0.185
23	0.101	0.1	-0.2	0.159	0.009	-0.02	0	0	-0.05	0.101	0.1	-0.2	0.159
24	0.082	0.093	-0.2	0.133	0.009	-0.02	-0.04	0.021	-0.04	0.082	0.077	-0.19	0.134

VALUES (Contd.)

	AA	AB	AC	AD	AE	AF	AG	AH	AI	AJ	AK	AL	AM
18	2									3			
19	kxa	kxb	kva	kvb	xa	xb	va	vb	r	kxa	kxb	kva	kvb
20													
21	0.009	-0.02	0	0	-0.06	0.128	0.1	-0.2	0.198	0.009	-0.02	0	0
22	0.009	-0.02	0	0	-0.05	0.11	0.1	-0.2	0.172	0.009	-0.02	0	0
23	0.009	-0.02	0	0	-0.05	0.091	0.1	-0.2	0.146	0.009	-0.02	-0.01	0.005
24	0.007	-0.02	-0.04	0.021	-0.04	0.074	0.057	-0.18	0.123	0.005	-0.02	-0.07	0.033

Model 4.12 (continued)

Formula Set I	Prototype Cell is A21	=0.5*(**B4***F21^2+**B5***G21^2)
Formula Set II	Prototype Cell is B21	=**B4***F21+**B5***G21
Formula Set III	Prototype Cell is C22	=C21+**B16**
Formula Set IV	Prototype Cell is D22	=D21+(I21+2*R21+2*AA21+AJ21)/6
Formula Set V	Prototype Cell is E22	=E21+(J21+2*S21+2*AB21+AK21)/6
Formula Set VI	Prototype Cell is F22	=F21+(K21+2*T21+2*AC21+AL21)/6
Formula Set VII	Prototype Cell is G22	=G21+(L21+2*U21+2*AD21+AM21)/6
Formula Set VIII	Prototype Cell is H21	=SQRT((E21-D21)^2+**F4** ^2)
Formula Set IX	Prototype Cell is I21	=**B16** *F21
Formula Set X	Prototype Cell is J21	=**B16** *G21
Formula Set XI	Prototype Cell is K21	=IF(H21>**F6** ,0,**F5** *(H21-**F6**)*(E21-D21)/H21-**F7** *(F21-G21)*((E21-D21)/H21)^2)***B16** /**B4**
Formula Set XII	Prototype Cell is L21	=-K21***B4/B5**
Formula Set XIII	Prototype Cell is M21	=D21+I21/2
Formula Set XIV	Prototype Cell is N21	=E21+J21/2
Formula Set XV	Prototype Cell is O21	=F21+K21/2
Formula Set XVI	Prototype Cell is P21	=G21+L21/2
Formula Set XVII	Prototype Cell is Q21	=SQRT((N21-M21)^2+**F4** ^2)
Formula Set XVIII	Prototype Cell is R21	=**B16** *O21
Formula Set XIX	Prototype Cell is S21	=**B16** *P21
Formula Set XX	Prototype Cell is T21	=IF(Q21>**F6** ,0,**F5** *(Q21-**F6**)*(N21-M21)/Q21-**F7** *(O21-P21)*((N21-M21)/Q21)^2)***B16/B4**
Formula Set XXI	Prototype Cell is U21	=-T21***B4/B5**
Formula Set XXII	Prototype Cell is V21	=D21+R21/2
Formula Set XXIII	Prototype Cell is W21	=E21+S21/2
Formula Set XXIV	Prototype Cell is X21	=F21+T21/2
Formula Set XXV	Prototype Cell is Y21	=G21+U21/2
Formula Set XXVI	Prototype Cell is Z21	=SQRT((W21-V21)^2+**F4** ^2)
Formula Set XXVII	Prototype Cell is AA21	=**B16** *X21
Formula Set XXVIII	Prototype Cell is AB21	=**B16** *Y21
Formula Set XXIX	Prototype Cell is AC21	=IF(Z21>**F6** ,0,**F5** *(Z21-**F6**)*(W21-V21)/Z21-**F7** *(X21-Y21)*((W21-V21)/Z21)^2)***B16/B4**
Formula Set XXX	Prototype Cell is AD21	=-AC21***B4/B5**
Formula Set XXXI	Prototype Cell is AE21	=D21+AA21
Formula Set XXXII	Prototype Cell is AF21	=E21+AB21
Formula Set XXXIII	Prototype Cell is AG21	=F21+AC21
Formula Set XXXIV	Prototype Cell is AH21	=G21+AD21
Formula Set XXXV	Prototype Cell is AI21	=SQRT((AF21-AE21)^2+**F4** ^2)
Formula Set XXXVI	Prototype Cell is AJ21	=**B16** *AG21
Formula Set XXXVII	Prototype Cell is AK21	=**B16** *AH21
Formula Set XXXVIII	Prototype Cell is AL21	=IF(AI21>**F6** ,0,**F5** *(AI21-**F6**)*(AF21-AE21)/AI21-**F7** *(AG21-AH21)*((AF21-AE21)/AI21)^2)***B16/B4**
Formula Set XXXIX	Prototype Cell is AM21	=-AL21***B4/B5**

FORMULAS

=(E14+B5*E13)/B7
=(E14-B5*E13)/B7
=(E14+B5*E13)/B7
=(E14+B4*E13)/B7

	A	B	C	D	E	F	G	H	I	J	K	L	M
1	Gliders in an air trough interacting via 2 springs, one precompressed												
2													
3		PUCKS				SPRINGS							
4	ma =	0.5	kg	<-initially	d =	0.05	m						
5	mb =	1	kg	0.5kg	k1 =	15	N/m						
6	m =	=B4*B5/B7			k2 =	100	N/m						
7	M =	=B4+B5			r0 =	0.15	m						
8					rtrig =	0.1	m						
9	vel. =>	va	vb										
10		(m/sec)	(m/sec)										
11	initial =>	0.1	-0.2	Krel =>	=0.5*B6*(B11-C11)^2								
12	final =>	=H62	=I62	Q =>	=IF(B62=F5+F6,-0.5*F6*(F7-F8)^2,0)								
13	fin1 =>			Δ =>	=SQRT(2/B6*(E11-E12))								
14	fin2 =>			P =>	=D21								
15													
16	Δt =	=F16/B17			tcol =	=SQRT(2.25*F7^2-F4^2)/ABS(B11-C11)							
17	cnt =	25						=-F16*B11					
18								=-F16*C11		0			
19	Δt	k	K(tot)	P	t	xa	xb	va	vb	r	kxa	kxb	kva
20	(sec)		(J)		(sec)	(m)	(m)	(m/sec)	(m/sec)				
21	=B16	=F5			0			=B11	=C11				
22–24	Form Set I	Form Set II	Form Set III	Form Set IV	Form Set V	Form Set VI	Form Set VII	Form Set VIII	Form Set IX	Form Set X	Form Set XI	Form Set XII	Form Set XIII

FORMULAS (Contd.)

	N	O	P	Q	R	S	T	U	V	W	X	Y	Z	AA
18						1								
19	kvb	xa	xb	va	vb	r	kxa	kxb	kva	kvb	xa	xb	va	vb
20														
21–24	Form Set XIV	Form Set XV	Form Set XVI	Form Set XVII	Form Set XVIII	Form Set XIX	Form Set XX	Form Set XXI	Form Set XXII	Form Set XXIII	Form Set XXIV	Form Set XXV	Form Set XXVI	Form Set XXVII

FORMULAS (Contd.)

	AB	AC	AD	AE	AF	AG	AH	AI	AJ	AK	AL	AM	AN	AO
18	2									3				
19	r	kxa	kxb	kva	kvb	xa	xb	va	vb	r	kxa	kxb	kva	kvb
20														
21–24	Form Set XXVIII	Form Set XXIX	Form Set XXX	Form Set XXXI	Form Set XXXII	Form Set XXXIII	Form Set XXXI	Form Set XXX	Form Set XXXVI	Form Set XXXVII	Form Set XXXVIII	Form Set XXXIX	Form Set XXX	Form Set XXXXI

Model 4.13

BUILD and verify Model 4.13.

ENTER data using B4 and B5 (mass of pucks a and b), B11 - C11 (initial velocities of pucks a and b), and F4 - F8 (separation between air troughs, spring constants for both springs, equilibrium length of spring, triggering distance for the precompressed spring).

READ output from the model at cells B12 - C14 (row 12 displays the final velocities of the two pucks obtained by solving the equations of motion, rows 13 and 14 display the two alternate final velocities obtained from the conservation laws for energy, taking into account the *Q*-value of the reaction, and momentum). Cell F11 calculates the kinetic energy of *relative* motion between the two pucks (i.e., total kinetic energy less the kinetic energy of the center of mass). Columns C and D show the total kinetic energy and total momentum of the system while columns E - J yield the time, positions, and velocities of the two pucks and the distance between them.

NOTICE that the formulas in cells B10 and C10 assume that the model extends to row 62.

Model 4.13 (continued)

VALUES

	A	B	C	D	E	F	G	H	I	J	K	L	M
1	Gliders in an air trough interacting via 2 springs, one precompressed												
2													
3		PUCKS				SPRINGS							
4	ma =	0.5	kg	<-initially	d =	0.05	m						
5	mb =	1	kg	0.5kg	k1 =	15	N/m						
6	m =	0.3333	kg		k2 =	100	N/m						
7	M =	1.5	kg		r0 =	0.15	m						
8					rtrig =	0.1 m							
9	vel. =>	va	vb										
10		(m/sec)	(m/sec)										
11	initial =>	0.1	-0.2	Krel =>	0.015	J							
12	final =>	-0.3	-2E-04	Q =>	0	J							
13	fin1 =>	0.1	0.1	Δ =>	0.3								
14	fin2 =>	-0.3	0	P =>	-0.15								
15													
16	Δt =	0.0292	sec		tcol =	0.7312	sec						
17	cnt =	25											
18										0			
19	Δt	k	K(tot)	P	t	xa	xb	va	vb	r	kxa	kxb	kva
20	(sec)		(J)		(sec)	(m)	(m)	(m/sec)	(m/sec)				
21	0.02925	15	0.0225	-0.15	0	-0.073	0.1462	0.1	-0.2	0.225	0.003	-0.01	0
22	0.02925	15	0.0225	-0.15	0.0292	-0.07	0.1404	0.1	-0.2	0.216	0.003	-0.01	0
23	0.02925	15	0.0225	-0.15	0.0585	-0.067	0.1345	0.1	-0.2	0.208	0.003	-0.01	0
24	0.02925	15	0.0225	-0.15	0.0877	-0.064	0.1287	0.1	-0.2	0.199	0.003	-0.01	0

VALUES (Contd.)

	N	O	P	Q	R	S	T	U	V	W	X	Y	Z	AA
18						1								
19	kvb	xa	xb	va	vb	r	kxa	kxb	kva	kvb	xa	xb	va	vb
20														
21	0	-0.07	0.143	0.1	-0.2	0.221	0.003	-0.01	0	0	-0.07	0.143	0.1	-0.2
22	0	-0.07	0.137	0.1	-0.2	0.212	0.003	-0.01	0	0	-0.07	0.137	0.1	-0.2
23	0	-0.07	0.132	0.1	-0.2	0.204	0.003	-0.01	0	0	-0.07	0.132	0.1	-0.2
24	0	-0.06	0.126	0.1	-0.2	0.195	0.003	-0.01	0	0	-0.06	0.126	0.1	-0.2

VALUES (Contd.)

	AB	AC	AD	AE	AF	AG	AH	AI	AJ	AK	AL	AM	AN	AO
18	2									3				
19	r	kxa	kxb	kva	kvb	xa	xb	va	vb	r	kxa	kxb	kva	kvb
20														
21	0.221	0.003	-0.01	0	0	-0.07	0.14	0.1	-0.2	0.216	0.003	-0.01	0	0
22	0.212	0.003	-0.01	0	0	-0.07	0.135	0.1	-0.2	0.208	0.003	-0.01	0	0
23	0.204	0.003	-0.01	0	0	-0.06	0.129	0.1	-0.2	0.199	0.003	-0.01	0	0
24	0.195	0.003	-0.01	0	0	-0.06	0.123	0.1	-0.2	0.191	0.003	-0.01	0	0

Model 4.13 (continued)

Formula Set I
Prototype Cell is A22

=IF(AND(B22=F5,(2*J22-J21)<F8),B16*(J22-F8)/(J21-J22),B16)

Formula Set II
Prototype Cell is B22

=IF(OR(A21<>B16,B21=(F5+F6)),F5+F6,F5)

Formula Set III
Prototype Cell is C21

=0.5*(B4*H21^2+B5*I21^2)

Formula Set IV
Prototype Cell is D21

=B4*H21+B5*I21

Formula Set V
Prototype Cell is E22

=E21+A21

Formula Set VI
Prototype Cell is F22

=F21+(K21+2*T21+2*AC21+AL21)/6

Formula Set VII
Prototype Cell is G22

=G21+(L21+2*U21+2*AD21+AM21)/6

Formula Set VIII
Prototype Cell is H22

=H21+(M21+2*V21+2*AE21+AN21)/6

Formula Set IX
Prototype Cell is I22

=I21+(N21+2*W21+2*AF21+AO21)/6

Formula Set X
Prototype Cell is J21

=SQRT((G21-F21)^2+F4^2)

Formula Set XI
Prototype Cell is K21

=A21*H21

Formula Set XII
Prototype Cell is L21

=A21*I21

Formula Set XIII
Prototype Cell is M21

=IF(J21>F7,0,B21*(J21-F7)*(G21-F21)/J21)*A21/B4

Formula Set XIV
Prototype Cell is N21

=-M21*B4/B5

Formula Set XV
Prototype Cell is O21

=F21+K21/2

Formula Set XVI
Prototype Cell is P21

=G21+L21/2

Formula Set XVII
Prototype Cell is Q21

=H21+M21/2

Formula Set XVIII
Prototype Cell is R21

=I21+N21/2

Formula Set XIX
Prototype Cell is S21

=SQRT((P21-O21)^2+F4^2)

Formula Set XX
Prototype Cell is T21

=A21*Q21

Formula Set XXI
Prototype Cell is U21

=A21*R21

Formula Set XXII
Prototype Cell is V21

=IF(S21>F7,0,B21*(S21-F7)*(P21-O21)/S21)*A21/B4

Formula Set XXIII
Prototype Cell is W21

=-V21*B4/B5

Formula Set XXIV
Prototype Cell is X21

=F21+T21/2

Formula Set XXV
Prototype Cell is Y21

=G21+U21/2

Formula Set XXVI
Prototype Cell is Z21

=H21+V21/2

Formula Set XXVII
Prototype Cell is AA21

=I21+W21/2

Formula Set XXVIII
Prototype Cell is AB21

=SQRT((Y21-X21)^2+F4^2)

Formula Set XXIX
Prototype Cell is AC21

=A21*Z21

Formula Set XXX
Prototype Cell is AD21

=A21*AA21

Formula Set XXXI
Prototype Cell is AE21

=IF(AB21>F7,0,B21*(AB21-F7)*(Y21-X21)/AB21)*A21/B4

Formula Set XXXII
Prototype Cell is AF21

=-AE21*B4/B5

Formula Set XXXIII
Prototype Cell is AG21

=F21+AC21

Formula Set XXXIV
Prototype Cell is AH21

=G21+AD21

Formula Set XXXV
Prototype Cell is AI21

=H21+AE21

Formula Set XXXVI
Prototype Cell is AJ21

=I21+AF21

Formula Set XXXVII
Prototype Cell is AK21

=SQRT((AH21-AG21)^2+F4^2)

Formula Set XXXVIII
Prototype Cell is AL21

=A21*AI21

Formula Set XXXIX
Prototype Cell is AM21

=A21*AJ21

Formula Set XXXX
Prototype Cell is AN21

=IF(AK21>F7,0,B21*(AK21-F7)*(AH21-AG21)/AK21)*A21/B4

Formula Set XXXXI
Prototype Cell is AO21

=-AN21*B4/B5

EXERCISE **4.6**

THE CONSERVATION OF MOMENTUM: COLLISIONS IN MULTIPLE DIMENSIONS

TO GET THE MOST OUT OF THIS EXERCISE, YOU SHOULD ALREADY KNOW

The definitions of the terms: conservation of energy and momentum

How to solve simultaneous equations.

How to solve differential equations using the Runge-Kutta method.

Elastic collisions can be analyzed without any detailed knowledge of the forces of interaction by noting that momentum and energy are conserved during the collision. In Exercise 4.5 you discovered that, for collisions in one dimension, the two conservation equations (Equations 4.83 and 4.84) have only two unknowns (the final velocities) when the starting velocities and masses are known.

In the case of two objects capable of movement in two or three dimensions, the situation is more complex. If Equation 4.83 has velocity components along both the x- and y-axes (Equations 4.96 and 4.97), then there are four unknowns [$(v_{a2})_x$, $(v_{a2})_y$, $(v_{b2})_x$, and $(v_{b2})_y$] but only three equations (Equations 4.99 - 4.101):

$$m_a(v_{a1})_x + m_b(v_{b1})_x = m_a(v_{a2})_x + m_b(v_{b2})_x \qquad (4.99)$$

$$m_a(v_{a1})_y + m_b(v_{b1})_y = m_a(v_{a2})_y + m_b(v_{b2})_y \qquad (4.100)$$

$$\frac{m_a v_{a1}^2}{2} + \frac{m_b v_{b1}^2}{2} = \frac{m_a v_{a2}^2}{2} + \frac{m_b v_{b2}^2}{2} \qquad (4.101)$$

If Equation 4.83 has components along x-, y-, and z-axes, then there are six unknowns and only four equations. *For collisions that occur in the real world of three dimensions, the principles of conservation of energy and momentum are insufficient to predict the final configuration of the*

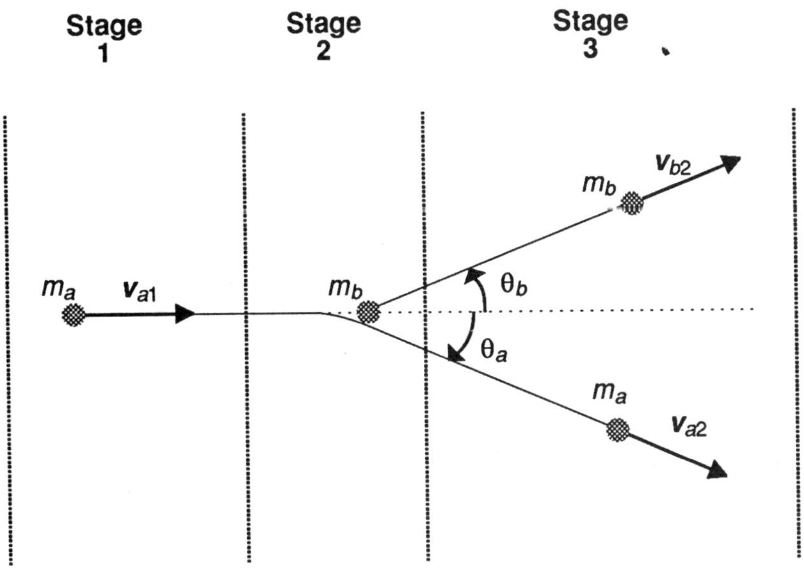

Figure 4.35

Two particles of mass m_a and m_b are capable of movement in two dimensions. Particle b is at rest and particle a is about to collide with it. A reference frame is chosen with its origin centered on m_b and its x-axis aligned with the straight-line trajectory of m_a. After the collision, m_a and m_b move away from each other at angles θ_a and θ_b with respect to the x-axis. For motions in two dimensions, knowledge of m_a, m_b, v_{a1}, and v_{b1} is insufficient to specify the behavior of the particles after collision. Other knowledge (say, the value of θ_a) is required to fully specify the system.

two objects, even if the initial velocities are known. Some information about the final state of the system must be supplied (usually the post-collision directions of motion).[1]

In most real situations, the force of interaction between two objects is not only equal and opposite (as required by Newton's third law) but also the force of interaction acts along the line joining them. When this is true, it can be shown that the final velocities of the colliding objects lie in the same plane as the initial velocities. For this reason, you only need to work out the details of two-dimensional collisions to understand the essence of the full, three-dimensional situation.

LESSON

Consider two objects on a frictionless surface (an air table for example) such that they can freely move in the x-, y-plane. One object, of mass m_a, initially moves parallel to the x-axis at a velocity v_{a1} toward a second object of mass m_b, *which is initially at rest.*[2] In general, the line of motion of object a will not produce a "head-on" collision, and so after the collision, the velocities v_{a2} and v_{b2} will make nonzero angles θ_a and θ_b with the x-axis (Figure 4.35). Because there are no external forces acting, both components of the vectorial momentum are conserved in the collision. This conservation is expressed in Equations 4.102 and 4.104. Furthermore, if no energy is lost in the collision (the collision is elastic), there is conservation of kinetic energy, given in Equation 4.101:

$$m_a v_{a1} = m_a v_{a2} \cos \theta_a + m_b v_{b2} \cos \theta_b \tag{4.102}$$

$$0 = m_a v_{a2} \sin \theta_a + m_b v_{b2} \sin \theta_b \tag{4.103}$$

$$\frac{m_a v_{a1}^2}{2} = \frac{m_a v_{a2}^2}{2} - \frac{m_b v_{b2}^2}{2} \tag{4.104}$$

It is important to note in these equations that v_{a1}, v_{a2}, and v_{b2} are the magnitudes of vectors and so are positive quantities by definition. Furthermore, Equation 4.103 cannot be satisfied unless the two angles θ_a and θ_b have the same sign. Because we take θ_a as positive, θ_b must be positive also.

In general, the values of the scattering angles, and so the final velocities, depend on both the details of the interaction and how far the initial line of motion is from the head-on condition. However, if the initial velocity v_{a1} and the final angle θ_a are known, then solutions for Equations 4.102 - 4.104 can be found, and the values for all the other quantities calculated. These solutions are

$$v_{a2} = \begin{cases} v_{a1}(W + \Delta) & \text{when} \quad m_b > m_a \\ v_{a1}(W \pm \Delta) & \text{when} \quad m_b < m_a \end{cases} \tag{4.105}$$

$$v_{b2} = \sqrt{\frac{m_a(v_{a1}^2 - v_{a2}^2)}{m_b}} \tag{4.106}$$

$$\tan \theta_b = \frac{v_{a2} \sin \theta_a}{v_{a1} - v_{a2} \cos \theta_a} \tag{4.107}$$

where $W = \dfrac{m_a \cos\theta_a}{m_a + m_b}$ and $\Delta = \sqrt{W^2 + \dfrac{m_b - m_a}{m_a + m_b}}$.

Equation 4.105 has two important features:

- If the initially stationary object is heavier than the moving object ($m_b > m_a$), then only one of the two roots of the quadratic equation for v_{a2} is physically acceptable. It is also true that θ_a can have any value in the range 0 to π.

- If the initially stationary object is lighter ($m_b < m_a$), then both roots of the quadratic equation for v_{a2} are acceptable — as indicated by the ± sign in that branch of the solution (the actual solution depends on the initial line of motion and the nature of the interaction — conservation laws cannot select between the two possibilities).

In this second case, not all values of θ_a are possible; the laws of motion require cos θ_a to be greater than or equal to the square root of $(m_a^2 - m_b^2)/m_a^2$ so that Δ does not become imaginary.

Build and verify Model 4.14 — a representation of an elastic collision of two objects in two dimensions under the circumstances described above. Explore the structure and notice that columns E - G give provisional values of the after-collision quantities of velocity and angle that are valid only if θ_b comes out positive. Similarly, the provisional values in columns K - M are valid only if θ_b is positive in these calculations.

Because angles of exactly 0° or 180° can cause problems for inverse trigonometric functions, the first value of θ_a is set to 0.01°, and 0.02° is subtracted from the last entry in the table.

Prepare plots of v_{a2}, v_{b2}, and θ_b vs θ_a. To plot θ_b vs θ_a, for example, use column A for the horizontal axis and both columns J and P for the vertical axis (Figure 4.36).

a. The model as initially described represents a collision between two identical masses.

Describe the relation between θ_b and θ_a in words. [HINT: Consider their sum.] What are the minimum and maximum values of θ_b? Describe the conditions corresponding to these angles.

b. Now let m_a decrease from 1 kg to .9 kg in steps of 0.02 kg.

Once m_a and m_b are no longer equal, what are the minimum and maximum values of θ_b and θ_a? Describe the character of the motion when the angles are near their extreme values.

c. Now consider a very small value for m_a, say 0.1 kg. Describe how θ_b is related to θ_a. [HINT: Consider the sum of twice θ_b and θ_a.] How are the postcollision velocities related?

d. What happens to your two-dimensional collision model as m_a is increased from 1.0 kg to 1.1 kg in steps of 0.02 kg? Are all values of θ_a physically possible?

For m_a = 1.1 kg, estimate the maximum value of θ_a. Notice that two values of the final configuration are possible for each allowed value of θ_a. Of the several possible mathematical solutions, which are

Figure 4.36

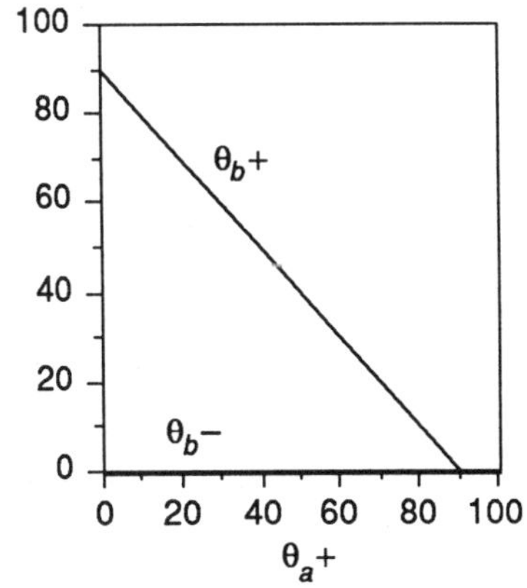

XY-PLOT

Data
Horizontal (x) axis: cells A12 - A102 (θ_a)
Vertical (y) axis: cells J12 - J102 (θ_b)

physically meaningful? Tune the value of the θ_a step size to make the two parts of the curves join (0.76°).

e. Finally, set m_a very large, say 10 kg. Suitably modify the step size of θ_a and estimate the largest possible value for θ_a. Why is it so small?

What is the allowable range for θ_b? Does your intuition agree that a light object is limited in the degree to which it can deflect a heavy object?

PROBLEM

The lesson considered the collision between a moving and a motionless object. However, because any inertial frame of reference is as good as any other, you really did not lose any generality by choosing a frame of reference centered on the "motionless" ball. Another reference frame, often used by physicists tackling collision problems is called the *center of momentum* (c.m.) frame. Within this reference frame, colliding objects move toward each other along a common line and then separate along another line making an angle θ with respect to the initial line (Figure 4.37). Laboratory observations are transferred into this frame for ease of calculation and then transferred back to the *laboratory frame* only after the calculations have been completed.

If the laboratory reference frame shows two objects of mass m_a and m_b approaching each other with velocities v_{a1} and v_{b1}, then the velocities in the c.m. frame are given by

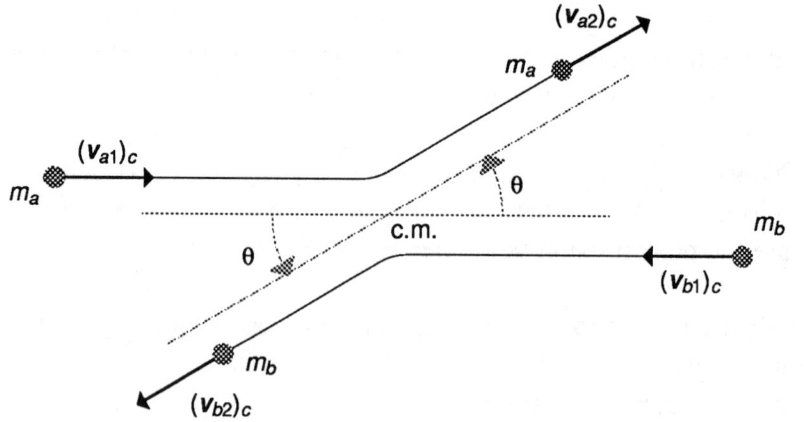

Figure 4.37

Collisions can be analyzed from numerous vantage points, and the choice of a suitable reference frame can significantly simplify the calculations. If the origin of the reference frame is the center of momentum between two colliding bodies, the bodies approach the collision point (the center of momentum) along lines parallel to a common axis. After collision the bodies separate along lines parallel to another axis that makes an angle θ with respect to the first. The angle θ can assume any value.

$$(v_{a1})_c = v_{a1} - V \qquad (4.108)$$

$$(v_{b1})_c = v_{b1} - V \qquad (4.109)$$

where $V = \dfrac{m_a v_a + m_b v_b}{m_a + m_b}$.

The total momentum of a system as seen from the c.m. frame equals zero, and momentum is conserved according to the equation

$$m_a(v_{a1})_c + m_b(v_{b1})_c = m_a(v_{a2})_c + m_b(v_{b2})_c = 0 \qquad (4.110)$$

For an elastic collision it can be shown that individual kinetic energies remain constant, and so the velocity components for the two particles after the collision are simply

$$x\text{-component of } (v_{a2})_c = |v_{a1}|_c \cos\theta \qquad (4.111)$$

$$y\text{-component of } (v_{a2})_c = -|v_{a1}|_c \sin\theta \qquad (4.112)$$

$$x\text{-component of } (v_{b2})_c = -|v_{b1}|_c \cos\theta \qquad (4.113)$$

$$y\text{-component of } (v_{b2})_c = |v_{b1}|_c \sin\theta \qquad (4.114)$$

The departing velocities in the laboratory frame can be found by adding the appropriate component of V to the above vectors.

a. Build a model based on a c.m. reference frame. Assuming that object a is initially moving along the x-axis in your laboratory frame and object b is initially at rest in the lab, you should reproduce the results of Model 4.14. Use the scattering angle θ in the c.m. frame as the independent variable.

   Increase θ from 0.01° to 179.99° in steps of 2°. Plot θ_b, the angle in the laboratory frame of particle b, as a function of θ_a. You should get the same results you found with Model 4.14, but the calculation is cleaner and more transparent. This the reason physicists use this reference frame.

b. Plot θ_a and θ_b vs θ, the c.m. angle, while m_b varies from less than to greater than 1 kg.

   Use your spreadsheet's "what-if table" or "data table" capability to verify the following:

m_a	θ_a max	θ_b max
0.1	179.989	89.995
0.5	179.980	89.995
0.9	179.900	89.995
0.99	179.000	89.995
0.999	170.099	89.995
1	89.995	89.995
1.001	87.357	89.995
1.01	81.930	89.995
1.1	65.372	89.995
1.5	41.810	89.995
2	30.000	89.995
5	11.536	89.995
10	5.739	89.995

Explain in simple English the meaning of these results.

MODELS

FORMULAS

	A	B	C	D	E	F	G	H	I	J	K	L	M	N	O	P
1	Elastic collision in two dimensions - one particle initially at rest															
2																
3	ma =	1	kg	M =	=B3+B4											
4	mb =	1	kg	P =	=B3*B5											
5	va1 =	2	m/sec	Ktot =	=0.5*B3*B5^2											
6																
7	Δ theta =	2	°													
8					***Only one set of the "Actual" values below can be true at any one time***											
9		cos			|------Provisional------			|---------Actual---------			|------Provisional------			|---------Actual---------|		
10	tha	tha	W	Δ	va2+	vb2+	thb+	va2+	vb2+	thb+	va2-	vb2-	thb-	va2-	vb2-	thb-
11	(°)				(m/s)	(m/s)	(deg)	(m/s)	(m/s)	(deg)	(m/s)	(m/s)	(deg)	(m/s)	(m/s)	(deg)
12	0.01															
13	Form	Form	Form	Form	Form	Form	Form	Form	Form	Form	Form	Form	Form	Form	Form	Form
14	Set	Set	Set	Set	Set	Set	Set	Set	Set	Set	Set	Set	Set	Set	Set	Set
15	I	II	III	IV	V	VI	VII	VIII	IX	X	XI	XII	XIII	XIV	XV	XVI

VALUES

	A	B	C	D	E	F	G	H	I	J	K	L	M	N	O	P
1	Elastic collision in two dimensions - one particle initially at rest															
2																
3	ma =	1	kg	M =	2	kg										
4	mb =	1	kg	P =	2	kg m/sec										
5	va1 =	2	m/sec	Ktot =	2	J										
6																
7	Δ theta =	2	°													
8					***Only one set of the "Actual" values below can be true at any one time***											
9		cos			|------Provisional------			|---------Actual---------			|------Provisional------			|---------Actual---------|		
10	tha	tha	W	Δ	va2+	vb2+	thb+	va2+	vb2+	thb+	va2-	vb2-	thb-	va2-	vb2-	thb-
11	(°)				(m/s)	(m/s)	(deg)	(m/s)	(m/s)	(deg)	(m/s)	(m/s)	(deg)	(m/s)	(m/s)	(deg)
12	0.01	1	0.5	0.5	2	3E-04	89.99	2	3E-04	89.99	0	2	0	ERR	ERR	ERR
13	2.01	0.9994	0.5	0.5	1.999	0.07	87.99	1.999	0.07	87.99	0	2	0	ERR	ERR	ERR
14	4.01	0.9976	0.499	0.499	1.995	0.14	85.99	1.995	0.14	85.99	0	2	0	ERR	ERR	ERR
15	6.01	0.9945	0.497	0.497	1.989	0.209	83.99	1.989	0.209	83.99	0	2	0	ERR	ERR	ERR

Formula Set I
Prototype Cell is A13

=A12+**B7**

Formula Set II
Prototype Cell is B12

=COS(A12*PI()/180)

Formula Set III
Prototype Cell is C12

=**B3***B12/**E3**

Formula Set IV
Prototype Cell is D12

=SQRT(C12^2+(**B4-B3**)/**E3**)

Formula Set V
Prototype Cell is E12

=**B5***(C12+D12)

Formula Set VI
Prototype Cell is F12

=SQRT(**B3***(**B5**^2-E12^2)/**B4**)

Formula Set VII
Prototype Cell is G12

=ATAN2(**B5**-E12*B12,E12*SIN(A12*PI()/180))*180/PI()

Formula Set VIII
Prototype Cell is H12

=IF(G12>0,E12,"ERR")

Formula Set IX
Prototype Cell is I12

=IF(G12>0,F12,"ERR")

Formula Set X
Prototype Cell is J12

=IF(G12>0,G12,"ERR")

Formula Set XI
Prototype Cell is K12

=**B5***(C12-D12)

Formula Set XII
Prototype Cell is L12

=SQRT(**B3***(**B5**^2-K12^2)/**B4**)

Formula Set XIII
Prototype Cell is M12

=ATAN2(**B5**-K12*B12,K12*SIN(A12*PI()/180))*180/PI()

Formula Set XIV
Prototype Cell is N12

=IF(M12>0,K12,"ERR")

Formula Set XV
Prototype Cell is O12

=IF(M12>0,L12,"ERR")

Formula Set XVI
Prototype Cell is P12

=IF(M12>0,M12,"ERR")

Model 4.14

BUILD and verify Model 4.14.

ENTER data using cells B3 - B5 and B7 [masses of pucks a and b, initial velocity of puck a (puck b is assumed to be initially stationary), increment in θ_a for each row of the table].

READ output from the model from column A (θ_a) and from either columns H - J or N - P (final velocities and θ_b). Cells in each trio of columns that have invalid data will yield an ERROR message.

NOTICE that the model should be extended such that θ_a can range from 0° to 180°. Change the last angle to 179.99° to avoid problems with trigonometric fructions at exactly 180°.

CHAPTER **5**

ROTATIONAL MOTION

A child's top — rotating, precessing, and dipping a bit while it performs a small nutation — Anon.

In the Netherlands in the 1600s, Dutchman Christiaan Huygens extended to the rotational motions of rigid bodies the simple but profound ideas of Galileo and Descartes. Shortly thereafter, Newton espoused his laws of motion, and the way was clear for others to build on Huygens' work in a formal setting.

In this chapter, you learn to apply the methods developed in earlier chapters to the general motion of a rigid body subject to combinations of applied forces. You first study the uniform rotation of a rigid body about a fixed axle and then graduate to the case where the axle remains fixed but the rotation speed changes in response to the dynamics of the system. Finally, you will be introduced to *general rotational motion* in which neither the axis of rotation nor speed is fixed.

When dealing with general rotational motion, it is most convenient to approach the problem by dividing it into three distinct parts:

- Find the translational motion of the center of mass
- Find the instantaneous rotational motion about coordinate axes through the center of mass and *fixed in the body*
- Find the rotational motion of these body-fixed coordinate axes with respect to an inertial reference frame

The first part of this problem has already been discussed in early chapters — the translational motion of the center of mass of a rigid body is treated identically to the translational motion of a particle. The remaining parts of this problem are similar to the first. You need to find rotational counterparts to the concepts of position, velocity, acceleration, force, and momentum. There are two main conceptual differences between the kinematics of translational motion and rotational motion. First, the rotational analog of position is not a vector. The second difference is more subtle. Although the rotational analog of velocity is a kind of vector *it is not the time derivative of a vector!*[1] Because of these differences, the kinematic task of working out rotational position from a known rotational velocity in the general case of rotational motion is somewhat more difficult than the corresponding problem in translational motion.

Figure 5.1

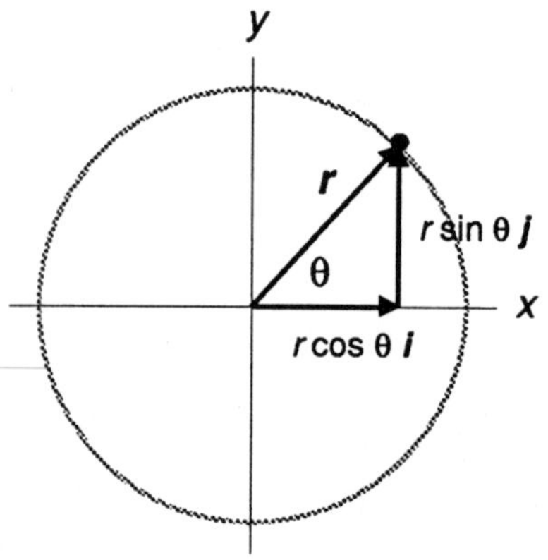

When a mass *m* moves such that its distance from some point *O* is fixed, the position of *m* can be given by vector ***r***. The Cartesian coordinates of *m* may be calculated as a vector function of θ (Equation 5.1).

When a mass moves so that its distance from some point in space is fixed, its ordinary Cartesian position can be given as a vector function of several angles. For example, if the mass moves in a circle on a plane (a particularly simple but important case), its position can be expressed in terms of the angle its position vector makes with the x-axis. If ***r*** is the particle's position vector with respect to the center of the circle and r is the circle's radius, then you have

$$\boldsymbol{r} = r(\cos\theta\, \boldsymbol{i} + \sin\theta\, \boldsymbol{j}) \tag{5.1}$$

This is shown in Figure 5.1. Differentiating Equation 5.1 to get the particle's velocity is easy:

$$\frac{d\boldsymbol{r}}{dt} = \boldsymbol{v} = r\frac{d\theta}{dt} \cdot (-\sin\theta\, \boldsymbol{i} + \cos\theta\, \boldsymbol{j}) \tag{5.2}$$

It is interesting and important to notice that the position vector and the velocity vector are perpendicular to one another (i.e., $\boldsymbol{r} \cdot \boldsymbol{v} = 0$). This means that $\boldsymbol{v}$ can be written as a cross product of some other vector with the position vector itself. In fact, physicists define the *angular velocity vector* $\boldsymbol{\omega}$ in such a way that

$$\boldsymbol{v} = \boldsymbol{\omega} \times \boldsymbol{r} \tag{5.3}$$

For a mass moving in a circle in the x-, y-plane, the angular velocity vector is

$$\boldsymbol{\omega} = \frac{d\theta}{dt}\, \boldsymbol{k} \tag{5.4}$$

where $\boldsymbol{k}$ is the unit vector along the z-axis.

This is the first time since Exercise 2.3 that you've encountered a vector cross product, and so you should take a moment to recall its properties. You should verify that

$$\boldsymbol{v} = \frac{d\theta}{dt}(\boldsymbol{k} \times \boldsymbol{r})$$

$$= \frac{d\theta}{dt}(\boldsymbol{k} \times \boldsymbol{i} \cos\theta + \boldsymbol{k} \times \boldsymbol{j} \sin\theta)$$

$$= \frac{d\theta}{dt}(\boldsymbol{j} \cos\theta - \boldsymbol{i} \sin\theta) \tag{5.5}$$

In general, $\boldsymbol{\omega}$ can be defined so that Equation 5.3 holds for any mass whose distance from the origin is a constant even if the motion is not a simple planar circle.

For a system of many masses whose distances are rigidly fixed with respect to each other, there is just one angular velocity for all of them. It is this fact that makes the description of the motion of a rigid body far simpler than that of a swarm of disconnected masses. The direction of

Figure 5.2

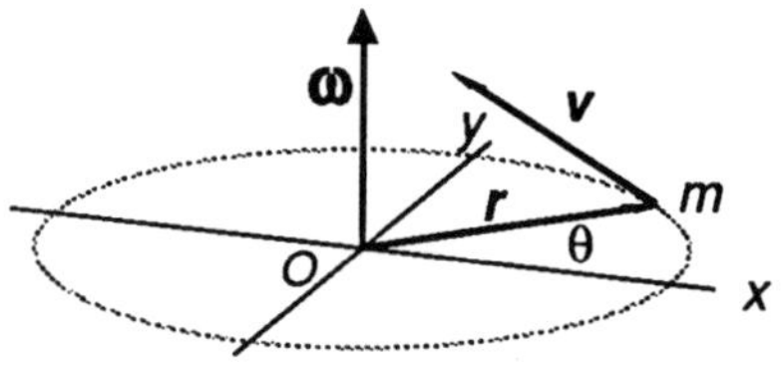

The angular velocity of a particle moving such that its distance from some point O remains fixed is a vector perpendicular to its translational velocity (Equation 5.3). When the particle's motion is a planar circle with the origin at its center as shown in the figure, the angular velocity is also perpendicular to the radius vector itself, but this is not generally the case. For example, if the origin were displaced below the plane of the circle, the angular velocity and radius vector would not be orthogonal.

this angular velocity is the instantaneous axis of rotation of the body (Figure 5.2). Both the direction and magnitude of $\boldsymbol{\omega}$ can change with time, of course, just as any vector in dynamics.

One last point. In sharp contrast to translational dynamics, angular acceleration $\boldsymbol{\alpha}$ (the time derivative of $\boldsymbol{\omega}$) plays almost no role in rotational dynamics because the general equations for rotational motion are more easily expressed using other variables. The reason for this big difference from translational motion is that the rotational analog of mass is often a function of time![2] Consequently, because "mass" can change, the concept of *angular momentum* Γ holds center stage in a way that linear momentum did not.

To get to the heart of the matter as quickly as possible, we'll jump over much interesting discussion about further relationships between translational and rotational kinematics and plunge into rotational dynamics (i.e., the rotational counterparts of force and momentum and the relationship that exists between them). You should refer to your introductory physics textbook to round out your understanding of rotational kinematics.

Let's develop the rotational analogs of force and momentum by starting with the dynamics of point masses. Consider a particle moving in three-dimensional space. The linear momentum $\boldsymbol{p}$ of this particle is given by Equation 5.6:

$$\boldsymbol{p} = m\boldsymbol{v} \tag{5.6}$$

where m is the mass of the particle and $\boldsymbol{v}$ is the velocity of the particle. $\boldsymbol{p}$ and $\boldsymbol{v}$, of course, are vectors.

Figure 5.3

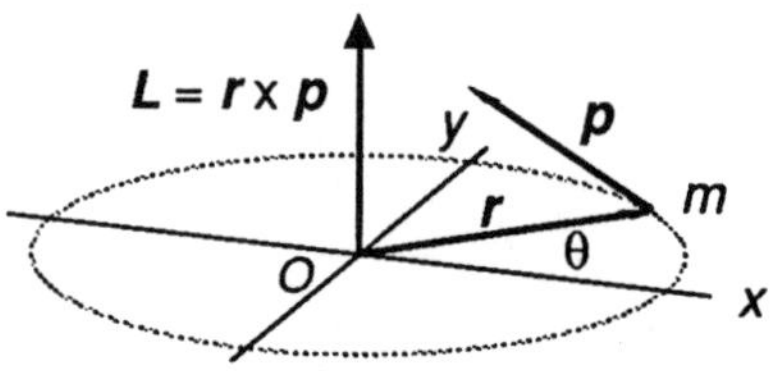

A particle of mass m and momentum $\boldsymbol{p}$ moving with respect to a point O exhibits an angular momentum $\boldsymbol{L}$ (with respect to point O) equal to $\boldsymbol{r} \times \boldsymbol{p}$, where $\boldsymbol{r}$ is the radius vector extending from point O to the particle.

The rotational counterpart to linear momentum $\boldsymbol{p}$ is angular momentum $\boldsymbol{L}$. For a single particle, $\boldsymbol{L}$ is defined as

$$\boldsymbol{L} = \boldsymbol{r} \times \boldsymbol{p} = \boldsymbol{r} \times m\boldsymbol{v} \tag{5.7}$$

where $\boldsymbol{r}$ is the vector from the origin of the coordinate system to the particle.

For the case that the particle is moving in a planar circle with its center at the origin of the coordinate system, $\boldsymbol{L}$ is directed at right angles to the plane defined by $\boldsymbol{r}$ and $\boldsymbol{v}$ (Figure 5.3). In this special case, $\boldsymbol{L}$ is just proportional to $\boldsymbol{\omega}$ (and the proportionality constant is mr^2). This can be shown from Equation 5.3.

The rotational counterpart to force $\boldsymbol{F}$ is torque Γ. In a manner exactly analogous to the definition of $\boldsymbol{L}$, Γ is the cross product of the radius vector and the force (Equation 5.8, Figure 5.4).

$$\Gamma = \boldsymbol{r} \times \boldsymbol{F} \tag{5.8}$$

When $\boldsymbol{r}$ and $\boldsymbol{F}$ define the x-, y-plane

$$\Gamma = \boldsymbol{r} \times \boldsymbol{F} = (xF_y - yF_x)\boldsymbol{k} \tag{5.9}$$

Angular momentum $\boldsymbol{L}$ and torque Γ are related via Equation 5.10, which follows directly from Newton's second law:

$$\Gamma = \frac{d\boldsymbol{L}}{dt} \tag{5.10}$$

To prove this, simply take the cross product of $\boldsymbol{r}$ on both sides of Newton's second law of motion and notice that

$$\frac{d\boldsymbol{L}}{dt} = \frac{d}{dt}(\boldsymbol{r} \times \boldsymbol{p}) = \frac{d\boldsymbol{r}}{dt} \times m\boldsymbol{v} + \boldsymbol{r} \times \frac{d\boldsymbol{p}}{dt} = \boldsymbol{r} \times \boldsymbol{F} = \Gamma \tag{5.11}$$

Notice the similarity between the torque-angular momentum relationship and Newton's second law of motion. In words, Equation 5.10 states that the torque on an object equals the time rate of change of its angular momentum. If net torque equals zero, then angular momentum is conserved.

Once the law governing the rotational motion of a single point object has been determined, it is easy to apply it to an extended object consisting of many such points by simply adding together many copies, one for each mass.

For an extended object, the angular momentum $\boldsymbol{L}_j$ of the jth particle of mass m_j and velocity $\boldsymbol{v}_j$ is

$$\boldsymbol{L}_j = \boldsymbol{r}_j \times m_j\boldsymbol{v}_j \qquad j = 1, n \tag{5.12}$$

where $\boldsymbol{r}_j$ equals the position of the jth particle and n is the number of particles in the system.

The total angular momentum of the object is defined as the sum of the angular momenta of all of its parts:

$$\boldsymbol{L} = \sum_{j=1}^{n} \boldsymbol{r}_j \times m_j\boldsymbol{v}_j \tag{5.13}$$

Figure 5.4

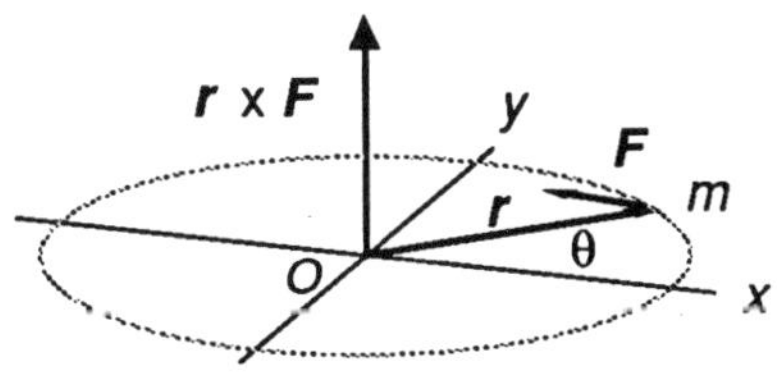

If a particle of mass m is subject to a force $\boldsymbol{F}$, the torque Γ (with respect to point O), equals $\boldsymbol{r} \times \boldsymbol{F}$, where $\boldsymbol{r}$ is the radius vector extending from point O to the particle.

If the object is a rigid body, then each velocity $\boldsymbol{v}_j$ can be expressed in terms of a common angular velocity and you obtain

$$\boldsymbol{L} = \sum_{j=1}^{n} \boldsymbol{r}_j \times m_j(\boldsymbol{\omega} \times \boldsymbol{r}_j) \tag{5.14}$$

If you expand the two cross products in this equation into components, you can reexpress Equation 5.14 as

$$L_x = I_{xx}\omega_x + I_{xy}\omega_y + I_{xz}\omega_z \tag{5.15a}$$

$$L_y = I_{yx}\omega_x + I_{yy}\omega_y + I_{yz}\omega_z \tag{5.15b}$$

$$L_z = I_{zx}\omega_x + I_{zy}\omega_y + I_{zz}\omega_z \tag{5.15c}$$

where

$$I_{xx} = \sum_{j=1}^{n} m_j(y_j^2 + z_j^2) \tag{5.16a}$$

$$I_{yy} = \sum_{j=1}^{n} m_j(z_j^2 + x_j^2) \tag{5.16b}$$

$$I_{zz} = \sum_{j=1}^{n} m_j(x_j^2 + y_j^2) \tag{5.16c}$$

$$I_{xy} = I_{yx} = -\sum_{j=1}^{n} m_j x_j y_j \tag{5.16d}$$

$$I_{yz} = I_{zy} = -\sum_{j=1}^{n} m_j y_j z_j \tag{5.16e}$$

$$I_{zx} = I_{xz} = -\sum_{j=1}^{n} m_j z_j x_j \tag{5.16f}$$

Sets of relations such as those above occur very often in physics and many other sciences, and an elegant and concise notation has been invented to make it easier to think about them. Just as you have used a single symbol to stand for a vector (which can have several components) so also a single symbol is often used to signify the nine-component object represented by Equations 5.16a - f above. In this book a bold, sans serif font (e.g., **I**) will be used to identify this kind of object (Equation 5.17):

$$\mathbf{I} = \begin{pmatrix} I_{xx} & I_{xy} & I_{xz} \\ I_{yx} & I_{yy} & I_{yz} \\ I_{zx} & I_{zy} & I_{zz} \end{pmatrix} \qquad (5.17)$$

Depending on exactly what people want to emphasize about this kind of thing, it is variously called a *matrix*, *dyad* (or *dyadic*), or *tensor*. The reason for the proliferation of names for the same type of mathematical object comes from the fact that as they arose in different disciplines and different people named them differently. For our purposes, the three names can be used interchangeably.

Using the dyadic notation, Equations 5.15a - c can be abbreviated

$$\boldsymbol{L} = \mathbf{I} \cdot \boldsymbol{\omega} \qquad (5.18)$$

where $\mathbf{I}$ is called the *inertia tensor*. (Some people experience a strange, almost tingling, sensation in their brains the first time they see Equation 5.18.)

Like vector algebra, dyadic algebra has a consistent set of rules that need be learned only once. Because this book is not about such algebras, however, we will translate equations involving dyadics into their component forms as they appear, and we will show you other commonly used notations and the conventions related to them.

Equation 5.18 is the rotational analog to Equation 4.79 (the definition of linear momentum) in Exercise 4.5 — $\mathbf{I}$ replaces m and $\boldsymbol{\omega}$ replaces $\boldsymbol{v}$. $\mathbf{I}$ describes not only the mass of the rotating object but also its distribution (which may be a function of time).

Working from Equations 5.10 and 5.18, an expression for the relationship between torque and angular acceleration becomes possible, and the analogy to Newton's second law of motion is complete (Equations 5.19 and 5.20):

$$\boldsymbol{\Gamma} = \frac{d\boldsymbol{L}}{dt} = \frac{d(\mathbf{I} \cdot \boldsymbol{\omega})}{dt} = \frac{d\mathbf{I}}{dt} \cdot \boldsymbol{\omega} + \mathbf{I} \cdot \frac{d\boldsymbol{\omega}}{dt} \qquad (5.19)$$

in general, and when the inertia tensor components are constant,

$$\boldsymbol{\Gamma} = \frac{d\boldsymbol{L}}{dt} = \mathbf{I} \cdot \frac{d\boldsymbol{\omega}}{dt} \qquad (5.20)$$

(In thinking about the difference between Equations 5.19 and 5.20, recall our earlier discussion that, because the rotational analog of mass

I can be a function of time, angular momentum plays a larger role in rotational motion than linear momentum plays in translational motion.)

We're now where we were for translational motion at the end of Chapter 3, and we can use both analytic and numeric methods to solve Equation 5.19 (or, in the simple case, Equation 5.20), for the angular velocity **ω**. Only a brief discussion of energy remains.

In contrast to the case of momentum and force, there is no "angular energy" associated with rotational motion. Ordinary kinetic energy is simply reexpressed using the variables of angular motion. The result is

$$K_{rot} = \frac{I_{xx}\omega_x^2 + I_{yy}\omega_y^2 + I_{zz}\omega_z^2}{2} + \frac{2I_{xy}\omega_x\omega_y + 2I_{yz}\omega_y\omega_z + 2I_{zx}\omega_z\omega_x}{2} \tag{5.21}$$

Using the concise notation of dyads, Equation 5.21 becomes

$$K_{rot} = \frac{\boldsymbol{\omega}\cdot \mathbf{L}}{2} = \frac{\boldsymbol{\omega}\cdot \mathbf{I}\cdot \boldsymbol{\omega}}{2} \tag{5.22}$$

When Christiaan Huygens first investigated rotational motion in the 1600s, he could not have known that angular momentum would have profound implications for twentieth-century physics and our understanding of Nature. Research in this century has shown that angular momentum is *quantized.* That is, angular momentum can have only certain discrete or *quantum* values that are integer or half-odd-integer multiples of the *quantum of angular momentum* (which has a value of 1.054×10^{-34} kg m^2 s^{-1}). Because this is a very small quantity when expressed in the macroscopic units of the *mks* system, the "graininess" of angular momentum can usually be ignored in macroscopic mechanics.

In the microscopic world, however, this graininess is both noticeable and consequential. Furthermore, all elementary particles in Nature, such as electrons, protons, or neutrons (and collections of them such as nuclei, atoms, and molecules), have their own *intrinsic* angular momentum called *spin.* Spin arises from the rotational symmetry properties of elementary particles, and so all particles of the same type have exactly the same spin. Furthermore, because spin is an angular momentum, a particular type of particle can have only integral or half-odd-integral value of the quantum.

- Particles with integral values are called *bosons*, named after Satyandra Nath Bose, who first described their properties. One example of a boson is the photon — the elementary particle of visible light.

- Particles with half-odd-integral values are called *fermions*, named after Enrico Fermi, who first recognized their properties. Onc cxample of a fermion is the electron   an elementary particle responsible for much of the chemistry of atoms and molecules.

Despite the very small value of the angular momentum quantum (in macroscopic units), the spin of elementary particles has dramatic macroscopic consequences.

- Quantum theory requires and experiment confirms that any number of bosons can be in exactly the same configuration or *quantum state* at the same time. It is this fact that allows enormous numbers of photons to clump together into electromagnetic waves and bring radio and television into your home.

- No two fermions, on the other hand, can be in the same quantum state at the same time. Fermions must therefore remain non-congenial and "spread around" in space. Electrons are fermions and ordinary macroscopic matter contains enormous numbers of them. Thus, the spatial extent of all the stuff around you originates in a fundamental property of fermions.

All of these wonderful things in Nature, and more, will be discussed in later volumes.

EXERCISE **5.1**

AN INTRODUCTION TO UNIFORM ROTARY MOTION

TO GET THE MOST OUT OF THIS EXERCISE, YOU SHOULD ALREADY KNOW

The definitions of the terms:
force
linear momentum
angular position
angular velocity
angular momentum
torque

How to calculate the cross product of a vector.

Many aspects of rotational motion may initially violate your intuition and so be difficult to understand. When you studied translational motions earlier in this book, you almost certainly were able to connect your studies to personal experiences such as throwing a ball or dragging equipment up an incline. Your experiences with rotational motion may not have been as complete. This exercise is designed to offer you a chance to get a feel for the behavior of rotating systems prior to your immersion in their formalities.

- For simple rigid bodies, you will work out the forces supplied by two axle bearings that assure uniform circular motion. In general, these forces add up to produce a nonzero net force (and net torque) on the body.

- You will see how these forces depend on the axial positions of the bearings.

  You will discover that if the axle passes through a special place in a rigid body (called the *center of mass*) no net force is required to keep it spinning uniformly. Furthermore, you will find that if the body rotates about certain special directions through the center of mass (called *principal axes*) no torque is required either. You will also find that principal axes are always at right angles to each other.

Previous exercises in this book generally have supplied a force and requested a motion. In this exercise, the scheme is reversed— uniform rotary motion is specified, and the forces necessary to produce it are requested.

Uniform rotary motions about fixed axles are very common in your everyday world. Most machines— from the flywheel in your automobile to the disk drive in your computer — involve such motions. In contrast to linear motion, uniform rotary motion does not always happen "for free." Unless the body rotates about an axle that passes through its center of mass in the direction of one of its principal axes, various pushes and pulls must be supplied by expensive bearings to assure uniform motion. In the real world, the wear and tear on these axle bearings is minimized whenever possible by judicious placement of the axle in the body.

Such special axle positions exist even for irregularly shaped bodies (although this may not be intuitively obvious). In this exercise, you perform numeric "experiments" to find these positions; in the next exercise, you learn an elegant technique for finding them algorithmicly. Except in Exercise 5.5, your task is simplified by using masses that rotate in a vacuum about perfectly oiled (frictionless) bearings.

Let's begin.

From your earlier studies of dynamics, you know that the total force $\boldsymbol{F}$ that acts on a body consisting of several masses ($m_1, m_2, ..., m_j$) is the time rate of change of its total momentum $\boldsymbol{P}$:

$$\boldsymbol{F} = \frac{d\boldsymbol{P}}{dt} \tag{5.23}$$

where $\boldsymbol{P} = \sum_{j=1}^{n} m_j \boldsymbol{v}_j$

There is a similar equation relating the total torque Γ on the system and the rate of change of its total angular momentum $\boldsymbol{L}$ (Equation 5.24 and Equation 5.10 in this chapter's introduction):

$$\Gamma = \frac{d\boldsymbol{L}}{dt} \tag{5.24}$$

where $L = \sum_{j=1}^{n} (r_j \times m_j v_j)$.

One important difference between these two equations is that the values for total torque and total angular momentum usually depend on the choice of origin for the coordinate system. This is never the case for the total force and the total momentum. You will see this by exploring the behavior of simple, rigid frameworks consisting of just a few point masses (Figure 5.5). The framework is built out of rigid rods of negligible mass that are attached to a fixed axle. The fixed axle is supported by two bearings through which forces and torques are applied.

To produce a uniform, rotary motion, the bearings must supply forces F_1 and F_2. These are related to the total force and total torque of Equations 5.23 and 5.24 by

$$F_1 + F_2 = F \quad (5.25)$$

$$r_1 \times F_1 + r_2 \times F_2 = \Gamma \quad (5.26)$$

where r_1 and r_2 are the vector locations of the ideal "point" bearings (Figure 5.5). You can use these equations to learn something about the bearing forces. By using Equation 5.25 to successively eliminate F_1 and F_2 in Equation 5.26, you get

$$(r_1 - r_2) \times F_1 = \Gamma - r_2 \times F \quad (5.27)$$

$$(r_1 - r_2) \times F_2 = -\Gamma + r_1 \times F \quad (5.28)$$

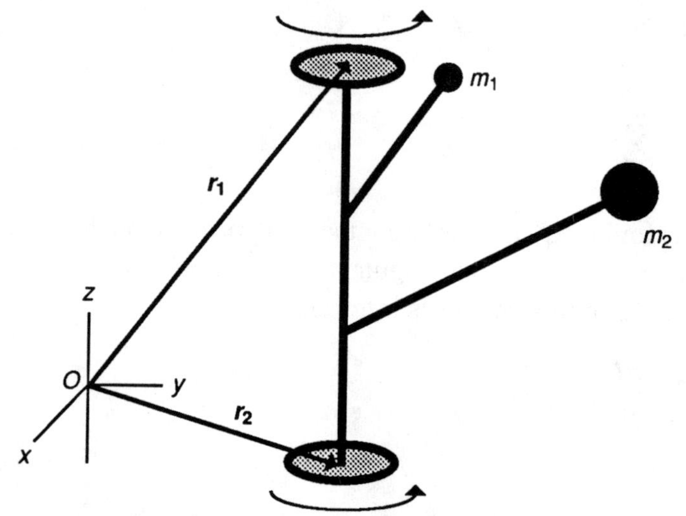

Figure 5.5

Masses attached to a rotating axle via rigid, massless rods make an excellent device with which to gain insight into rotational motion. In the figure, two masses m_1 and m_2 are attached to an axle whose bearings deliver suitable forces so that the device always exhibits its uniform rotary motion. The locations of the axle bearings are identified by the vectors r_1 and r_2 extending from the coordinate frame origin. In your idealized model, "point" bearings are assumed that can only supply force to one point on the axle. Real bearings always have some axial extent, and so they normally supply forces to a range of points along the axle.

Because the bearings are both on the axle, the vector $\boldsymbol{r}_1 - \boldsymbol{r}_2$ is along the axle itself. By remembering the properties of the cross product, you can see that these equations do not allow you to determine the component of the bearing forces *along* the axle. The axial component of $\boldsymbol{F}_1$ can be anything. Given it, you can calculate the axial component of $\boldsymbol{F}_2$ from Equation 5.25, but the laws of dynamics say no more; they do not specify how strongly the bearings push against each other along the axle. We will set these undetermined forces to zero, although in real machinery they needn't be.

If the axle is taken along the z-axis (e.g., Figure 5.6) and the bearings are at z_1 and z_2, then you can write Equations 5.27 and 5.28 in components to get

$$(F_1)_x = \frac{\Gamma_y - z_2 F_x}{z_1 - z_2} \tag{5.29}$$

$$(F_2)_x = \frac{-\Gamma_y + z_1 F_x}{z_1 - z_2} \tag{5.30}$$

$$(F_1)_y = \frac{-\Gamma_x - z_2 F_y}{z_1 - z_2} \tag{5.31}$$

$$(F_2)_y = \frac{\Gamma_x + z_1 F_y}{z_1 - z_2} \tag{5.32}$$

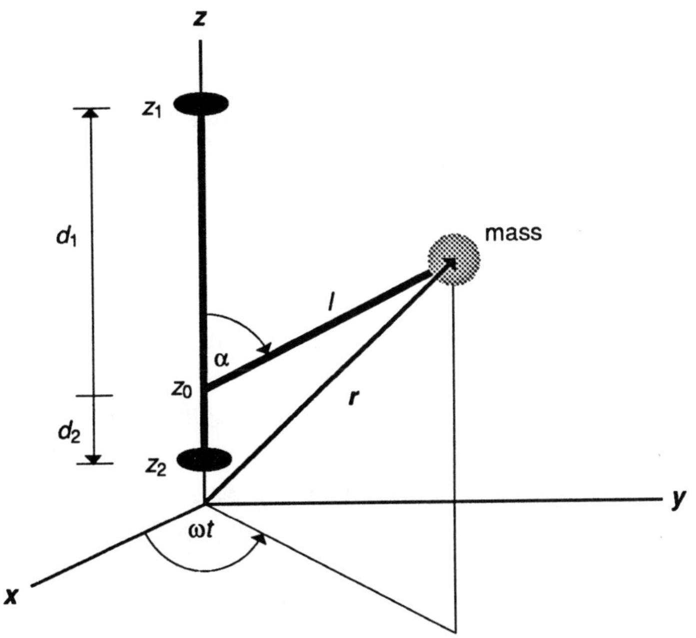

Figure 5.6

A simple arrangement of one mass attached to an axle. The axle is spun with angular velocity ω. By changing the angle α, you can study what happens as you change the geometry of the system. By changing the distance z_0, you can study the dependence of the dynamical quantities on the choice of origin. By changing the distances d_1 and d_2, you can see how the placement of the bearings affects the actual forces they must supply to maintain uniform rotary motion. Notice that, in this sketch, the distance d_2 has a negative value.

In this exercise the axle is always arranged along the z-axis for convenience, but the relationship of the axle to the rigid body will change.

The first order of business then is to select suitable parameters with which to describe the internal geometry of the body, the relative location of the axle with respect to it, and the relative location of the origin.

You work primarily with two simple, rigid bodies in this exercise:

- Figure 5.6 shows the simplest example of a rigid body, one with just a single point mass. The mass is attached to the (massless) axle with a (massless) rod of length l that makes an angle α with the axle as shown. During any one "spin," you hold this angle fixed but sometimes change it between "spins." By changing this angle, you can adjust the relative positions of the axle and the body and study what happens.

  By changing the distance z_0, you can study the dependence of the dynamical quantities on the relative location of the origin of the coordinate system.

  The positions of the bearings are given by d_1 and d_2, their distances from where the rod holding the mass is attached to the axle.[1] Thus, the formulas for z_1 and z_2 are $z_1 = z_0 + d_1$ and $z_2 = z_0 + d_2$ (Figure 5.6).

  Now everything follows from Equation 5.33 — the formula for the position vector $\boldsymbol{r}$ of the mass.

  $$\boldsymbol{r} = l \sin\alpha \cos(\omega t)\,\boldsymbol{i} + l \sin\alpha \sin(\omega t)\boldsymbol{j} + (z_0 + l\cos\alpha)\boldsymbol{k} \quad (5.33)$$

  where ωt identifies the rotation angle of the rotating mass and is the angle that the projection of the mass onto the x-, y-plane makes with the x-axis (Figure 5.6).

- Figure 5.7 shows a second version of the simple, rigid framework of masses that you will use. Now there are two masses m_1 and m_2 at the ends of two rods of lengths l_1 and l_2. The masses and their connecting rods lie in a single plane with the axle, and (when the masses are on the same side of the axle) the opposite ends of the rods are connected together at a perpendicular distance z_0 from the x-axis. The angle between the two rods is θ.

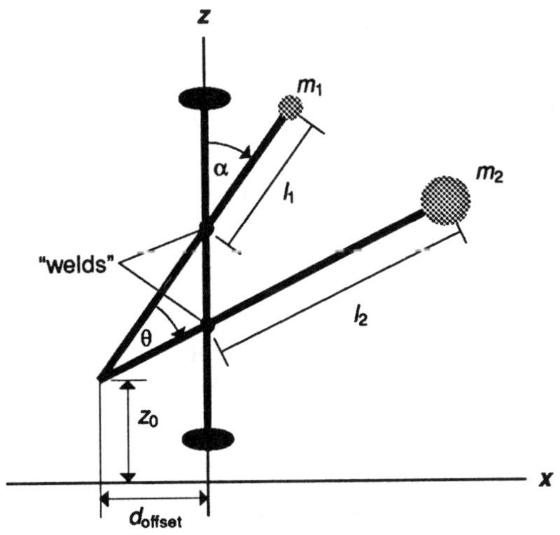

Figure 5.7

Consider a rigid body consisting of two masses. The two masses, their supporting rods, and the axle are shown at a moment when the two masses are in the $y = 0$ plane. The y-axis is normal to the page of the book, going into it. A moment later, the two masses will have rotated to just outside the page. The location of the axle (the z-axis) relative to the object itself is defined by the two parameters d_{offset} and α. Finally, the origin is always on the axle and is located a distance z_0 from the place where the two rods are joined (as shown).

The device can be thought of as consisting of two parts:

(a) A rigid body consisting of two point masses supported by two rods. The rods are of length l_1 and l_2, respectively, and the angle between them is θ. Notice that this body can be mounted on the axle in many ways.

(b) An axle with two bearings.

The rigid body is then mounted on the axle by "welding" the rods to the axle. Exactly where they are welded is defined by two parameters d_{offset} and α, the specification of which uniquely defines where the one or two welds are to be made. Note that if the masses are on opposite sides of the axle, there is just one weld.

Finally, on top of this completed system or rigid body and axle, impose a coordinate system. The origin of the coordinate system is always on the axle, and the z-distance from the join point to the origin is defined by the parameter z_0.

The axle need not pass through the attachment point of the two rods but can be offset a distance d_{offset} from the intersection of the two rods (Figure 5.7). Finally, the rod holding the first mass makes an angle α with the axle. For a particular rigid body (i.e., a body with specified values of m_1, m_2, l_1, l_2, and θ), you will study what happens as the origin of coordinates is moved (changing z_0), the location of the axle is modified (changing d_{offset}), and the axle's orientation relative to the body is changed (α). The bearings are taken just the same as in the one-mass rigid body.

With this geometry, the position vectors of the two masses are

$$\boldsymbol{r}_1 = (l_1 \sin\alpha - d_{offset})[\cos(\omega t)\boldsymbol{i} + \sin(\omega t)\boldsymbol{j}] + (l_1 \cos\alpha + z_0)\boldsymbol{k} \quad (5.34)$$

$$\boldsymbol{r}_2 = [l_2 \sin(\alpha+\theta) - d_{offset}][\cos(\omega t)\boldsymbol{i} + \sin(\omega t)\boldsymbol{j}] + [l_2 \cos(\alpha+\theta) + z_0]\boldsymbol{k} \quad (5.35)$$

As you will see from Model 5.1, the quantity

$$x_{cm}(t=0) = \frac{m_1 x_1(t=0) + m_2 x_2(t=0)}{m_1 + m_2} \quad (5.36)$$

is very important. It is the distance of the body's *center of mass* from the axle. As you will see, when this quantity is zero, *there is no net force acting on the body* when it is in steady rotation. There may be a torque required, but (in contrast to when the center of mass is not on the axle) *its value does not depend on where the origin of coordinates is located.* Finally, when the axle passes through the center of mass, you will find that even the torque vanishes for certain values of α.

SPREADSHEET TECHNIQUE

Building Models Using Existing Components

As you begin to use your personal computer regularly to think about and to solve problems, you will rapidly develop a library of spreadsheet formulas and systems of formulas that have been carefully verified in numerous previous situations.

Rather than going through the tedium of rewriting and verifying these formulas every time you encounter a situation that requires them, it is far better to reuse previously verified formula systems. Different spreadsheet programs and computer operating systems permit you to extract the "mechanisms" of previously built spreadsheet files in different ways.

Regardless of your hardware-software combination, you can always make a copy of a previous file and then edit it to its new requirements.

If you use a computer that operates in a graphical environment that supports windows, you may need only CUT the material you need from one spreadsheet and PASTE it into its new environment. If you use a program such as Lotus 1-2-3 version 2.0 you can use the EXTRACT and COMBINE commands to transfer the material from one spreadsheet to another and then the MOVE commands to position the retrieved material properly.

Models 5.1 and 5.2 of this exercise, for example, were originally constructed by first retrieving the formulas for a vector cross product from Model 2.6 of Exercise 2.3.

LESSONS

1. Build and verify Model 5.1. This model calculates the position of the single, point mass in Figure 5.6 as a function of time and then numerically differentiates that position to obtain velocity.[2] The velocity is multiplied by the mass to obtain the momentum, and the angular momentum is then calculated. Finally, both the momentum and angular momentum are differentiated numerically to find the three components of the force and the torque necessary to produce the motion. The angular velocity vector ω plays no direct role in this lesson; you should explore it in the problems if you wish.

   We attempted to avoid mistakes in constructing this model by retrieving the verified, vector cross-product formulas from Model 2.6 in Exercise 2.3. We used Spreadsheet Technique "Building Models Using Existing Components" to do this.

   Begin your explorations using the initial values described for Model 5.1. The angular velocity of 3600 rpm (revolutions per minute) is typical of the crankshaft of an automobile going about 60 miles per hour. It is also the rate at which the hard disk platter in your personal computer rotates.

a. Explore the structure of Model 5.1 until you understand it and how it implements Equation 5.33.

b. Plot y vs x (Figure 5.8) and verify the uniform circular motion of the mass attached to the axle (if the aspect ratio of your plot permits). Because there is a z-component of the position vector, the full position vector is not represented but only its projection into the x-, y-plane.

Notice Figure 5.8 uses points rather than lines to represent the motion so that you can verify the equal angular spacing between the points. Furthermore, the motion begins on the $+x$-axis, and ends just beyond the $-x$-axis. This slight asymmetry was included to help you remember which end is the starting point.

c. Next, plot P_y vs P_x (Figure 5.9). Because the momentum has no component along the z-axis, you are plotting the full momentum vector.

The slight asymmetry deliberately built into Figure 5.8 is also present in Figure 5.9. Thus, you can see that the plot begins at the $+y$-axis and ends just beyond the $-y$-axis. Study your plots of y vs x and P_y vs P_x and assure yourself that, in circular motion, the velocity vector is always perpendicular to the position vector.

d. Plot L_y vs L_x. Because $\boldsymbol{L}$ has a z-component, you are plotting the projection of $\boldsymbol{L}$ into the x-, y-plane. What is the geometric relationship that exists between position, linear momentum, and angular momentum?

e. Plot F_y vs F_x. There is no z-component of the force, and so you are seeing the actual force vector $\boldsymbol{F}$. In which direction does the force always point?

f. Plot Γ_y vs Γ_x. Is this the full torque vector? How is Γ related to the earlier vectors?

Make a sketch (at a single moment in time) that shows the relative directions in the x-, y-plane of the five vectors just explored. How will your diagram change as time increases?

Figure 5.8

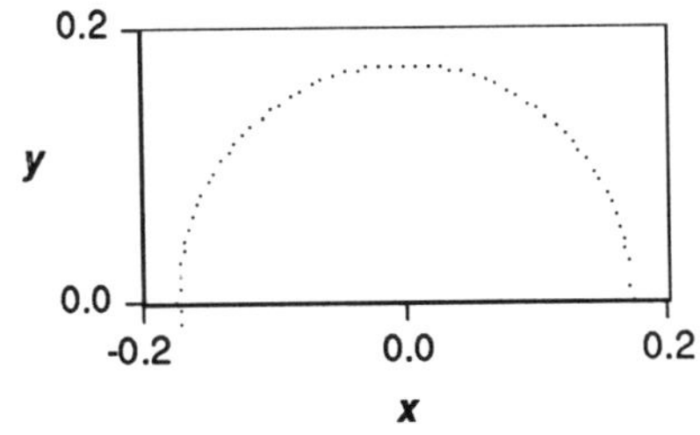

XY-PLOT

Data
Horizontal (x) axis: cells H14 - H75 (x-position)
Vertical (y) axis: cells I14 - I75 (y-position)

Notice that the plot points are not connected with lines. This makes it easier to see the constant spacing

Figure 5.9

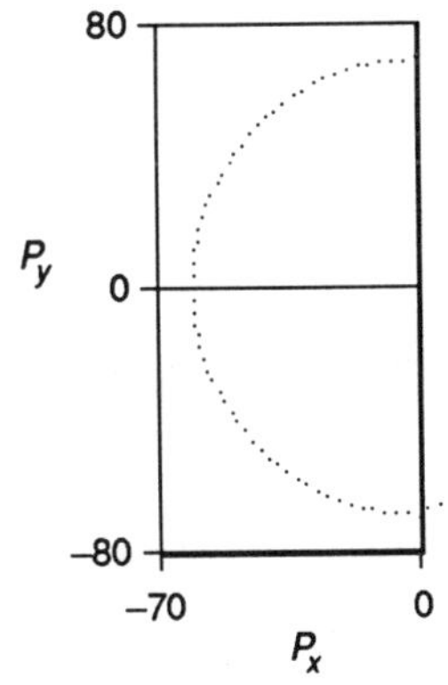

XY-PLOT

Data
Horizontal (x) axis: cells O14 - O75 (x-component of momentum)
Vertical (y) axis: cells P14 - P75 (y-component of momentum)

Notice that the plot points are not connected with lines. This makes it easier to see the constant spacing

2. You've been told that if the axle passes through a "special place" in rigid bodies (called the *center of mass*) no net force is required to keep a rigid body spinning uniformly.

   a. Plot L_x, L_y, and L_z vs time. Notice that L_z is constant and L_x and L_y have correlated sinuisoidal variations that are the hallmark of circular motion.

   b. Now plot L_x, L_y, and L_z vs time for each of the values of z_0 in the left-hand column of the table in Lesson 2c. Don't attempt to complete the remainder of the table just yet.

   What happens to the angular momentum vector as the origin of your coordinate system moves along the axle? Why?

   Where is "most" of the angular momentum when the origin is far away from the rod's insertion point on the axle?

   What value of z_0 causes the angular momentum to become constant in time?

   Why does the angular momentum become constant for this special value of z_0?

   c. Summarize your observations by completing the following table. Take time t_0 to be 0 and time t_1 to be that moment at which the mass is on the y-axis. (t= 0.004166 s at this point when the initial conditions described for Model 5.1 are in place.)

   Plot L_x vs L_y for each value of z_0 (once for $t = 0$ and again for $t = t_1$).

   If you wish, automate the creation of this table by using "data table" or "what-if" table commands on your spreadsheet. This will make subsequent plotting of these tables easier.

	$t = t_0$			$t = t_1$		
z_0	L_x	L_y	L_z	L_x	L_y	L_z
100	____	____	____	____	____	____
10	____	____	____	____	____	____
5	____	____	____	____	____	____
.2	____	____	____	____	____	____
.1	____	____	____	____	____	____
0	____	____	____	____	____	____
–.1	____	____	____	____	____	____
–.2	____	____	____	____	____	____
–.3	____	____	____	____	____	____
–5	____	____	____	____	____	____
–10	____	____	____	____	____	____
–100	____	____	____	____	____	____

d. Sketch the direction of the projection of the angular momentum in the x-, y-plane relative to the position vector for $z_0 > -0.1$ m, $z_0 = -0.1$ m, and $z_0 < -0.1$ m. What do you suppose would happen if the origin of coordinates were taken off the axle?

3. Now plot the three components of force necessary to maintain the motion and study their dependence on z_0 by filling in a table similar to the previous one. Take t_0 as the first time step and t_1 as previously done. Does the necessary force depend on the location of the origin along the axle? What if the origin were off the axle? Notice that the units of force are Newtons. Would you consider this force negligible, moderate, or quite large? (You can get a feeling for the size of a force by comparing it to the weight of a specific volume of water.)

	$t = t_0$			$t = t_1$		
z_0	F_x	F_y	F_z	F_x	F_y	F_z
100	____	____	____	____	____	____
10	____	____	____	____	____	____
5	____	____	____	____	____	____
.2	____	____	____	____	____	____
.1	____	____	____	____	____	____
0	____	____	____	____	____	____
−.1	____	____	____	____	____	____
−.2	____	____	____	____	____	____
−.3	____	____	____	____	____	____
−5	____	____	____	____	____	____
−10	____	____	____	____	____	____
−100	____	____	____	____	____	____

4. Repeat explorations similar to those of Lessons 2 and 3 for the three components of the torque.

   Notice that the value of the torque depends on the location of the origin. Why? For what value of the origin does the torque vanish? Why? The units of the torques you have calculated are Newton-meters, that is, the torque created by a force of 1 N acting at a distance of 1 m. How would you characterize the magnitudes of the torques you have just calculated?

5. Finally, use appropriate tools to examine the four components of the bearing forces.

   Why are these independent of the location of the origin, although the calculated torque depends strongly on its location? Notice the magnitude of these forces in your model. Describe qualitatively the time dependence of these bearing forces. Which bearing must supply the greater force? Is it reasonable to you that one bearing must provide more force than the other? Do you consider these forces big, moderate, or negligible? How big a steel bearing do you suppose would be required to support them?

6. Modify Model 5.1 by placing the formula =E4+1 in cell E3. Now the distance d_1 is always 1 m greater then d_2, and you can study the effect of moving the location of the bearings along the axle while keeping the distance between them constant.

   Complete the table below to explore the relationship between the bearings' location and the magnitude of the bearing forces. (Automate the procedure with "data tables" if you wish.)

d_2	F_1	F_2
-5.0	______	______
-3.0	______	______
-1.0	______	______
-0.9	______	______
-0.8	______	______
-0.7	______	______
-0.6	______	______
-0.5	______	______
-0.4	______	______
-0.3	______	______
-0.2	______	______
-0.1	______	______
0.0	______	______
0.1	______	______
0.2	______	______
0.3	______	______
0.4	______	______
0.5	______	______
1.0	______	______
3.0	______	______
5.0	______	______

   a. If the two bearings had equal strength, where would you put them to minimize possible problems with them? Make a careful sketch showing the geometry for this configuration. Do you find it reasonable that the bearing forces become so large as the bearings are moved far away from the mass?

   b. Suppose one bearing were very much weaker than the other. How would you arrange them in this case?Again make a sketch to understand the geometry for this case. Notice that in this case a single bearing is sufficient to support the whole load.

7. To accelerate the mass, you know that a force must be applied directly to it. This is actually supplied by the rod to which the mass is attached. Although you have not worked out the exact details of how the rod does this, you have calculated just what this force must be. For circular motion, this force is directed from the mass directly toward the axle and is called the *centripetal force*. Centripetal force points to that place on the axle where a single bearing can support

the full load and may be called the center of force. The center of force is a natural location for the origin of coordinates used in calculations. The physics, of course, is never dependent on your choice of coordinate frame, but a shrewd choice can often greatly simplify the mathematics.

a. Set up the parameters in your model so that z_1 is at the center of the force and set z_0 properly so that $z_1 = 0$. Study the results. Describe the angular momentum with this set of parameters. What about the torque?

b. As you have just seen, you can arrange the angular momentum in this problem to be constant and the torque to be zero by putting the origin of your coordinates at a special place. Investigate what happens as you change the angle α from 0° to 180° in steps of 30°, arranging that the origin is always at the center of force.

   Do you understand the results when α is either 0° or 180°?

   What happens to the center of force as α is changed?

   How does the centripetal force depend on $l \sin \alpha$, the distance from the axle to the mass?

8. Your studies of Model 5.1 have shown you that such vectors as position, momentum, and angular momentum have a constant magnitude, a constant component along the axle, and a projection into the x-, y-plane that moves in circles.

   Now you will use Model 5.2 to study what happens with changes in the location and direction of the axle and what happens when additional masses are added to the axle. You can easily change the time point at which the vectors are calculated by changing the entry in cell I27.

   Build and verify Model 5.2 (using parts of Model 5.1 where appropriate). This model simulates the rigid body shown in Figure 5.7 but has not been designed to calculate the detailed time dependence of dynamical vectors such as position, momentum, or angular momentum. Rather, these vectors are calculated at just one moment ($t = 0$) using numerical differentiation.[3]

   You can use your spreadsheet graphics capabilities to build an on-screen graphic to represent the geometry of the rigid body using the instructions in Figure 5.10. In this plot, the vertical axis is the axle

Figure 5.10

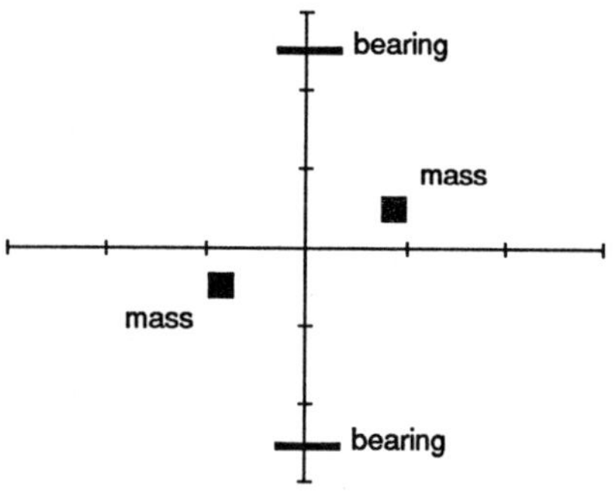

You can use your on-screen graphic capability to draw a simple picture of the rigid body represented by Model 5.2.

Data Plotted as Lines

Horizontal (*x*) axis: cells I34 and I35
Vertical (*y*) axis: cells L34 and L35
(assures sufficient length of *x*-axis)

Horizontal (*x*) axis: cells I37 and I38
Vertical (*y*) axis: cells L37 and L38
(assures sufficient length of *y*-axis)

Horizontal (*x*) axis: cells I40 and I41
Vertical (*y*) axis: cells L40 and L41
(first bearing location)

Horizontal (*x*) axis: cells I43 and I44
Vertical (*y*) axis: cells L43 and L44
(second bearing location)

Data Plotted As Symbols

Horizontal (*x*) axis: cells I30 and I31
Vertical (*y*) axis: cells K30 and K31
(location of masses)

Some spreadsheets will allow you to plot a single, continuous stretch of *x*-axis. We have broken up that stretch here to guarantee that inappropriate connections between lines are discarded. In Lotus products, the *x*-axis range is I30 - I44, and the two *y*-axis ranges are K30 - K44, L30 - L44.

(z-axis), the horizontal axis is the x-axis, and you are looking at the system from a point on the y-axis. The picture is drawn at $t = 0$, and the mass on the right is moving into the screen at this instant, while that on the left is moving out of it. The bearings are indicated on the axle as short lines, and the masses are indicated as symbols. The various connecting rods are not shown (although you could add them if you like). You will use this plot in the exercises that follow to visualize the relation of the two masses to the axle and the location of the bearings.[4]

a. Verify that Model 5.2 is initialized with values such that the net force that must be supplied by the bearings to maintain uniform rotary motion is zero. Notice, however, that the bearings must supply a torque. What is the geometric relationship of the two masses to each other (see Figure 5.10)?

   Compare this situation with the case of the single mass in Model 5.1. (You can get the same result in this model by temporarily setting the mass in the second row of the table to zero [cell A31].)

   Also compare this situation to a case in which two equal masses are present at the end of equal-length rods, but the masses are not "in line" with each other. What happens when the angle θ for the second mass is not 180°?

b. To accelerate a single mass under conditions of uniform rotary motion, external force must be delivered by the axle bearings. Curiously, however, no external force is required to accelerate the two masses as initially aligned in Model 5.2! Under these conditions, the two masses experience centripetal forces of equal magnitude but opposite direction, and *they balance each other out*. Thus, all forces necessary for acceleration are within the rigid body — no external forces are required.

   Set the angle θ for the second mass to 90° so that the system is unbalanced and external force is required to maintain uniform rotary motion. Your on-screen plot of Figure 5.10 should show that both masses are now on the same side of the axle (as seen from the y-axis).

   Now complete the following table:

d_{offset}	F_x	X_{cm}
0.05	______	______
0.06	______	______
0.07	______	______
0.08	______	______
0.09	______	______
0.10	______	______
0.11	______	______
0.12	______	______
0.13	______	______
0.14	______	______
0.15	______	______
0.16	______	______
0.17	______	______
0.18	______	______
0.19	______	______
0.20	______	______

Plot the F_x vs X_{cm}. What offset value "balances" the system and permits uniform rotary motion without a net external force? Refine your estimate to three decimal places by entering new offset values into cell B8. Examine your on-line graphic corresponding to Figure 5.10 and describe the new position of axle when the offset balances the system.

c. The value of d_{offset} that balances the system can always be calculated using the formula

$$d_{\text{offset}} = \frac{m_1 l_1 \sin\alpha + m_2 l_2 \sin(\alpha + \theta)}{m_1 + m_2} \tag{5.37}$$

Develop this formula independently and verify its correctness by inserting it into cell B9 (Don't forget to convert degrees to radians in the formula!). Changing the parameters of your model should now always yield a balanced system because the axle will always offset itself to accommodate any inbalance.

d. The balance condition you have just seen occurs when the the parameter X_{cm} is zero, or more geometrically, when the axle passes through the center of mass of the rigid body. When it is satisfied, the system is said to be statically balanced. Vary different parameters such as the angles, masses, and lengths and look at the resulting pictures of your system until you get a feel for where the axle should be placed to have static balance.

In a uniform gravitational field, you can find the center of mass of a rigid body by finding the support point that "balances" it (i.e., the body is statically supported at the center of mass when the body is free to move but doesn't). The mechanic at your local tire store does this when she starts the process of balancing the tires on your car.

e. With the formula developed for the value of d_{offset} in place, study the effect of the parameter z_0 on the angular momenta and torque. Fill out the following table:

z_0	L_x	L_y	L_z	Γ_x
-100	____	____	____	____
-20	____	____	____	____
-1	____	____	____	____
0	____	____	____	____
1	____	____	____	____
20	____	____	____	____
100	____	____	____	____

Describe in words what you find. Why do you get this result? When two forces equal in magnitude and opposite in direction act along different lines, people call the force system a *couple.* Although, in general, torque is dependent on the position of the origin of coordinates as you saw in Model 5.1, this is not the case when you have a couple.

f. As you have seen, a torque is usually required to rotate the body uniformly, and so people say that it is not "dynamically balanced." If the bearings are not perfectly rigid, the body will wobble as it rotates because real bearings always "give" at least a little when they supply forces to an axle. You can study the more subtle issue of dynamic balancing by setting θ (for the second mass, of course) to 180° so that $d_{offset} = 0$ for static balance. Now complete the following table:

α	Γ_y	$(F_1)_x$	$(F_2)_x$	L_x	L_y	L_z
0	____	____	____	____	____	____
10	____	____	____	____	____	____
20	____	____	____	____	____	____
30	____	____	____	____	____	____
40	____	____	____	____	____	____
50	____	____	____	____	____	____
60	____	____	____	____	____	____
70	____	____	____	____	____	____
80	____	____	____	____	____	____
90	____	____	____	____	____	____
100	____	____	____	____	____	____
110	____	____	____	____	____	____
120	____	____	____	____	____	____
130	____	____	____	____	____	____
140	____	____	____	____	____	____
150	____	____	____	____	____	____
160	____	____	____	____	____	____
170	____	____	____	____	____	____
180	____	____	____	____	____	____

Notice how the bearing forces and the angular momentum are related to the torque. Plot Γ_y vs α. For which values of α does the torque vanish? Explain qualitatively why this is the case. What sort of function does $\Gamma_y(\alpha)$ look like? Develop an analytic formula for this function and compare your result with the numeric calculation.

Study the angular momentum in the previous table. What direction does it have when the torque vanishes?

Notice that there are two axle directions (0° and 180° really are the same axle direction and so are not counted as different) for which no torque is necessary to maintain steady rotation. When the system is statically balanced and an axle has either of these very special directions, the system is said to be in *dynamic balance*. These directions are called the *principal axes* of the body, and they pass through the body's center of mass. What is the angle between the two principal axes you have just found for the simple two-mass rigid body? Plot pictures of your system for these two special values of α.

As you will see in the next exercise, an axle through the center of mass and perpendicular to *both* of the principal axes you have just found is also a principal axis. *In general, there are always three principal axes, and they are always at right angles to each other.*

PROBLEMS

1. In the lessons of this exercise, you discovered some important properties about steady circular motion for a rigid body made up of just two masses. These properties, however, are in fact general in the sense that all rigid bodies

   - Have a center of mass

   - Have three orthogonal principal axes
     (Thus far, you've found two.)

   Model 5.2 was constructed so that you could extend your explorations to rigid bodies containing more than three discrete masses in a plane. Notice that row 32 of Model 5.2 contains nothing more than a sequence of hyphens. This row marks the location where additional masses can be added by simply inserting a new row immediately above (i.e., between row 31 and row 32). Consult your spreadsheet manual for an INSERT ROW command if you've forgotten how to do this.

Insert a new row and then "mount" a mass of 2 kg on a rod of radius 0.3 m with an angle θ of 90°. Copy the remainder of the preceding row into the new one. Your spreadsheet program is designed to automatically adjust all formulas in the matrix to accommodate the new row. (However, if you still have a formula for automatically calculating d_{offset} in cell B9, this formula is no longer valid.) You can also get rid of masses by just DELETING the row. Be sure not to delete the row containing the sequence of hyphens! That's really the reason it is specially marked.

a. Find two different axes for which the system is both statically and dynamically balanced by varying d_{offset} and α.

   You may wish to develop a formula for d_{offset} that ensures that the center of mass is always on the axle as you vary α.

   What are the two values of α that define the principal axes? What is the angle between them?

   What are the corresponding values of d_{offset}?

b. What are the two values of the angular momentum when the axle is along each of the two principal axes?

   The ratios of the angular momentum to the angular velocity for these two cases are called the two principal moments of inertia. What are their values?

   Notice that the mass is more concentrated near the axle for the principal axis with the smaller principal moment of inertia.

2. Using Model 5.2, arrange four masses of 1 kg each on the four corners of a square by taking all four radii values to be 0.1 m and the four θ angles to be 0°, 90°, 180°, and 270°. Where is the center of mass of this mass configuration? What are its principal axes? What are its principal moments of inertia about these axes? Can you understand this result qualitatively?

   Suppose the mass at $\theta = 0°$ were slightly heavier than the rest (say, 1.010 kg). Where would you put a 10 g counterweight to keep the center of mass at the center of the square? Where are the principal axes of this new configuration, and what are the principal moments of inertia?

3. Use Model 5.2 to arrange three equal masses of 1 kg each in the shape of an equilateral triangle. The thee radii are 0.3 m, and the three θ angles are 0°, 120°, and 240°, respectively.

   Find the center of mass, the principal axes, and the principal moments of inertia. Discuss.

   What happens if the symmetry of the object is broken by letting one mass equal 2 kg?

4. Thus far you have not explored the angular velocity vector $\boldsymbol{\omega}$. It is nevertheless a crucial notion in rigid-body motion. For each mass in a rigid body, you can write

$$\boldsymbol{v} = \boldsymbol{\omega} \times \boldsymbol{r} \qquad (5.38)$$

   with *the same angular velocity vector* $\boldsymbol{\omega}$ *for every mass.*

   Show that you can calculate the velocity of each mass at $t = 0$ from Equation 5.38 by choosing $\boldsymbol{\omega} = \omega\boldsymbol{k}$ and comparing the result of this formula with what you get from averaging the velocites in Model 5.2 at $t = -\Delta t/2$ and $t = +\Delta t/2$. This clearly is a much simpler way to calculate velocity for circular motion than numeric differentiation!

5. In Problem 4, the angular velocity for Model 5.2 was given. Calculate the magnitude of the angular momentum for various values of the magnitude of the angular velocity for any particular rigid body and axle location in Model 5.2. Plot $|\boldsymbol{L}|$ versus $|\boldsymbol{\omega}|$ and describe the relation in words. Is the direction of the angular momentum parallel to that of the angular velocity for all locations of the axle? Under what special conditions is the angular momentum vector in the same direction as the angular velocity vector? (In the next exercise, we define the idea of a "dyad," which expresses the relation between these two vectors.)

6. Add a suitable column to Model 5.2 that allows you to calculate the total kinetic energy of the rotating body, namely,

$$\sum_{i=1}^{n} \frac{m_i v_i^2}{2} \qquad (5.39)$$

By numeric calculation show that the kinetic energy is always equal to $(\omega \cdot \boldsymbol{L})/2$. You may wish to reduce the size of Δt to improve the accuracy of your numeric calculation. Does the kinetic energy depend on time? [HINT: Change the value in cell I27.]

7. Construct a body of several masses by choosing values for the masses, angles θ, and lengths l and verify that you can always find values of d_{offset} and α that define two principal axes and that these are perpendicular to each other. Verify the various relations developed in the previous problems for your particular rigid body.

MODELS

Model 5.1

BUILD and verify Model 5.1.

ENTER data using cells B3 and B4 (mass, angular speed), B7 and B9 (location of rod insertion point, rod length, angle between rod and axle), and E3 and E4 (bearing locations).

READ output from the model from columns A - F (bearing forces), G (time), H - J (*x*-, *y*-, *z*-components of position vector), O - Q (*x*-, *y*-, *z*-components of linear momentum vector), R - T (*x*-, *y*-, *z*-components of angular momentum vector), U-W (*x*-, *y*-, *z*-components of force vector), and X-Z (*x*-, *y*-, *z*-components of torque vector).

NOTICE that the model uses vector operations that can be imported directly from Model 2.6 (Exercise 3.2) using Spreadsheet Technique "Building Models Using Existing Components."

FORMULAS

	A	B	C	D	E	F	G	H	I	J	K	L
1	Single mass attached to an axle											
2												
3	m =	1	kg	d1 =	0.5	m	z1 =	=B7+E3				
4	ang.speed =	3600	rpm	d2 =	-0.5	m	z2 =	=B7+E4				
5	omega =	=2*PI()*B4/60					z1-z2 =	=H3-H4				
6												
7	z0 =	0	m	Δt =	=PI()/60/B5							
8	l =	0.2	m	l cos al =	=COS(B9*PI()/180)*B8							
9	alpha =	60	deg	l sin al =	=SIN(B9*PI()/180)*B8							
10												
11	Bearing 1 Force			Bearing 2 Force				position1 (t)				position1 (t+dt/2)
12	Fx1	Fy1	F1	Fx2	Fy2	F2	t	x	y	z	t+dt/2	x
13	(N)	(N)	(N)	(N)	(N)	(N)	(s)	(m)			(sec)	(m)
14							0					
15	Form	Form	Form	Form	Form	Form	Form	Form	Form	Form	Form	Form
16	Set	Set	Set	Set	Set	Set	Set	Set	Set	Set	Set	Set
17	I	II	III	IV	V	VI	VII	VIII	IX	X	XI	XII

FORMULAS (Contd.)

	M	N	O	P	Q	R	S	T	U	V	W	X	Y	Z
11			momentum1 (t)			angular momentum1 (t)			dP/dt = Force			dL/dt = GAM = Torque		
12	y	z	px	py	pz	Lx	Ly	Lz	Fx	Fy	Fz	GAMx	GAMy	GAMz
13			(kg m/s)			(kg m^2/s)			(Newton)			(Newton meter)		
14														
15	Form	Form	Form	Form	Form	Form	Form	Form	Form	Form	Form	Form	Form	Form
16	Set	Set	Set	Set	Set	Set	Set	Set	Set	Set	Set	Set	Set	Set
17	XIII	XIV	XV	XVI	XVII	XVIII	XIX	XX	XXI	XXII	XXIII	XXIV	XXV	XXVI

VALUES

	A	B	C	D	E	F	G	H	I	J	K	L
1	Single mass attached to an axle											
2												
3	m =	1	kg	d1 =	0.5	m	z1 =	0.5	m			
4	ang.speed =	3600	rpm	d2 =	-0.5	m	z2 =	-0.5	m			
5	omega =	376.991	rad/sec				z1-z2 =	1	m			
6												
7	z0 =	0	m	Δt =	0.00014	sec						
8	l =	0.2	m	l cos al =	0.1	m						
9	alpha =	60	deg	l sin al =	0.17321	m						
10												
11	Bearing 1 Force			Bearing 2 Force				position1 (t)				position1 (t+dt/2)
12	Fx1	Fy1	F1	Fx2	Fy2	F2	t	x	y	z	t+dt/2	x
13	(N)	(N)	(N)	(N)	(N)	(N)	(s)	(m)			(sec)	(m)
14							0	0.1732	0	0.1	7E-05	0.1731
15	-14756.289	-579.78	14767.7	-9837.5	-386.52	9845.12	0.00014	0.173	0.0091	0.1	0.0002	0.1727
16	-14705.723	-1351.3	14767.7	-9803.8	-900.84	9845.12	0.00028	0.1723	0.0181	0.1	0.0003	0.1717
17	-14614.849	-2119.1	14767.7	-9743.2	-1412.7	9845.12	0.00042	0.1711	0.0271	0.1	0.0005	0.1703

VALUES (Contd.)

	M	N	O	P	Q	R	S	T	U	V	W	X	Y	Z
11			momentum1 (t)			angular momentum1 (t)			dP/dt = Force			dL/dt = GAM = Torque		
12	y	z	px	py	pz	Lx	Ly	Lz	Fx	Fy	Fz	GAMx	GAMy	GAMz
13			(kg m/s)			(kg m^2/s)			(Newton)			(Newton meter)		
14	0.0045	0.1	-0.855	65.289	0	-6.529	-0.085	11.308						
15	0.0136	0.1	-4.27	65.155	0	-6.516	-0.427	11.308	-24594	-966.3	0	96.629	-2459	-4E-11
16	0.0226	0.1	-7.675	64.842	0	-6.484	-0.767	11.308	-24510	-2252	0	225.21	-2451	9E-11
17	0.0316	0.1	-11.06	64.352	0	-6.435	-1.106	11.308	-24358	-3532	0	353.18	-2436	-6E-11

Model 5.1 (continued)

Formula Set I	Prototype Cell is A15	=(Y15-**H4** *U15)/**H5**
Formula Set II	Prototype Cell is B15	=(-X15-**H4** *V15)/**H5**
Formula Set III	Prototype Cell is C15	=SQRT(A15^2+B15^2)
Formula Set IV	Prototype Cell is D15	=(-Y15+**H3** *U15)/**H5**
Formula Set V	Prototype Cell is E15	=(X15+**H3** *V15)/**H5**
Formula Set VI	Prototype Cell is F15	=SQRT(D15^2+E15^2)
Formula Set VII	Prototype Cell is G15	=G14+**E7**
Formula Set VIII	Prototype Cell is H14	=**E9** *COS(**B5** *G14)
Formula Set IX	Prototype Cell is I14	=**E9** *SIN(**B5** *G14)
Formula Set X	Prototype Cell is J14	=**E8** +**B7**
Formula Set XI	Prototype Cell is K14	=G14+**E7** /2
Formula Set XII	Prototype Cell is L14	=**E9** *COS(**B5** *K14)
Formula Set XIII	Prototype Cell is M14	=**E9** *SIN(**B5** *K14)
Formula Set XIV	Prototype Cell is N14	=**E8** +**B7**
Formula Set XV	Prototype Cell is O14	=(L14-H14)*2/ **E7** *__B3__
Formula Set XVI	Prototype Cell is P14	=(M14-I14)*2/**E7** *__B3__
Formula Set XVII	Prototype Cell is Q14	=(N14-J14)*2/ **E7** *__B3__
Formula Set XVIII	Prototype Cell is R14	=I14*Q14-J14*P14
Formula Set XIX	Prototype Cell is S14	=-Q14*H14+O14*J14
Formula Set XX	Prototype Cell is T14	=H14*P14-I14*O14
Formula Set XXI	Prototype Cell is U15	=(O15-O14)/ **E7**
Formula Set XXII	Prototype Cell is V15	=(P15-P14)/ **E7**
Formula Set XXIII	Prototype Cell is W15	=(Q15-Q14)/ **E7**
Formula Set XXIV	Prototype Cell is X15	=(R15-R14)/**E7**
Formula Set XXV	Prototype Cell is Y15	=(S15-S14)/ **E7**
Formula Set XXVI	Prototype Cell is Z15	=(T15-T14)/**E7**

Model 5.2

BUILD and verify Model 5.2.

ENTER data using cell B4 (angular speed), cells B8 - B10 z-location of the point where the two rigid rods are joined, axle offset from the same point), B13 and B14 (bearing locations), and columns A - C starting at row 30 (mass, radius, and θ of attached masses). Additional masses can be added by inserting new rows immediately above row 32 (the row of hyphens).

READ output from the model at cells B21 - D21 (components of net force delivered by bearings), B22 - D23 (components of angular momentum), and B24 - D24 (components of torque delivered by bearings). You an also use cells I30 - L44 to build an on-screen graphic of the geometric relationships of axle, axle bearings, and masses.

NOTICE that column L has no entries in the main table but several entries below it. The purpose of this peculiar arrangement is to allow you to easily make plots showing the geometry of the system you are investigating. *However, you should build your model initially without this gap.* By so doing, you can use COPY and MOVE commands to build the model efficiently and quickly. Finally, after all of your formulas are in place and working, create the gap by MOVING column L in the three rows of the main table (including the row of hyphens!) and all to its right over one column. It is not really necessary to have the gap, but it it tends to be easier to have all plot data close together.

FORMULAS

	A	B	C	D	E	F	G	H	I	J	K	L	M	N
1	Two masses attached to an axle and forced to rotate with angular velocity omega.													
2														
3	ROTATION													
4	ang.speed =	3600	rpm	w =	=2*PI()*B4/60									
5				Δt =	=PI()/60/E4									
6														
7	ORIGIN & AXLE CHARACTERISTICS													
8	z0 =	0	m											
9	offset =	0	m											
10	alpha =	60	deg											
11														
12	BEARING LOCATIONS						BEARING FORCES							
13	d1 =	0.5	m	z1 =	=B8+B13		F1 =	=(C24-E14*B21)/E15	=(-B24-E14*C21)/E15	N				
14	d2 =	-0.5	m	z2 =	=B8+B14		F2 =	=(-C24+E13*B21)/E15	=(B24+E13*C21)/E15	N				
15				z1-z2 =	=E13-E14									
16														
17	SYSTEM DYNAMICS						CENTER OF MASS							
18	Axes ->	x	y	z										
19	P(-Δt) =	Formula Set I			kg m/s		Xcm =	=SUM(AB30:AB32)/SUM(A30:A32)						
20	P(Δt) =	Formula Set II			kg m/s									
21	F(0) =	Formula Set III			N									
22	L(-Δt) =	Formula Set IV			kg m^2/s									
23	L(Δt) =	Formula Set V			kg m^2/s									
24	GAM(0) =	Formula Set VI			N m									
25														
26	t->					-Δt			zero				Δt	
27	t->					=I27-E5			0				=I27+E5	
28	mass	length	theta	x0	z0	x	y	z	x	y	z		x	y
29	(kg)					(m)	(m)	(m)	(m)	(m)	(m)		(m)	(m)
30	1	0.2		F.S. VII	F.S. VIII	F.S. IX	F.S. X	F.S. XI	F.S. XII	F.S. XIII	F.S. XIV		F.S. XV	F.S. XVI
31	1	0.2	180											
32	------	------	------	------	------	------	------	------	------	------	------	------	------	------

FORMULAS (Contd.)

	O	P	Q	R	S	T	U	V	W	X	Y	Z	AA	AB
26		-Δt/2			Δt/2			-Δt/2			Δt/2			
27														
28	z	px	py	pz	px	py	pz	Lx	Ly	Lz	Lx	Ly	Lz	mx
29	(m)													
30	F.S. XVII	F.S. XVIII	F.S. XIX	F.S. XX	F.S. XXI	F.S. XXII	F.S. XXIII	F.S. XXIV	F.S. XXV	F.S. XXVI	F.S. XXVII	F.S. XXVIII	F.S. XXIX	F.S. XXX
31														
32	------	------	------	------	------	------	------	------	------	------	------	------	------	------

FORMULAS (Contd.)

	I	J	K	L
33				
34	=1.2*MAX(MAX(F30:F32,H30:H32,E13:E14),-MIN(F30:F32,H30:H32,E13:E14))			0
35	=-I34			0
36				
37	0			=I34
38	0			=-L37
39				
40	=-I34/10			=E13
41	=-I40			=L40
42				
43	=I40			=E14
44	=I41			=L43

Model 5.2 (continued)

VALUES

	A	B	C	D	E	F	G	H	I	J	K	L	M	N
1	Two masses attached to an axle and forced to rotate with angular velocity omega.													
2														
3	ROTATION													
4	ang.speed =	3600	rpm	w =	376.99	rad/sec								
5				Δt =	0.0001	sec								
6														
7	ORIGIN & AXLE CHARACTERISTICS													
8	z0 =	0	m											
9	offset =	0	m											
10	alpha =	60	deg											
11														
12	BEARING LOCATIONS						BEARING FORCES							
13	d1 =	0.5	m	z1 =	0.5	m	F1 =	-4922	0	N				
14	d2 =	-0.5	m	z2 =	-0.5	m	F2 =	4922.1	0	N				
15				z1-z2 =	1	m								
16														
17	SYSTEM DYNAMICS						CENTER OF MASS							
18	Axes ->	x	y	z										
19	P(-Δt) =	0	3E-14	0	kg m/s		Xcm =	6E-17						
20	P(Δt) =	0	3E-14	0	kg m/s									
21	F(0) =	0	0	0	N									
22	L(-Δt) =	-13.05	0.3418	22.609	kg m^2/s									
23	L(Δt) =	-13.05	-0.342	22.609	kg m^2/s									
24	GAM(0) =	0	-4922	0	N m									
25														
26	t->					-Δt			zero				Δt	
27	t->					-1E-04			0				0.0001	
28	mass	length	theta	x0	z0	x	y	z	x	y	z		x	y
29	(kg)					(m)	(m)	(m)	(m)	(m)	(m)		(m)	(m)
30	1	0.2	0	0.1732	0.1	0.173	-0.009	0.1	0.1732	0	0.1		0.173	0.0091
31	1	0.2	180	-0.173	-0.1	-0.173	0.0091	-0.1	-0.173	0	-0.1		-0.173	-0.009
32														

VALUES (Contd.)

	O	P	Q	R	S	T	U	V	W	X	Y	Z	AA	AB
26		-Δt/2			Δt/2			-Δt/2			Δt/2			
27														
28	z	px	py	pz	px	py	pz	Lx	Ly	Lz	Lx	Ly	Lz	mx
29	(m)													
30	0.1	1.7091	65.267	0	-1.709	65.267	0	-6.527	0.1709	11.305	-6.527	-0.171	11.305	0.1732
31	-0.1	-1.709	-65.27	0	1.7091	-65.27	0	-6.527	0.1709	11.305	-6.527	-0.171	11.305	-0.173
32														

VALUES (Contd.)

	I	J	K	L
33				
34	0.6			0
35	-0.6			0
36				
37	0			0.6
38	0			-0.6
39				
40	-0.06			0.5
41	0.06			0.5
42				
43	-0.06			-0.5
44	0.06			-0.5

Model 5.2 (continued)

Formula Set	Prototype Cell	Formula
I	B19	=SUM(P30:P32)
II	B20	=SUM(S30:S32)
III	B21	=(B20-B19)/ **E5**
IV	B22	=SUM(V30:V32)
V	B23	=SUM(Y30:Y32)
VI	B24	=(B23-B22)/ **E5**
VII	D30	=B30*SIN((C30+ **B10**)*PI()/180)-**B9**
VIII	E30	=B30*COS((C30+ **B10**)*PI()/180)
IX	F30	=D30*COS(**F27** *E4)
X	G30	=D30*SIN(**F27** *E4)
XI	H30	=E30+ **B8**
XII	I30	=D30*COS(**I27** *E4)
XIII	J30	=D30*SIN(**I27** *E4)
XIV	K30	=E30+ **B8**
XV	M30	=D30*COS(**M27** *E4)
XVI	N30	=D30*SIN(**M27** *E4)
XVII	O30	=E30+ **B8**
XVIII	P30	=A30*(I30-F30)/ **E5**
XIX	Q30	=A30*(J30-G30)/ **E5**
XX	R30	=A30*(K30-H30)/ **E5**
XXI	S30	=A30*(M30-I30)/ **E5**
XXII	T30	=A30*(N30-J30)/ **E5**
XXIII	U30	=A30*(O30-K30)/ **E5**
XXIV	V30	=G30*R30-H30*Q30
XXV	W30	=-R30*F30+P30*H30
XXVI	X30	=F30*Q30-G30*P30
XXVII	Y30	=J30*U30-K30*T30
XXVIII	Z30	=-U30*I30+S30*K30
XXIX	AA30	=I30*T30-J30*S30
XXX	AB30	=A30*I30

EXERCISE **5.2**

THE CENTER OF MASS AND THE INERTIA TENSOR

The general motion of a rigid body can be determined by first resolving the motion into two components:

- The translational motion of the center of mass
- The rotational motion about the center of mass

The translational motion of the center of mass of a solid body can be analyzed using Newton's three laws of motion and the techniques developed in earlier chapters. The rotational motion about the center of mass of a solid body can be analyzed using the rotational analog to Newton's second law of motion (Equation 5.40):

$$\boldsymbol{\Gamma} = \frac{d\boldsymbol{L}}{dt} \qquad (5.40)$$

where $\boldsymbol{\Gamma}$ is torque and $\boldsymbol{L}$ is angular momentum.

In this exercise, you learn how to calculate a body's center of mass and how to express its angular momentum about its center of mass in terms of its angular velocity.

In linear dynamics the momentum vector $\boldsymbol{p}$ is proportional to the velocity vector $\boldsymbol{v}$, but in rotational dynamics the angular momentum

TO GET THE MOST OUT OF THIS EXERCISE, YOU SHOULD ALREADY KNOW

The definitions of the terms:
center of mass
moment of inertia

vector $\boldsymbol{L}$ is, in general, not proportional to the angular velocity vector $\boldsymbol{\omega}$. Indeed, you discovered in Exercise 5.1 that while the *magnitudes* of $\boldsymbol{L}$ and $\boldsymbol{\omega}$ are proportional their *directions* are usually different.[1] The relationships between the components of $\boldsymbol{L}$ and the components of $\boldsymbol{\omega}$ are

$$L_x = I_{xx}\omega_x + I_{xy}\omega_y + I_{xz}\omega_z \quad (5.41a)$$

$$L_y = I_{yx}\omega_x + I_{yy}\omega_y + I_{yz}\omega_z \quad (5.41b)$$

$$L_z = I_{zx}\omega_x + I_{zy}\omega_y + I_{zz}\omega_z \quad (5.41c)$$

These are the same equations you saw as Equations 5.15a - 5.15c in this chapter's introduction. The nine numbers I_{ij} are components of the *inertia tensor* (or *inertia dyad*).

Just as the notation $\boldsymbol{L}$ is a convenient shorthand for the three components of a vector, so also people use a concise, suggestive notation for the three Equations 5.41a - 5.41c (Equation 5.42):

$$\boldsymbol{L} = \mathbf{I}\cdot\omega \quad (5.42)$$

where $\mathbf{I}$ is the inertia tensor. In the elegant mathematical notation of matrices, you can write the nine components of the $\mathbf{I}$ as

$$\mathbf{I} = \begin{pmatrix} I_{xx} & I_{xy} & I_{xz} \\ I_{yx} & I_{yy} & I_{yz} \\ I_{zx} & I_{zy} & I_{zz} \end{pmatrix} \quad (5.43)$$

You learn how to calculate $\mathbf{I}$ for a rigid body in this exercise and acquire a sense of some of its properties in Exercise 5.3. A few points about $\mathbf{I}$, however, should be made now. In the last exercise, where the angular velocity ω had just one component along the z-axis and was constant with time, you saw that in general the three components of angular momentum $\boldsymbol{L}$ varied with time. A brief examination of Equation 5.42 should indicate to you that, under these conditions, the inertia tensor $\mathbf{I}$ must vary with time. Furthermore, just as the components of a vector depend on the coordinate system that you choose to use, so also the components of $\mathbf{I}$ depend on the coordinate system.

The pathway to the solution of a rotational motion about the center of mass involves the following steps:

- Find the coordinates of the center of mass and the components of the inertia tensor at a fixed moment of time in a convenient coordinate system. You will study this process in this exercise.

  It is useful to think of this coordinate system as being rigidly fixed to the body so that in this *noninertial* coordinate system the components of both the center of mass and the inertia tensor are independent of time. Such noninertial reference frames are called *body coordinates*. It is important to realize that there is no unique set of body coordinates and people typically use several different sets in the course of a calculation.

- Find the principal axes through the center of mass in body coordinates. You will study this process in Exercise 5.3.

- Solve Equation 5.40 for the angular velocity ω and integrate the angular velocity to find the actual orientation of the body as a function of time.[2] The actual orientation of the body must be expressed in terms of an inertial frame of reference.[3]

  This last step is analogous to integrating the velocity of an object to find its position in the dynamics of point masses. You will explore this process using numeric methods in Exercises 5.5 and 5.6 for various kinds of rotational motion.

Let's begin then with the methods required to calculate the center of mass and the inertia tensor in one particular set of body coordinates. We will confine ourselves to masses represented as discrete chunks of matter at specific locations. The extension of this concept to a continuous mass is, of course, made by reducing the size of the chunks to the limit in which sums become integrals.[4] You should consult your primary physics textbook on this matter.

In this exercise, you consider only bodies that can be approximated as thin plates. These are called two-dimensional masses. Let such a mass be referenced to an appropriate rectangular coordinate system fixed in the body (body coordinates) with the z-axis perpendicular to the plate and the x-, y-plane centered in it. Partition the plate into equal area elements (m_{ij}) that can be made as small as desired (Figure 5.11).[5] The first of the two indices of m_{ij} refers to the x-direction and the second to the y-direction. Once this is done, the coordinates of the center of mass (x_{cm}, y_{cm}) are

Figure 5.11

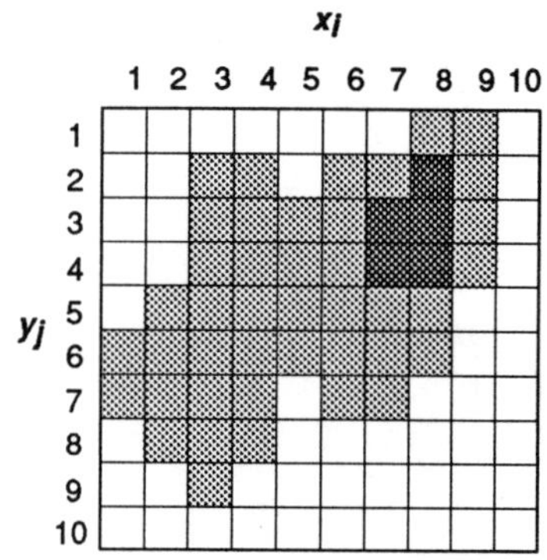

The coordinates (x_{cm}, y_{cm}) for the center of mass of a two-dimensional object can be found by dividing the object into equal-size area elements in the framework of a Cartesian coordinate system. The dark region of the mass reflects a region of greater mass density.

To find x_{cm}, the products of each mass element and its x-coordinate are added together and the total is divided by the total mass. The y-coordinate of the center of mass is found in a similar fashion (Equation 5.44).

$$x_{cm} = \frac{\sum_i \left(\sum_j m_{ij}\right) x_i}{M}, \qquad y_{cm} = \frac{\sum_j \left(\sum_i m_{ij}\right) y_j}{M} \tag{5.44}$$

where M is the total mass of the body and x_i and y_j are the coordinates of each area element. Figure 5.12 shows how this process might be represented in an electronic spreadsheet. How would you calculate the z-coordinate of the center of mass of a three-dimensional object?

The choice of reference frame in Equation 5.44 was in large part arbitrary. When dealing with rotational motion, it is often a very good idea (but not an absolutely necessary one) to choose a second set of body coordinates with its origin located at the center of mass of the object. The axes of this second frame remain parallel to those of the original reference frame (Figure 5.13). In this exercise the capital letters (X, Y) will identify locations in this coordinate system. Derived quantities such as velocity $\boldsymbol{V}$ and position vectors $\boldsymbol{R}$ will also be capitalized when referring to this frame.

The relationship between the coordinates of the masses in the two frames is, of course

$$X_i = x_i - x_{cm} \tag{5.45a}$$

$$Y_j = y_j - y_{cm} \tag{5.45b}$$

Inserting the important relation for rotational motion

$$\boldsymbol{V}_{ij} = \boldsymbol{\omega} \times \boldsymbol{R}_{ij} \tag{5.46}$$

into the definition of angular momentum about the center of mass

$$\boldsymbol{L} = \sum_i \sum_j (\boldsymbol{R}_{ij} \times m_{ij}\boldsymbol{V}_{ij}) \tag{5.47}$$

Figure 5.12

The mass distribution of a two-dimensional object may be conveniently represented in the two-dimensional array of an electronic spreadsheet. The x- and y-coordinates of the center of mass are calculated in cells B1 and B2. $\Sigma m_{ij}x_i$ is the sum of the contents of cells E1 - N1. How is $\Sigma m_{ij}y_j$ calculated?

Take a moment to explore this figure and try to build the model shown. Model 5.3 contains the answer.

	A	B	C	D	E	F	G	H	I	J	K	L	M	N
1	x(cm) =	5.563		m(xi)x(i) =	4	16	48	56	40	72	147	168	72	0
2	y(cm) =	4.509		m(xi) =	4	8	16	14	8	12	21	21	8	0
3					x values									
4	m(yi)y(i)	m(yi)		Index (i)	1	2	3	4	5	6	7	8	9	10
5	4	4	y	1								2	2	
6	30	15		2			2	2		2	2	5	2	
7	60	20	v	3			2	2	2	2	5	5	2	
8	80	20	a	4			2	2	2	2	5	5	2	
9	85	17	l	5		2	2	2	2	2	5	2		
10	96	16	u	6	2	2	2	2	2	2	2	2		
11	84	12	e	7	2	2	2	2		2	2			
12	48	6	s	8		2	2	2						
13	18	2		9			2							
14	0	0		10										

it can be shown that the components of the inertia tensor referenced to the center of mass of a thin plate are

$$I_{xx} = \sum_j \left(\sum_i m_{ij}\right) Y_j^2 \tag{5.48a}$$

$$I_{yy} = \sum_i \left(\sum_j m_{ij}\right) X_i^2 \tag{5.48b}$$

$$I_{zz} = \sum_i \sum_j m_{ij} \left(X_i^2 + Y_j^2\right) \tag{5.48c}$$

$$I_{xy} = -\sum_i \sum_j m_{ij} X_i Y_j \tag{5.48d}$$

$$I_{yx} = I_{xy} \tag{5.48e}$$

$$I_{xz} = I_{zx} = 0 \tag{5.48f}$$

$$I_{yz} = I_{zy} = 0 \tag{5.48g}$$

Notice that the z-axis must be taken into account even for a two-dimensional object.[6] The reason is simply to accommodate the possibility that the angular velocity of a flat object can have a component perpendicular to its plane (or even have only this component, as is the usual case with wheels on axles). Finally, notice that the inertia tensor has units of kg m².

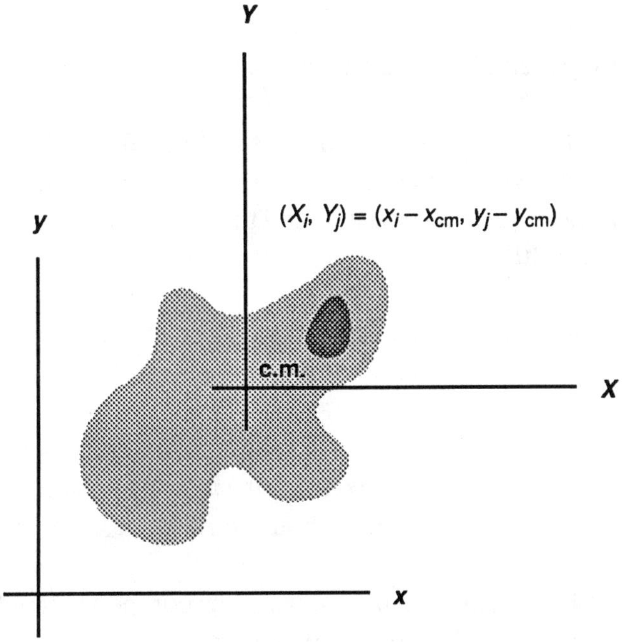

Figure 5.13

Rotational motions are often best represented in a reference frame with an origin at the center of mass. The inertia tensor **I**, for example, is usually represented in such a reference frame.

In this figure (x, y) are coordinates in the original reference frame and (X, Y) are coordinates in the "center of mass" (c.m.) frame.

LESSONS

1. Build and verify Model 5.3. The model calculates the coordinates of the center of mass and the components of the inertia tensor for a two-dimensional solid object. Explore the model carefully until you understand its structure and its relationship to the equations above. (Notice that the spreadsheet labels Mx^2 and My^2 refer to the terms of Equations 5.48a - 5.48g).

   You can, at any time, extend the model to a larger object (or an object defined with greater resolution) by making the grid larger.[7]

2. The object in Model 5.3 is meant to represent a table-tennis paddle (and so the units of the model might be inches and grams).

   a. What happens to the center of mass of the paddle if the handle is given more weight by inserting a metal rod into its center (e.g., change the values in cells M16 - P16 from 8 to 20)?

   b. Remove the metal rod from the handle (thus resetting your model to its original condition) and then add enough mass at the very end of the handle ($x = 10$, $y = 8$) so that the center of mass is moved to precisely ($x = 5$, $y = 8$). How much mass did you add?

   c. After you have a paddle with a center of mass at (5,8), add 200 g to the area element at (5,8). What happens to the center of mass? Why?

      What happens if the 200 units of mass are added to area elements progressively farther away from the center of mass?

3. Change the table-tennis paddle to a racketball paddle by eliminating mass everywhere except in the handle and outer rim. (You should also remove the extra mass added to the end of the handle in Lesson 2b, if it's still there.)

   Gradually reduce the mass in the handle until the center of mass is located at a place where no mass is present. Explain.

4. Return your model to its initial settings but add 1.4 g to the end of the handle (cell P16) and examine the inertia tensor for a table-tennis paddle. The center of mass should now be at $x = 5$, $y = 8$.

   Cells B4 - D6 represent the values of the components of **I**. Notice that only the diagonal elements of the matrix (I_{xx}, I_{yy}, I_{zz}) have values appreciably different from zero. Matrices with zero off-diagonal elements are said to be *diagonalized.* For diagonalized matrices, the three orthogonal coordinate axes used to compute the components of the inertia tensor are also the principal axes of the body. Clearly, a diagonalized inertia tensor is very important! You will study these matters in greater depth in Exercise 5.3.

a. Add 100 g to cell L13 ($x = 6, y = 5$). What are the coordinates of this point in the X- , Y-coordinate system (i.e., the coordinate system whose origin is located at the center of mass prior to adding the mass)?

   Are the diagonal elements of **I** affected by this addition? (Notice that the center of mass has moved with the addition of the 100 g and the inertia tensor is being computed about a new place in the paddle.)

   If the paddle were spun about a fixed axis (say, through the length of the handle of the paddle) with a constant angular velocity, do you suppose this mass addition would increase, decrease, or leave unchanged the magnitude of the angular momentum? Of the kinetic energy? Why?

b. Add another 100 g of mass to a suitable cell so that the center of mass is returned to ($x = 5, y = 8$). What is the cell and what are its (x, y) and (X, Y) coordinates? What happens to the off-diagonal components of the inertia tensor? The diagonal components?

c. Add 100 g of mass at two other locations so that a diagonal inertia tensor is generated and the center of mass remains at ($x = 5, y = 8$). Comment on the symmetry of the resulting geometric distribution of the added mass.

d. Remove the two masses that you added in Lesson 4c and instead put equal added masses at ($X = 1, Y = 4$) and ($X = -1, X = -4$). (Notice that you can use a simple formula in one of these cells so that it tracks the value of the other.)

   Do these additional masses change the location of the center of mass?

   Adjust the mass values so that the inertia tensor is again diagonal. What masses are required? Comment on the symmetry of the added masses. Comment on the changes to the *diagonal* components of **I** and explain their behavior.

e. Examine the values of

$$\sum_j m_{ij}(y_j - y_{cm})^2 \quad \text{and} \quad \sum_i m_{ij}(x_i - x_{cm})^2$$

in your model (labeled My^2 and Mx^2 in your model) and notice the large contributions from masses well off the axes.

Because **I** depends on the *square* of the distance from the origin to the mass location, the more distant masses have a larger contribution.

PROBLEMS

1. Using Model 5.3, build a 7 x 7 square planar object of uniform mass density of 1 g per cell and let the distance units be centimeters (Figure 5.14). You can get a sense of the meaning of the terms of the inertia tensor by changing the mass distribution of an object and seeing what happens to the tensor.

   a. Where is the center of mass? What are the initial values in the inertia tensor **I**? Why do I_{xx} and I_{yy} have the same value? Comment on the fact that **I** is diagonal.

   b. Erase the line of values along the line $x = 4$. What happens to **I** and the center of mass?

   c. Reset the $x = 4$ line and erase the line of values along $y = 4$. What happens to **I** and the center of mass?

   d. Reset the $y = 4$ line and erase the mass at $x = y = 4$. What happens to **I** and the center of mass?

   e. Add an enormous mass at the center of mass position. What happens to **I**?

   f. Erase all but the outer edge of values. What happens to **I** and the center of mass? Is **I** reduced in the same ratio as the mass itself?

   g. Increase the mass values on the outer edges of Problem 1f until **I** has the same values as your original uniform planar object. By what factor did mass per cell increase? Interpret this result.

Figure 5.14

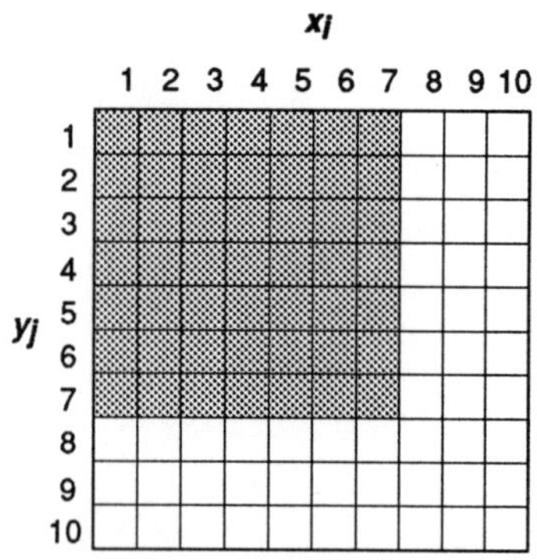

Problem 1 asks you to examine the center of mass and inertia tensor of a square, planar object of uniform mass distribution.

h. Erase the four corners and report the value of **I**. Compare with the results of Problem 1f when all the interior of the figure is removed.

i. Let each corner have a mass of 100 g. Report on **I**.

j. Put different masses in the four corners. Is the center of mass still at the center of the square? Is **I** still diagonal?

k. Place several different masses at other locations at random and carefully observe what happens to the center of mass and **I**. Can you arrange asymmetric mass distributions that result in a diagonal inertia tensor?

2. Use Model 5.3 to study a 7 ft x 11 ft rectangular steel plate (Figure 5.15). Assume that the plate is 1 in. thick and the steel has a weight density of 0.28 lbs/in.^{3}.

   a. Where is the center of mass, and what is **I**? Comment on the structure of **I**. Give **I** in units of lb ft^2/g where g is the acceleration of gravity.

   b. If a square 1 ft x 1 ft is cut off one corner, what other changes could you make to keep the center of mass at the geometric center of the plate and keep **I** diagonal? Suppose you didn't care if the center of mass were at the center of the plate or not. Then, give another solution.

   c. Starting with a full plate again, suppose a 3 ft x 3 ft hole is cut out of the very center of the plate. By what fraction is the mass of the plate reduced? By what fraction are the three diagonal components of **I** reduced? Comment on the results.

   d. Finally, it is desired to weld four lumps of steel, each as small as possible, to the plate so that its inertia tensor is the same as it was before the central hole was cut and the center of mass stays at the geometric center of the plate. What masses would you use and where would you put them?

3. Use Model 5.3 to build an L-shaped object of uniform unit mass density (Figure 5.16). Notice that the center of mass of the object is located outside the mass of the object.

Figure 5.15

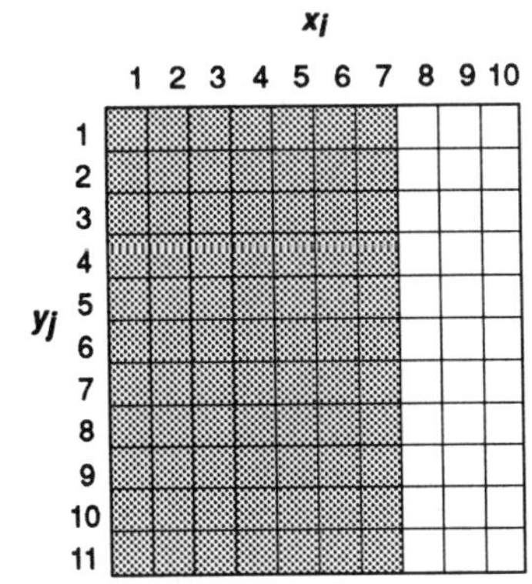

Problem 2 asks you to examine the center of mass and inertia tensor of a rectangular, planar object of uniform mass distribution.

Figure 5.16

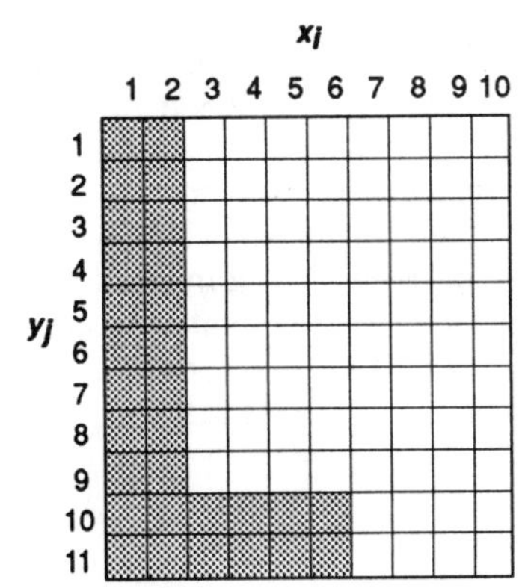

Problem 3 asks you to examine the center of mass and inertia tensor of a L-shaped planar object of uniform mass distribution.

a. Where on the body would you add a single lump of 9 units of mass to move the center of mass as far from the lower left corner of the L as possible?

b. Place a second 9 units of mass to move the center of mass as far away as possible. Where should it be put?

c. Add more masses to the L shape until $I_{xx} = I_{yy}$. Can this be accomplished without moving the center of mass from where it was in Problem 3b?

4. Build and verify a model that calculates the center of mass for a three-dimensional object.

MODELS

Model 5.3

BUILD and verify Model 5.3.

ENTER data using the rectangular grid of cells G9 - P22 (the mass of each area element).

READ output from the model at cells G2, (the mass of the object), J1 and J2 (the *x*- , *y*-coordinates of the center of mass) and cells B4 - D6 (the components of the inertia tensor).

NOTICE that the model assumes that your spreadsheet treats blank cells as though they contained a value of zero. If your spreadsheet treats blanks differently, you must insert zeros into those parts of the grid G9 - P22 that contain zero mass. Also, you can expand your model by inserting new rows between rows 22 and 23 (just above the row of hyphens) and new columns between columns P - Q and AD and AE (to the left of the column of vertical lines). The formulas in your spreadsheet will automatically adjust to the new size if you place your row and column insertions in these locations.

FORMULAS

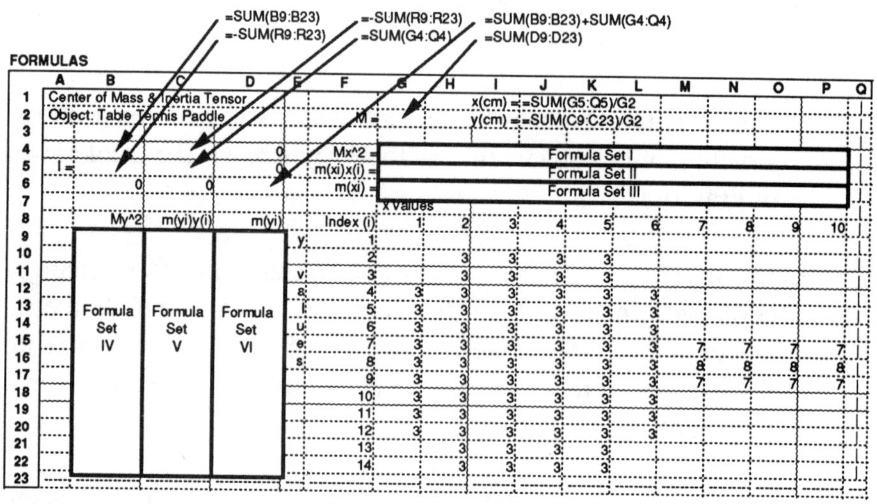

FORMULAS (Contd.)

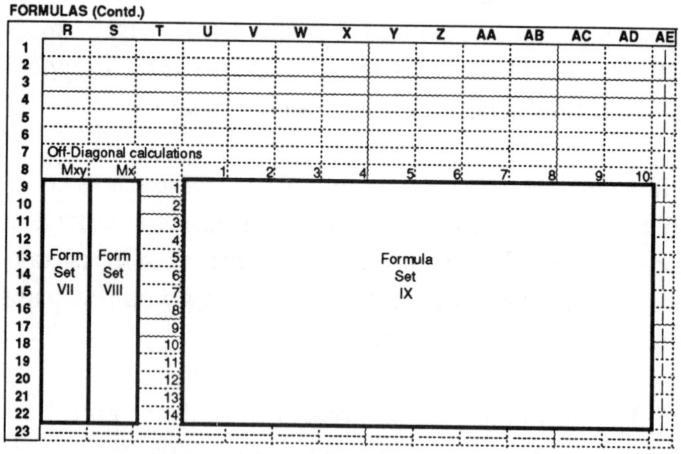

Model 5.3 (continued)

VALUES

	A	B	C	D	E	F	G	H	I	J	K	L	M	N	O	P
1		Center of Mass & Inertia Tensor							x(cm) =	4.977						
2		Object: Table Tennis Paddle				M =	298		y(cm) =	8						
3																
4		2600	2.84E-14	0		Mx^2 =	426.9	345.5	152.4	37.19	0.022	28.28	90.08	201.1	356.1	555.2
5	I =	2.84E-14	2192.836	0		m(xi)x(i) =	27	78	117	156	195	162	154	176	198	220
6		0	0	4792.836		m(xi) =	27	39	39	39	39	27	22	22	22	22
7							x values									
8		My^2	m(yi)y(i)	m(yi)		Index (i)	1	2	3	4	5	6	7	8	9	10
9		0	0	0	y	1										
10		432	24	12		2		3	3	3	3					
11		300	36	12	v	3		3	3	3	3					
12		288	72	18	a	4	3	3	3	3	3	3				
13		162	90	18	l	5	3	3	3	3	3	3				
14		72	108	18	u	6	3	3	3	3	3	3				
15		46	322	46	e	7	3	3	3	3	3	3	7	7	7	7
16		0	400	50	s	8	3	3	3	3	3	3	8	8	8	8
17		46	414	46		9	3	3	3	3	3	3	7	7	7	7
18		72	180	18		10	3	3	3	3	3	3				
19		162	198	18		11	3	3	3	3	3	3				
20		288	216	18		12	3	3	3	3	3	3				
21		300	156	12		13		3	3	3	3					
22		432	168	12		14		3	3	3	3					
23																

VALUES (Contd.)

	R	S	T	U	V	W	X	Y	Z	AA	AB	AC	AD
1													
2													
3													
4													
5													
6													
7	Off-Diagonal calculations												
8	Mxy	Mx		1	2	3	4	5	6	7	8	9	10
9	0	0	1	0	0	0	0	0	0	0	0	0	0
10	-106	-17.7	2	0	-8.93	-5.93	-2.93	0.07	0	0	0	0	0
11	-88.6	-17.7	3	0	-8.93	-5.93	-2.93	0.07	0	0	0	0	0
12	-106	-26.6	4	-11.9	-8.93	-5.93	-2.93	0.07	3.07	0	0	0	0
13	-79.7	-26.6	5	-11.9	-8.93	-5.93	-2.93	0.07	3.07	0	0	0	0
14	-53.2	-26.6	6	-11.9	-8.93	-5.93	-2.93	0.07	3.07	0	0	0	0
15	72.08	72.08	7	-11.9	-8.93	-5.93	-2.93	0.07	3.07	14.16	21.16	28.16	35.16
16	0	86.17	8	-11.9	-8.93	-5.93	-2.93	0.07	3.07	16.19	24.19	32.19	40.19
17	-72.1	72.08	9	-11.9	-8.93	-5.93	-2.93	0.07	3.07	14.16	21.16	28.16	35.16
18	53.15	-26.6	10	-11.9	-8.93	-5.93	-2.93	0.07	3.07	0	0	0	0
19	79.73	-26.6	11	-11.9	-8.93	-5.93	-2.93	0.07	3.07	0	0	0	0
20	106.3	-26.6	12	-11.9	-8.93	-5.93	-2.93	0.07	3.07	0	0	0	0
21	88.59	-17.7	13	0	-8.93	-5.93	-2.93	0.07	0	0	0	0	0
22	106.3	-17.7	14	0	-8.93	-5.93	-2.93	0.07	0	0	0	0	0
23													

Formula Set I Prototype Cell is G4	Formula Set IV Prototype Cell is B9	Formula Set VII Prototype Cell is R9
=G6*(G8-**J1**)^2	=D9*(F9-**J2**)^2	=S9*(**J2** -T9)

Formula Set II Prototype Cell is G5	Formula Set V Prototype Cell is C9	Formula Set VIII Prototype Cell is S9
=G8*G6	=F9*D9	=SUM(U9:AE9)

Formula Set III Prototype Cell is G6	Formula Set VI Prototype Cell is D9	Formula Set IX Prototype Cell is U9
=SUM(G9:G23)	=SUM(G9:Q9)	=G9*(U8-**J1**)

EXERCISE **5.3**

THE INERTIA TENSOR OF A UNIFORMLY ROTATING BODY

TO GET THE MOST OUT OF THIS EXERCISE, YOU SHOULD ALREADY KNOW

The definitions of the terms:
force
linear momentum
angular position
angular velocity
angular momentum
torque

How to calculate the cross product of a vector

In Exercise 5.1, you built and explored models of simple rigid bodies exhibiting uniform rotary motion about a fixed axle (Figure 5.17) and discovered that

- If the axle of a rigid body passes through a special place in the body (called the center of mass), uniform rotary motion can be maintained with no net external force ($\boldsymbol{F} = 0$).[1]

- If the axle of a rigid body is aligned along one of at least two special directions in the plane of the body (called the principal axes), uniform rotary motion can be maintained with no net external torque ($\boldsymbol{\Gamma} = 0$).[2]

You made these discoveries by exploiting the fundamental relationships that govern translational (Equation 5.49) and rotational motion (Equation 5.50):

$$\boldsymbol{F} = \frac{d\boldsymbol{P}}{dt} \tag{5.49}$$

$$\boldsymbol{\Gamma} = \frac{d\boldsymbol{L}}{dt} \tag{5.50}$$

where F and Γ are the total force and total torque (respectively) applied to the body and P and L are the total linear and angular momentum (respectively) of the body.

Total torque Γ and total angular momentum L are the central concepts of rotational motion, and it is important that you understand them. Total torque is the time rate of change of total angular momentum (Equation 5.50); total angular momentum is the sum of the angular momenta of each point mass in the rigid body (Equation 5.51). You should review these matters in your primary physics textbook if necessary.

$$L = \sum_i (r_i \times m_i v_i) \qquad (5.51)$$

where r_i is the position vector of mass m_i and v_i is the velocity of m_i.

In Exercise 5.1, you used Equation 5.49 and 5.51 to explore the nature of the center of mass and principal axes of a rigid body. Recall, however, that you made your discoveries in much the same way that an artillery officer finds a target — by trial and error. You learned in Exercise 5.2 how to directly calculate the center of mass of a rigid body. Is there a direct method for finding the principal axes? Yes — that's part of what the inertia tensor **I** is all about!

Recall from this chapter's introduction that L can be expressed with remarkable conciseness in terms of angular velocity ω and the inertia tensor **I** (Equation 5.52).

$$L = \mathbf{I} \cdot \omega \qquad (5.52)$$

where the nine components of **I** can be calculated using Equations 5.48a - 5.48g in Exercise 5.2.

In this exercise you use **I** to calculate L and reproduce the results of Exercise 5.1. In the following exercise, you will learn how **I** can be used to calculate the principal axes and principal moments of inertia directly.

In this exercise (as in Exercise 5.1), consider only uniform rotation of a planar object about various axles in the body's plane. Each axle you consider passes through the body's center of mass and is fixed rigidly to the body. The angular velocity vector ω is constant and points along the axle. Furthermore, as you saw in Exercise 5.1, the two rotationally important dynamical vectors L and Γ behave just as though they were fixed in the body itself. In other words, they have constant magnitudes and rotate about the axle with the object.

Figure 5.17

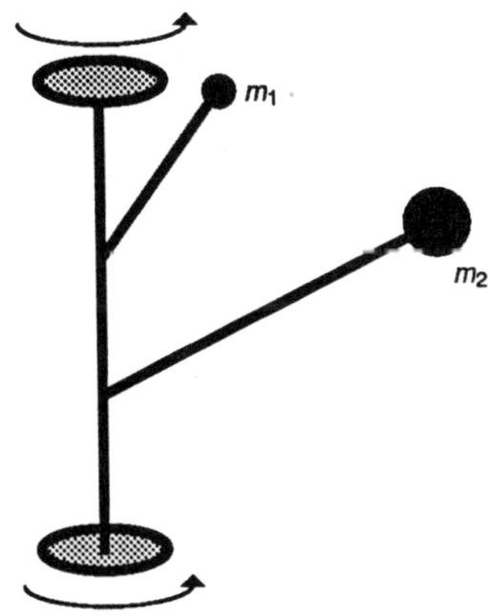

Masses m_1 and m_2 are attached to a rotating axle via rigid, massless rods such that the masses and the axle all lie in the same plane. The bearings of the axle deliver forces and torques so that the device always exhibits uniform rotary motion.

Consider the implications for the inertia tensor. As the body rotates, its parts are clearly at different places at different times. Therefore, all components of the inertia tensor will vary with time (except in cases of special, highly symmetric objects). For an object steadily rotating about a fixed axis, you know the position of each mass at every moment and so can calculate the inertia tensor at each moment by using the methods of Exercise 5.1. At this same moment then, you can calculate the angular momentum from the angular velocity using Equation 5.52. However, to get the torque at this moment from Equation 5.50, it would seem that you must actually work out all the components of **I** as functions of time and then differentiate each one of them (see Equation 5.19 at the beginning of this chapter). Although this task is not impossible, it is certainly tedious!

There is a wonderful shortcut when the angular velocity vector is constant. It can be shown, and you will verify this important relation in the lesson, that when $\boldsymbol{\omega}$ is constant

$$\boldsymbol{\Gamma} = \frac{d\boldsymbol{L}}{dt} = \boldsymbol{\omega} \times \boldsymbol{L} = \boldsymbol{\omega} \times (\mathbf{I} \cdot \boldsymbol{\omega}) \qquad (5.53)$$

[WARNING: Don't forget that $\boldsymbol{\omega}$ must be constant in Equation 5.53. The general equation of which this is a special example is discussed in Exercise 5.6.]

Knowing the angular momentum and torque at one moment when $\boldsymbol{\omega}$ is constant, you effectively know them always because in this special case these vectors just spin about the axle as the body does. So by using cross products *at a single moment of time that you can choose for convenience of calculation*, you can find out all important dynamical quantities about a uniformly rotating body without doing any time differentiations.

In these lessons, you calculate the time dependence of the inertia tensor in an inertial coordinate system. Note, however, that you really only need **I** at just one time to obtain all dynamical information about a uniformly spinning body. One purpose of this exercise is to emphasize how efficient calculation becomes using cross products and the inertia tensor at one moment of time. Much is going on behind the scenes nevertheless.

LESSONS

In these lessons you verify the very important relations

$$\boldsymbol{v} = \frac{d\boldsymbol{r}}{dt} = \boldsymbol{\omega} \times \boldsymbol{r} \qquad (5.54)$$

$$L = \mathbf{I} \cdot \boldsymbol{\omega} \tag{5.55}$$

$$\Gamma = \frac{dL}{dt} = \boldsymbol{\omega} \times L \tag{5.56}$$

when $\boldsymbol{\omega}$ is a constant and all masses in the body are in a plane. These relations underlie all mathematics of the inertia tensor!

Model 5.2 of Exercise 5.1 calculates such dynamical vectors as position, momentum, and angular momentum at a single moment in time using direct numeric differentiation. In Exercise 5.1, you calculated these vectors only at the time t = 0, but the model is constructed in such a way that any vector can be calculated at any time by simply changing the contents of cell I27. Model 5.4 in this exercise simply adds to Model 5.2 the ability to also calculate these vectors using $\boldsymbol{\omega}$ and **I**.

1. Build and verify Model 5.4 by adding new cells to Model 5.2 of Exercise 5.1.

   a. Cells AC28 - AD31 perform intermediate calculations for the *y*- and *z*-components of the center of mass of a simple rigid body of the same sort as that described in Figure 5.13 of Exercise 5.2.

   b. Cells H20 and H21 contain the *y*- and *z*-components of the center of mass. Labels identifying these cells are present in cells G20 and G21.

   c. Cells L16 - N16 compute the *x*-, *y*-, and *z*-components of the angular velocity vector.

   d. Cells AE28 - AG31 compute the *x*-, *y*-, and *z*-components of the velocity vector using the numeric differentiation formulas already present in the model.

      Cells AH28 - AJ31 compute these same velocity vector components by using Equation 5.54.

      Cells AK28 - AM31 show the difference between these two ways of calculating $\boldsymbol{v}$.

   e. The ingredients of the inertia tensor **I** are calculated in cells AN28 - AS31, and **I** is represented in matrix form in cells L11 - N13.

   f. The *x*-, *y*-, and *z*-components of the angular momentum (via the inertia tensor) are displayed in cells L18 - N18, while the same components computed via numeric differentiation are shown in cells L19 - N19.

The difference in the two calculations is in cells L20 - N20.

g. The x-, y-, and z-components of the torque (via the inertia tensor) are displayed in cells L22 - N22, while the same components computed via numeric differentiation are shown in cells L23 - N23.

The difference in the two calculations exists in cells L24 - N24.

2. Using the initial values in your model, notice that there are small differences between the results obtained by numeric differentiation and those obtained by cross products and the inertia tensor.

a. Verify that these differences are due to the inaccuracies of numeric differentiation by completing the following table (use a "what-if" or "data" table, if you wish).

Errors at Time $t = 0$ s

	Angular Momentum Differences			Torque Differences		
Δt	x-	y-	z-	x-	y-	z-
0.01	15.095500	0	-26.14610	0	3669.937000	0
0.001	0.307146	____	____	____	____	____
0.0001	0.003093	____	____	____	____	____
0.00001	0.000030	____	____	____	____	____
0.000001	0.000000	____	____	____	____	____

b. Although you have used Model 5.1 to calculate things only at $t = 0$, the model is designed in such a way that any time t can be explored by setting the new t value in cell I27. Verify that the efficient methods of calculating quantities with cross products and the inertia tensor give the same results as the direct differentiation methods by setting cell I27 to a time of 2ms (0.002 s), and then filling out the following table by changing Δt at cell E5:

Errors at Time $t = 0.002$ s

	Angular Momentum Differences			Torque Differences		
Δt	x-	y-	z-	x-	y-	z-
0.01	____	____	____	____	____	____
0.001	____	____	____	____	____	____
0.0001	____	____	____	____	____	____
0.00001	____	____	____	____	____	____
0.000001	____	____	____	____	____	____

3. It is pointed out in the introduction to this exercise that, when the inertia tensor is calculated in an inertial frame, it normally depends

on time. You can observe this important fact in this model by simply changing the contents of cell I27, which is the time at which the dynamical quantities are being calculated. To see the time dependence of the components of the inertia tensor, fill in the following table and confirm the plot of the components of the inertia tensor as functions of time (Figure 5.18). In the table, the time step corresponds to 1/20 of a revolution so the table covers one revolution of the body. (Your spreadsheet's data table commands can help, if you wish.)

t	I_{xx}	I_{xy}	I_{xz}	I_{yy}	I_{yz}	I_{zz}
0.000000	0.02	0	-0.03464	0.08	0	0.06
0.000833						
0.001666						
0.002500						
0.003333						
0.004166						
0.005000						
0.005833						
0.006666						
0.007500						
0.008333						
0.009166						
0.010000						
0.010833						
0.011666						
0.012500						
0.013333						
0.014166						
0.015000						
0.015833						
0.016666						

What frequencies are represented in these time dependences? In particular, what are the frequencies of oscillation of each of the six components of inertia tensor? By thinking about the definition of these tensor components, convince yourself that these results are reasonable. Remember that the two masses are in a straight line, i.e., $\theta = 180°$.

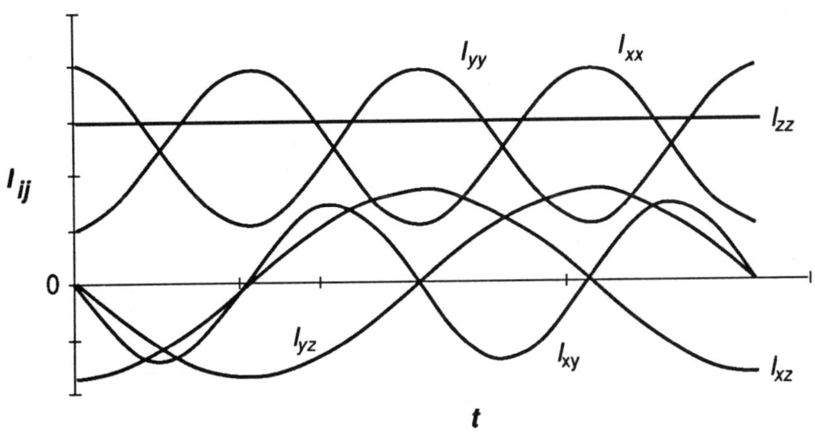

Figure 5.18

As many rigid bodies rotate, the components of the inertia tensor change. The plots shown belong to the table associated with Lesson 3.

Can you imagine a body in which the components of the inertia tensor do not change as the body rotates?

Exercise 5.1 identified several dynamical vectors including angular momentum L that (under the conditions of uniform rotational motion) behave as if they were geometric objects fixed to the body and rotating with it. Because L is obtained by operating on the angular velocity with the inertia tensor ($L = \mathbf{I} \cdot \omega$), you might expect the inertia tensor to also act this way. Does it? Discuss.

4. Change the angle α to 90° and then complete and plot the following table:

t	I_{xx}	I_{xy}	I_{xz}	I_{yy}	I_{yz}	I_{zz}
0.000000	___	___	___	___	___	___
0.000833	___	___	___	___	___	___
0.001666	___	___	___	___	___	___
0.002500	___	___	___	___	___	___
0.003333	___	___	___	___	___	___
0.004166	___	___	___	___	___	___
0.005000	___	___	___	___	___	___
0.005833	___	___	___	___	___	___
0.006666	___	___	___	___	___	___
0.007500	___	___	___	___	___	___
0.008333	___	___	___	___	___	___
0.009166	___	___	___	___	___	___
0.010000	___	___	___	___	___	___
0.010833	___	___	___	___	___	___
0.011666	___	___	___	___	___	___
0.012500	___	___	___	___	___	___
0.013333	___	___	___	___	___	___
0.014166	___	___	___	___	___	___
0.015000	___	___	___	___	___	___
0.015833	___	___	___	___	___	___
0.016666	___	___	___	___	___	___

Describe in words the behavior of the inertia tensor in this case. On the basis of this inertia tensor, what can you say about the angular momentum? The torque?

5. Finally, set angle α to 0° and study the inertia tensor by completing and plotting the table:

t	I_{xx}	I_{xy}	I_{xz}	I_{yy}	I_{yz}	I_{zz}
0.000000	___	___	___	___	___	___
0.000833	___	___	___	___	___	___
0.001666	___	___	___	___	___	___
0.002500	___	___	___	___	___	___
0.003333	___	___	___	___	___	___
0.004166	___	___	___	___	___	___
0.005000	___	___	___	___	___	___
0.005833	___	___	___	___	___	___
0.006666	___	___	___	___	___	___
0.007500	___	___	___	___	___	___
0.008333	___	___	___	___	___	___
0.009166	___	___	___	___	___	___
0.010000	___	___	___	___	___	___
0.010833	___	___	___	___	___	___
0.011666	___	___	___	___	___	___
0.012500	___	___	___	___	___	___
0.013333	___	___	___	___	___	___
0.014166	___	___	___	___	___	___
0.015000	___	___	___	___	___	___
0.015833	___	___	___	___	___	___
0.016666	___	___	___	___	___	___

Explain the result (which may have surprised you at first). Would any real dumbbell have an inertia tensor like this?

PROBLEMS

1. You know that the three diagonal components of the inertia tensor are given by the equations

$$I_{xx} = \sum_i m_i(Y_i^2 + Z_i^2) \tag{5.57a}$$

$$I_{yy} = \sum_i m_i(Z_i^2 + X_i^2) \tag{5.57b}$$

$$I_{zz} = \sum_i m_i(X_i^2 + Y_i^2) \tag{5.57c}$$

where X_i, Y_i, Z_i are the coordinates of mass m_i with respect to the center of mass of the body. The sum of these three components of the inertia tensor is called its *trace*. By remembering that the quantity $R_i^2 = X_i^2 + Y_i^2 + Z_i^2$ has the same value no matter what three orthogonal directions are used to set up a coordinate system in a body, show that the trace of the inertia tensor also has this property. Verify this property of the inertia tensor using the tables you generated in the lesson. The trace of the inertia tensor is called an *invariant* of the tensor because its value does not depend on the coordinate system used to calculate the tensor's components. You will see later that there are two other invariants.

2. Add further masses to your rigid body and verify that the method of calculating time derivatives by doing cross products did not depend on the fact that only two masses were used. Notice that at $t = 0$ the inertia tensor always has a simple form.

3. What kinds of rigid bodies have a diagonal inertia tensor at all times? Try to find an arrangement of masses so that the inertia tensor is diagonal at all times. Why is this not possible in this model?

4. Build a model which at $t = 0$ can have its masses not only in the x-, z-plane (as is the case in Models 5.1 and 5.2) but at any point in space. Let the axle of rotation always be along the z-axis. Calculate all time derivatives using the cross product method. In this more general model, can you arrange things so that the inertia tensor is always diagonal?

MODELS

Model 5.4

MODIFY Model 5.2 to create Model 5.4 and verify it. Only the changes are displayed in the figure.

ENTER data using cell B4 (angular speed), cells B8 - B10 z-location of the point where the two support rods are joined, axle offset from this same point, angle between first rod and axle), B13 and B14 (bearing locations), and columns A - C starting at row 30 (mass, radius, and θ of attached masses). Additional masses can be added by inserting new rows immediately above row 32 (the row of hyphens).

READ output from the model at cells B21 - D21 (components of net force delivered by bearings), B22 - D23 (components of angular momentum), and B24 - D24 (components of torque delivered by bearings). You can also use cells I30 - L44 to build an on-screen graphic of the geometric relationships of axle, axle bearings, and masses. Cells L11 - N13 show the inertia tensor of the masses entered in the table above row 32, while cells M18 - N24 compare and contrast the dynamic vectors calculated by differentiation and by vector cross product calculations.

NOTICE that column L has no entries in the main table but several entries below it. The purpose of this peculiar arrangement is to allow you to easily make plots showing the geometry of the system you are investigating. *However, you should build your model initially without this gap.* By so doing, you can use COPY and MOVE commands to build the model efficiently and quickly. Finally, after all formulas are in place and working, create the gap by MOVING column L in the five rows of the main table (including the titles and row of hyphens!) and all to its right over one column. It is not really necessary to have the gap, but it it tends to be easier to have all plot data close together.

FORMULAS (to be added to Model 5.2)

	A
17	SYSTEM DYNAMICS BY DIFFERENTIATION

	G	H
20	ycm	=SUM(AC30:AC32)/SUM(A30:A32)
21	zcm	=SUM(AD30:AD32)/SUM(A30:A32)

	J	K	L	M	N	O
9		COMPARATIVE SYSTEM DYNAMICS				
10						
11			=SUM(AN30:AN32)	=SUM(AO30:AO32)	=SUM(AP30:AP32)	
12		I =	=M11	=SUM(AQ30:AQ32)	=SUM(AR30:AR32)	
13			=N11	=N12	=SUM(AS30:AS32)	
14						
15			x	y	z	
16		w (vector) =	0	0	=E4	
17						
18		L (vector) =	=N16*N11	=N16*N12	=N16*N13	(by inertia tensor)
19		L (vector) =	Formula Set I			(by differentiation)
20		Difference =	Formula Set II			
21						
22		GAM (vector) =	=M16*N18-N16*M18	=N16*L18-L16*N18	=L16*M18-M16*L18	(by cross product)
23		GAM (vector) =	Formula Set III			(by differentiation)
24		Difference =	Formula Set IV			

	AC	AD	AE	AF	AG	AH	AI	AJ
24			Velocity by			Velocity by		
25			Differentiation			Cross Product		
26			0					
27								
28	my	mz	vx	vy	vz	vx	vy	vz
29			(m/s)					
30-31	F.S. V	F.S. VI	F.S. VII	F.S. VIII	F.S. IX	F.S. X	F.S. XI	F.S. XII
32	---------	---------	---------	---------	---------	---------	---------	---------

	AK	AL	AM	AN	AO	AP	AQ	AR	AS
24				Calculation of I					
25	Velocity Difference			(at t = 0)					
26									
27									
28	Δvx	Δvy	Δvz	I.XX	I.XY	I.XZ	I.YY	I.YZ	I.ZZ
29				(kg m^2)					
30-31	F.S. XIII	F.S. XIV	F.S. XV	F.S. XVI	F.S. XVII	F.S. XVIII	F.S. XIX	F.S. XX	F.S. XXI
32	---------	---------	---------	---------	---------	---------	---------	---------	---------

Model 5.4 (continued)

VALUES (to be added to Model 5.2)

	A
17	SYSTEM DYNAMICS BY DIFFERENTIATION

	G	H
20	ycm	0
21	zcm	-3E-17

	J	K	L	M	N	O
9		COMPARATIVE SYSTEM DYNAMICS				
10						
11			0.02	0	-0.035	
12		I =	0	0.08	0	
13	N		-0.035	0	0.06	
14	N					
15			x	y	z	
16		w (vector) =	0	0	376.99	
17						
18		L (vector) =	-13.06	0	22.619	(by inertia tensor)
19		L (vector) =	-13.05	0	22.609	(by differentiation)
20		Difference =	0.006	0	-0.01	
21						
22		GAM (vector) =	0	-4923	0	(by cross product)
23		GAM (vector) =	0	-4922	0	(by differentiation)
24		Difference =	0	1.1247	0	

	AC	AD	AE	AF	AG	AH	AI	AJ
24			Velocity by			Velocity by		
25			Differentiation			Cross Product		
26			0					
27								
28	my	mz	vx	vy	vz	vx	vy	vz
29			(m/s)					
30	0	0.1	0	65.267	0	0	65.297	0
31	0	-0.1	0	-65.27	0	0	-65.3	0
32								

	AK	AL	AM	AN	AO	AP	AQ	AR	AS
24				Calculation of I					
25	Velocity Difference			(at t = 0)					
26									
27									
28	Δvx	Δvy	Δvz	I.XX	I.XY	I.XZ	I.YY	I.YZ	I.ZZ
29				(kg m^2)					
30	0	-0.03	0	0.01	0	-0.017	0.04	0	0.03
31	0	0.0298	0	0.01	0	-0.017	0.04	0	0.03
32									

Model 5.4 (continued)

Formula Set I	Prototype Cell is L19	=(B22+B23)/2
Formula Set II	Prototype Cell is L20	=L19-L18
Formula Set III	Prototype Cell is L23	=B24
Formula Set IV	Prototype Cell is L24	=L23-L22
Formula Set V	Prototype Cell is AC30	=A30*J30
Formula Set VI	Prototype Cell is AD30	=A30*K30
Formula Set VII	Prototype Cell is AE30	=(P30+S30)/2/A30
Formula Set VIII	Prototype Cell is AF30	=(Q30+T30)/2/A30
Formula Set IX	Prototype Cell is AG30	=(R30+U30)/2/A30
Formula Set X	Prototype Cell is AH30	=**M16** *K30-**N16** *J30
Formula Set XI	Prototype Cell is AI30	=**N16** *I30-**L16** *K30
Formula Set XII	Prototype Cell is AJ30	=**L16** *J30-**M16** *I30
Formula Set XIII	Prototype Cell is AK30	=AE30-AH30
Formula Set XIV	Prototype Cell is AL30	=AF30-AI30
Formula Set XV	Prototype Cell is AM30	=AG30-AJ30
Formula Set XVI	Prototype Cell is AN30	=A30*((J30-**H20**)^2+(K30-**H21**)^2)
Formula Set XVII	Prototype Cell is AO30	=-A30*(I30-**H19**)*(J30-**H20**)
Formula Set XVIII	Prototype Cell is AP30	=-A30*(I30-**H19**)*(K30-**H21**)
Formula Set XIX	Prototype Cell is AQ30	=A30*((I30-**H19**)^2+(K30-**H21**)^2)
Formula Set XX	Prototype Cell is AR30	=-A30*(J30-**H20**)*(K30-**H21**)

Formula Set XXI
Prototype Cell is AS30

=A30*((I30-**H19**)^2+(J30-**H20**)^2)

EXERCISE **5.4**

DIAGONALIZING THE INERTIA TENSOR

Let's pause for a moment and summarize a few of this chapter's principal lessons. The rotational equivalent to Newton's second law of motion is

$$\boldsymbol{\Gamma} = \frac{d\boldsymbol{L}}{dt} \tag{5.58}$$

where $\boldsymbol{\Gamma}$ is the total torque applied to the body and $\boldsymbol{L}$ is the total angular momentum of the body. To see the analogy between Equation 5.58 and Newton's second law for translational motion, recall that $\boldsymbol{\Gamma}$ can also be expressed as shown in Equation 5.59 and $\boldsymbol{L}$ is defined by Equation 5.60:

$$\boldsymbol{\Gamma} = \sum_{j=1}^{n} \boldsymbol{r}_j \times \boldsymbol{F}_j \tag{5.59}$$

$$\boldsymbol{L} = \sum_{j=1}^{n} \boldsymbol{r}_j \times m_j \boldsymbol{v}_j \tag{5.60}$$

where $\boldsymbol{r}_j$, $\boldsymbol{F}_j$, m_j, and $\boldsymbol{v}_j$ are the the position, force, mass, and velocity, respectively, of the jth particle in a system of n particles.

Earlier exercises in this chapter have emphasized that, for a rigid body with its center of mass at rest, Equation 5.60 can also be expressed in

TO GET THE MOST OUT OF THIS EXERCISE, YOU SHOULD ALREADY KNOW

The definitions of the terms:
force
linear momentum
angular position
angular velocity
angular momentum
torque
inertia tensor

How to calculate the cross product of a vector

terms of the angular momentum **ω** and a multicomponent quantity called the inertia tensor **I**:

$$L = \mathbf{I} \cdot \omega \tag{5.61}$$

You know from Exercise 5.1 that, in general, L and w have different directions. You have also seen in Exercise 5.3 that for a rigid body in uniform rotation the torque can be calculated from a cross product (Equation 5.62).

$$\Gamma = \frac{dL}{dt} = \omega \times L \tag{5.62}$$

Finally, you saw in Exercise 5.1 that there are special directions through a body's center of mass for which no torque is required to maintain steady motion. These important directions are called the principal axes of the body and the "no-torque" condition can be taken as their operational definition.

You can use this definition of the principal axes and Equations 5.61 and 5.62 to find an efficient algebraic procedure for calculating them. This procedure is called *finding the eigenvectors of* ***I***. It also produces certain quantities called the *eigenvalues of* ***I*** and in the context of the inertia tensor, these eigenvalues are called the *principal moments of inertia* of the body (see Problems 1 and 2 of Exercise 5.1).

Substituting Equation 5.61 into Equation 5.62 gives

$$\Gamma = \omega \times (\mathbf{I} \cdot \omega) \tag{5.63}$$

Finding a principal axis of the body amounts to finding a direction of a non-zero angular velocity **ω** so that the torque vanishes. But, from Equation 5.63, you can see that this just means that $\mathbf{I} \cdot \omega$ must be parallel to **ω** itself. From the properties of the cross product, you can see that the *length* of $\mathbf{I} \cdot \omega$ is irrelevant for this consideration, and so you can write the algebraic condition

$$\mathbf{I} \cdot \omega = \lambda \omega \tag{5.64}$$

where λ is a constant. *Any vector* ***ω*** *satisfying this equation will be parallel to a principal axis of the body.* In general, different eigenvectors will be associated with different values of λ. This constant is the eigenvalue associated with the eigenvector or, in the inertia tensor context, the principal moment of inertia associated with the particular principal axis.

In Exercise 5.1, you saw that at least two principal axes exist for very simple mass distributions. In fact, solutions to Equation 5.64 show that, for any rigid body, there are always at least three principal axes and each is orthogonal to the other two.[2]

If you write Equation 5.64 as three equations, one for each component, you will see that you have three homogeneous equations in three unknowns, ω_x, ω_y, and ω_z (Equations 5.65 - 5.67):

$$(I_{xx} - \lambda)\omega_x + I_{xy}\omega_y + I_{xz}\omega_z = 0 \tag{5.65}$$

$$I_{yx}\omega_x + (I_{yy} - \lambda)\omega_y + I_{yz}\omega_z = 0 \tag{5.66}$$

$$I_{zx}\omega_x + I_{zy}\omega_y + (I_{zz} - \lambda)\omega_z = 0 \tag{5.67}$$

You know from your algebra that three such equations have only one solution, $\omega_x = 0, \omega_y = 0, \omega_z = 0$, unless the determinant of the coefficients is zero. Because we want a vector $\boldsymbol{\omega}$ that points in *some* direction (i.e., a nonzero vector $\boldsymbol{\omega}$), matters must be arranged so that the determinant of the coefficients is zero. This will be the case only for very special values of λ. In fact, λ must be a root of the cubic polynomial in λ obtained by evaluating the determinant of the coefficients in Equations 5.65 - 5.67.[3]

Because a cubic polynomial has a maximum of three real roots, λ can have a maximum of three different values. Let's look at the various cases:

First, suppose you evaluate $\mathbf{I}$ and discover that

$$\mathbf{I} = \begin{pmatrix} A & 0 & 0 \\ 0 & B & 0 \\ 0 & 0 & C \end{pmatrix} \tag{5.68}$$

In this case, the job is done! The coordinate axes used to calculate $\mathbf{I}$ are already the principal axes, and the three numbers A, B, C are the principal moments of inertia about these axes: A is the principal moment about the x-axis, B about the y-axis, and C about the z-axis. You can easily verify this by showing that

$$\mathbf{I} \cdot \boldsymbol{\omega} = \begin{pmatrix} A & 0 & 0 \\ 0 & B & 0 \\ 0 & 0 & C \end{pmatrix} \cdot (\omega_x \boldsymbol{i} + \omega_y \boldsymbol{j} + \omega_z \boldsymbol{k})$$

$$= A\omega_x \boldsymbol{i} + B\omega_y \boldsymbol{j} + C\omega_z \boldsymbol{k} \tag{5.69}$$

From this, you can see that if $\omega_y = \omega_z = 0$, then $\boldsymbol{\omega}$ is along the x-axis and it is an eigenvector with eigenvalue A, and similarly for the other two cases. You saw examples of such diagonal matrices in Exercise 5.2. The unit vectors $\boldsymbol{i}, \boldsymbol{j}, \boldsymbol{k}$ along the three axes of reference frame are said to be the *normalized eigenvectors* of the matrix ($\boldsymbol{e}_1$, $\boldsymbol{e}_2$, $\boldsymbol{e}_3$), and the three diagonal components are called the corresponding *eigenvalues* (λ_1, λ_2, λ_3).

Next, suppose that

$$\mathbf{I} = \begin{pmatrix} A & E & 0 \\ E & B & 0 \\ 0 & 0 & C \end{pmatrix} \tag{5.70}$$

Exercise 5.2 offered many examples of $\mathbf{I}$ in this form. In this case, you can easily verify that the z-axis is a principal axis and C is the associated principal moment of inertia (i.e., a unit vector along the z-axis is a normalized eigenvector ($\boldsymbol{e}_3$) with eigenvalue $\lambda_3 = C$). The other two eigenvalues can be found by solving the quadratic equations. The results are

$$\lambda_1 = \frac{A + B}{2} + \sqrt{\left(\frac{A - B}{2}\right)^2 + E^2} \tag{5.71}$$

$$\lambda_2 = \frac{AB - E^2}{\lambda_1} \tag{5.72}$$

λ_1, of course, will always be a positive, real number, and it can be shown that λ_2 is also always positive. If $E = 0$, you should notice that the two eigenvalues are just A and B, exactly as you saw in Equation 5.68. In this case, λ_1 will be the larger of A and B.

Unit vectors along the principal axes are

$$\boldsymbol{e}_1 = \alpha\,\boldsymbol{i} + \beta\,\boldsymbol{j} \tag{5.73}$$

$$\boldsymbol{e}_2 = -\beta\,\boldsymbol{i} + \alpha\,\boldsymbol{j} \tag{5.74}$$

where $\boldsymbol{e}_1$ and $\boldsymbol{e}_2$ are unit vectors and

$$\alpha = \frac{a}{N} \tag{5.75}$$

$$\beta = \frac{E}{N} \tag{5.76}$$

$$a = \lambda_1 - B \tag{5.77}$$

$$N = \sqrt{E^2 + a^2} \tag{5.78}$$

Assuming that $N \neq 0$, you should be able to easily see that $\boldsymbol{e}_1 \cdot \boldsymbol{e}_2 = 0$, or in other words, these vectors are perpendicular to each other. Further, they are obviously perpendicular to the third eigenvector $\boldsymbol{k}$.

Clearly, the really special situation in Equation 5.70 is that two of the off-diagonal components of **I** are zero. In that matrix they were I_{xz} and I_{yz}. A similar situation would arise if any other pair of off-diagonal components were to vanish, in particular, either I_{xy} and I_{xz} or I_{xy} and I_{yz}. You should be able to write explicit formulas analogous to Equations 5.73 - 5.78 to handle these cases.

These equations for the eigenvectors fail when $E = 0$ and $B \geq A$, or in other words, in some cases when the inertia tensor is really like Equation 5.68. This may seem so special that it would hardly ever happen, but quite the contrary: It is the usual case with very symmetric objects such as wheels and disks. One way to avoid this sort of problem is to first check to see if you really have Equation 5.68 before using the equations for cases for which Equation 5.70 is the inertia tensor. But because the two off-diagonal zeros can be in any of three combinations, this strategy can make for rather complicated logic formulas. However, there is an easier way to handle the problem in the context of a spreadsheet calculation. Simply make a physically insignificant but mathematically important artificial modification to the problem so that the *exact* condition $E = 0$ is no longer true. For example you could let $E = 0.00001\ B$. This is called *breaking the symmetry.*

The eigenvalue λ_1 is said *to belong* to $\boldsymbol{e}_1$ and similarly for the other two pairs.

A very important fact to appreciate is that, if you had started out using as your x-axis a direction along $\boldsymbol{e}_1$ and your y-axis along $\boldsymbol{e}_2$, you would have gotten for the inertia matrix

$$\mathbf{I} = \begin{pmatrix} \lambda_1 & 0 & 0 \\ 0 & \lambda_2 & 0 \\ 0 & 0 & C \end{pmatrix} \tag{5.79}$$

with no extra calculation.

You will see this in the lesson. For this reason, people refer to this process as *diagonalizing the inertia matrix* by which they mean finding a set of coordinate axes with respect to which the inertia matrix is diagonal.

Finally, for a body that lacks any obvious symmetry, no pair of off-diagonal components of the inertia tensor will be zero (if any one vanishes, nothing special happens, so we do not specially consider that case). Suppose

$$\mathbf{I} = \begin{pmatrix} A & E & F \\ E & B & G \\ F & G & C \end{pmatrix} \tag{5.80}$$

Considerably more arithmetic is required to find the eigenvalues in this case because a cubic equation must be solved.The procedure is as follows.

First evaluate the three *invariants*:

$$T = A + B + C \tag{5.81}$$

$$S = (AB - E^2) + (AC - F^2) + (BC - G^2) \tag{5.82}$$

$$D = A(BC - G^2) - E(EC - GF) + F(EG - BF) \tag{5.83}$$

T, S, and D are called the *trace*, *second invariant*, and *determinant* of **I**, respectively. They are called invariants because no matter which coordinate system you use to evaluate **I** (so long as its origin is the center of mass), these combinations of the components of the inertia matrix have the same value. Their values depend on the mass distribution of the body, not the coordinate system used to describe that distribution.

Next form the combinations

$$P = \frac{T^2}{9} - \frac{S}{3} \tag{5.84}$$

$$Q = \frac{D}{2} - \frac{TS}{6} + \left(\frac{T}{2}\right)^3 \tag{5.85}$$

$$\phi = \frac{1}{3}\cos^{-1}\left(\frac{Q}{\sqrt{P^3}}\right) \tag{5.86}$$

$$R = 2\sqrt{P} \tag{5.87}$$

The eigenvalues of **I** are

$$\lambda_1 = \frac{T}{3} + R\cos\phi \tag{5.88}$$

$$\lambda_2 = \frac{T}{3} + R\cos\left(\phi - \frac{2\pi}{3}\right) \tag{5.89}$$

$$\lambda_3 = \frac{T}{3} + R\cos\left(\phi + \frac{2\pi}{3}\right) \tag{5.90}$$

It can be shown, and you will see it in the lesson, that P is always positive so you will never be embarrassed by asking your spreadsheet program to find the square root of a negative quantity in Equations 5.86 and 5.87 (except possibly when round-off errors combine in just the wrong way — which can be corrected by making an extremely small change to the tensor). Furthermore, the argument of the inverse cosine in Equation 5.86 always has an absolute value less than 1 so that you will never get an error in evaluating ϕ (again, barring round-off error problems). Finally, all three eigenvalues will always turn out positive, with $\lambda_1 \geq \lambda_2 \geq \lambda_3$.

The eigenvectors associated with these eigenvalues are

$$\boldsymbol{e}_1 = \alpha_{11}\,\boldsymbol{i} + \alpha_{12}\boldsymbol{j} + \alpha_{13}\,\boldsymbol{k} \tag{5.91}$$

$$\boldsymbol{e}_2 = \alpha_{21}\,\boldsymbol{i} + \alpha_{22}\boldsymbol{j} + \alpha_{23}\,\boldsymbol{k} \tag{5.92}$$

$$\boldsymbol{e}_3 = \boldsymbol{e}_1 \times \boldsymbol{e}_2 \tag{5.93}$$

where

$$\alpha_{ij} = \frac{a_{ij}}{N_i} \tag{5.94}$$

$$a_{i1} = F(\lambda_i - B) + GE \tag{5.95}$$

$$a_{i2} = FE + G(\lambda_i - A) \tag{5.96}$$

$$a_{i3} = (\lambda_i - B)(\lambda_i - A) - E^2 \tag{5.97}$$

$$N_i = \sqrt{a_{i1}^2 + a_{i2}^2 + a_{i3}^2} \tag{5.98}$$

Just as in the simpler case treated above, it is possible that these equations can fail because either N_1, N_2, or both are zero. This will happen whenever the body being considered has a great deal of symmetry, as you will see in the lesson. If this happens, the simplest remedy is to break the symmetry by a very tiny amount. There do exist

Figure 5.19

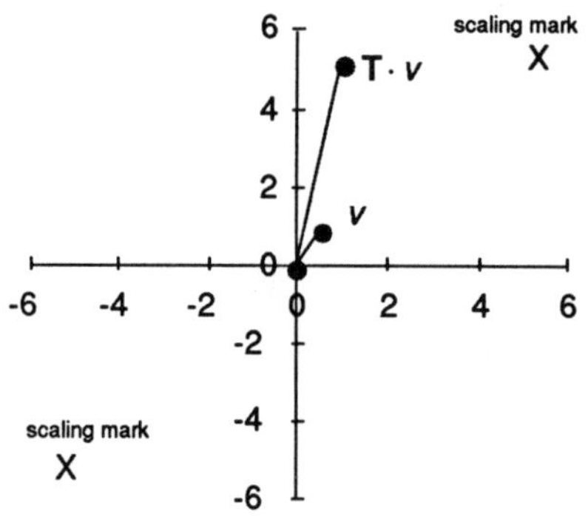

XY–PLOT

Data
Horizontal (*x*) axis: cells B7 and B8 (*x*-component of ***v***)
Vertical (*y*) axis: cells C7 and C8 (*y*-component of ***v***)

Horizontal (*x*) axis: cells B10 and B11 (*x*-component of **T** · ***v***)
Vertical (*y*) axis: cells C10 and C11 (*y*-component of **T** · ***v***)

Horizontal (*x*) axis: cells B13 and B14 (scaling device)
Vertical (*y*) axis: cells C13 and C14 (scaling device)

If possible, the first two components of the plot should contain both symbols and connecting lines (to represent the vectors ***v*** and **T** · ***v***), while the last component should contain only symbols (or even be rendered invisible because it is present only to assure a proper scale to the plot).

On some Lotus products it appears that, to plot the two vectors in different colors, one of the vectors must be moved to a blank column (e.g., MOVE cells C10 and C11 to D10 and D11).

algorithms for finding eigenvectors that do not suffer from this sort of problem, but they are more complicated than the formulas given here.

LESSONS

1. For a two-dimensional body, the inertia matrix has two of its off-diagonal components equal to zero (as in Equation 5.70). In this case, the angular momentum associated with an angular velocity in the *x*-, *y*-plane is also in the *x*-, *y*-plane. Therefore, if you limit yourself to such bodies, you can temporarily forget about the *z*-direction and explore the inertia tensor in a graphically simple environment. Because the 2 x 2 matrices you will study need not be inertia tensors, they will be called **T** (for tensor). Similarly, the vector it operates on need not be the angular velocity, so it will be called ***v*** (for vector).

   a. Build and verify that portion of Model 5.5 that exists in the block of cells A1 - G11. In this model, a two-dimensional inertia tensor **T** is entered into cells B4 - C5 and the magnitude $R(v)$ and angle $\theta(v)$ of a two-dimensional vector ***v*** is entered into cells F7 and F8. The model determines the vector that results from operating on ***v*** by **T** (**T**·***v***) and outputs the magnitude R (i.e., **T**·***v***) and angle θ(**T**·***v***) in cells F10 and F11. All angles are determined with respect to the *x*-axis.

   Also build an on-screen graphic to represent the vectors ***v*** and **T**·***v*** in an easily interpretable fashion (Figure 5.19). Notice that some Lotus products require a slight modification of the model to fully exploit a color screen (useful in distinguishing ***v*** from **T**).

   As noted above you can use Model 5.5 to study any two-dimensional matrix, not only inertia matrices. Inertia matrices are always symmetric, i.e., $I_{ij} = I_{ji}$, whereas that is not necessary here. Furthermore, the components of an inertia matrix also satisfy special conditions so that its eigenvalues, the principal moments of inertia, always end up being nonnegative, and it is never possible for only one of them to be nonzero. Again, these conditions are not imposed in Model 5.5. In this lesson you will study only *symmetric* matrices, as are all inertia matrices, leaving to the problems investigations of nonsymmetric tensor operators. These problems will help you appreciate the specialness of inertia tensors.

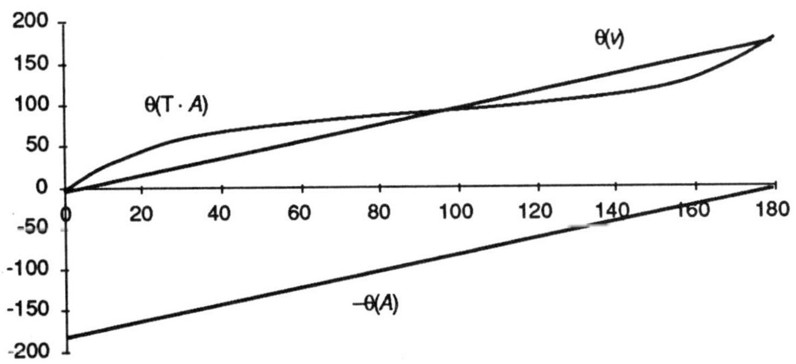

Figure 5.20

XY–PLOT

Data
Horizontal (*x*) axis: cells B17 - B107 [$-\theta(v)$]
Vertical (*y*) axis: cells A17 - A107 [$\theta(v)$], B17 - B107 [$-\theta(v)$], and H17 - H107 [$-\theta(\mathbf{T}\cdot v)$]

b. Initialize your model in cells B4 - C5 using the matrix

$$\mathbf{T} = \begin{pmatrix} 2 & 0 \\ 0 & 6 \end{pmatrix}$$

and a unit vector v such that $R(v) = 1$. Vary $\theta(v)$ from 0° to 180° (in increments of 10°) and notice that for most angles $\theta(v)$ the vectors v and $\mathbf{T}\cdot v$ point in different directions.

Verify that only the values of $\theta(v)$ = 0°, 90°, and 180° yield conditions where v and $\mathbf{T}\cdot v$ point in the same direction (your on-screen graphic of Figure 5.19 gets you close and an actual comparison of the numbers in cells F8 and F11 take you the rest of the way). If you had taken $\theta(v) = 270°$, you would have found the same thing. These four v vectors are eigenvectors of **T**. However, because the vectors with $\theta(v) = 0$ and $\theta(v) = 180°$ are proportional to each other, they are not considered distinct eigenvectors (and similarly for the pair at 90° and 270°). In this sense then, there are really only two eigenvectors.

Notice also that, because v has been taken as a unit vector [$R(v) = 1$], the value of $R(\mathbf{T}\cdot v)$ when v is an eigenvector is the magnitude of the associated eigenvalue. You can verify this by looking at the components of v and $\mathbf{T}\cdot v$.

What are the eigenvalues of **T**?

c. Now build and verify the remaining parts of Model 5.5 (i.e., the table that begins in row 16). This table simply automates the process of finding values of $\theta(v)$ that cause v and $\mathbf{T}\cdot v$ to point in the same direction. Build an on-screen graphic that plots $\theta(v)$ vs $\theta(\mathbf{T}\cdot v)$ (Figure 5.20). This graphic also includes plots of $\theta(v)$ vs $\theta(v)$ and $\theta(v)$ vs $-\theta(v)$ as useful orienting markers. Whenever $\theta(\mathbf{T}\cdot v)$ intersects the plot of $\theta(v)$, both v and $\mathbf{T}\cdot v$ are pointing in the same direction: the specified eigenvector of **T** has the direction $\theta(\nu)$, and the eigenvalue is positive. Whenever $\theta(\mathbf{T} \cdot \nu)$ intersects the plot of $-\theta(\nu)$, ν and $\mathbf{T} \cdot \nu$ point in opposite directions: the eigenvector of **T** has the direction $-\theta(\nu)$, and the eigenvalue is negative.

(Pause for a moment and confirm that you understand why the previous sentence is true.)

If **T** and ν have the same values as in Lesson 1b, then your on-screen graphic should give you an instantaneous readout of the eigenvectors 0°, 90°, and 180°. (Recall from Lesson 2b that only two distinct eigenvectors have been detected because 0° and 180° are not distinct.)

What is the eigenvalue associated with the eigenvector along the x-axis? With the one along the y-axis? Notice that the eigenvectors are orthogonal to each other. Can this matrix represent an inertia matrix?

d. Change the matrix under study to

$$\mathbf{T} = \begin{pmatrix} 2 & 0 \\ 0 & -6 \end{pmatrix}$$

Use on-screen graphics represented by both Figures 5.19 and 5.20 to determine the eigenvectors and eigenvalues of this matrix. What is the angle between the two eigenvectors? Can this be an inertia matrix?

e. Change the matrix under study to

$$\mathbf{T} = \begin{pmatrix} 2 & 0 \\ 0 & 2 \end{pmatrix}$$

and study its properties.How many *different* eigenvalues does this tensor have? How many *different* eigenvectors does it have? Can this tensor be an inertia tensor?

When a matrix has fewer *different* eigenvalues than its dimension, the matrix is said to have a *degeneracy*. Because the eigenvalues of a matrix are the roots of a polynomial, the case of degeneracy corresponds to the polynomial having multiple roots.

A *symmetric* degenerate matrix, such as the one under study, will have an infinite number of eigenvectors associated with a multiple root. Any two orthogonal vectors in the plane can be chosen from this infinite set and used as *the* eigenvectors. Any other eigenvector can be expressed as a sum of multiples of the two orthogonal ones chosen. Sums of vectors like this are called *linear combinations* of them.

Inertia matrices with degeneracy correspond to highly symmetric bodies, and so occur commonly for idealized man-made objects.In real life, however, no macroscopic body is perfectly symmetric, and its small imperfections, no matter how small, will cause some particular pair of orthogonal vectors to be the eigenvectors. People say that the degeneracy is "lifted" by the imperfection which breaks the symmetry.

f. Break the symmetry in the last example by setting the two off-diagonal components to 0.5. Study the effect of this matrix operator on various vectors.[4] What are its eigenvalues and eigenvectors? What is the angle between the eigenvectors? Can this tensor be an inertia tensor?

Notice that the degeneracy of Lesson 1e has been lifted in a particular way, and particular directions in the plane have been "selected" by the symmetry breaking. You will see later that (or how, if you wish), no matter how small this particular type of symmetry breaking is, the same pair of eigenvectors are selected.

g. Break the symmetry in Lesson 1e differently by taking

$$\mathbf{T} = \begin{pmatrix} 2 & 0 \\ 0 & 2.5 \end{pmatrix}$$

What do you conclude about the eigenvectors in this case? Be sure to notice the angle between them.

Again, as you can easily see in this case, this particular pattern of symmetry breaking always produces the same eigenvectors, no matter how small or large it may be.

h. Next break the symmetry of the matrix in Lesson 1e by setting

$$\mathbf{T} = \begin{pmatrix} 2 & 0.5 \\ 0.5 & 2.5 \end{pmatrix}$$

What are the directions of the eigenvectors now? What are the eigenvalues? Can this be an inertia matrix?

i. Consider next the matrix

$$\mathbf{T} = \begin{pmatrix} 2 & -0.5 \\ -0.5 & 2.5 \end{pmatrix}$$

How does changing the sign of the off-diagonal components change the eigenvalues? The eigenvectors? Can this be an inertia matrix?

j. Next consider the tensor

$$\mathbf{T} = \begin{pmatrix} 2 & 2 \\ 2 & 2 \end{pmatrix}$$

You may be surprised by the plots; to help you understand what is happening, plot $R(\mathbf{T} \cdot \nu)$ vs $\theta(\nu)$.What happens to $R(\mathbf{T} \cdot \nu)$ at $\theta(\nu)=135°$? This is the case that the tensor has a *zero* eigenvalue. A tensor operator with a zero eigenvalue is said to be *singular*. Look at the vector diagram for values of $\theta(\nu)$ near 135° to get a feeling for what happens near an eigenvector with zero eigenvalue. Can a matrix like this be an inertia matrix for any real object? What about for idealizations of real objects?

k. Suppose the symmetry in the last example is slightly broken so the matrix is changed to

$$\mathbf{T} = \begin{pmatrix} 2 & 1.99 \\ 1.99 & 2 \end{pmatrix}$$

Where are the eigenvectors and the associated eigenvalues? Can this matrix be an inertia matrix? To study in detail exactly what is happening near the critical angle of 135°, change cell B17 of your model to 134° and cell B18 to =B17+0.1 where B17 is a relative address. Copy the contents of cell B18 into cells B19 - B106. Your model should now permit you to complete and graph the table below:

$\theta(\nu)$	$\theta(\mathbf{T} \cdot \nu)$
134.0	______
134.1	______
134.2	______
134.3	______
134.4	______
134.5	______
134.6	______
134.7	______
134.8	______
134.9	______
135.0	______

135.1	________
135.2	________
135.3	________
135.4	________
135.5	________
135.6	________
135.7	________
135.8	________
135.9	________
136.0	________

l. Study what happens if the off-diagonal components in the last example are slightly greater than 2.00, for example, 2.01. Can this tensor be an inertia tensor?

m. Finally study the matrix

$$\mathbf{T} = \begin{pmatrix} 2 & 3 \\ 3 & 2 \end{pmatrix}$$

Can it be an inertia matrix?

2. Modify Model 5.5 by writing formulas to represent Equations 5.71 - 5.74 (Model 5.6). Your model can now directly calculate the eigenvalues and eigenvectors of a 2 x 2 inertia tensor. The values of the two eigenvectors are used to calculate the angles with respect to the x-axis, θ_1 and θ_2, of the eigenvectors.

   a. Using the initial values of Model 5.6, the angles of the two eigenvectors are directly computed as 45° (cell F20) and 135° (cell F21). Verify that these angles are correct by checking the angle of $\theta(\mathbf{T} \cdot \boldsymbol{v})$ when $\theta(\boldsymbol{v})$ has the angle predicted for the eigenvectors.

   Suppose the matrix is degenerate (as in Lesson 1e). What eigenvectors are calculated directly by your model? Are these the only eigenvectors? Verify your answer by entering other values of $\theta(\boldsymbol{v})$ into your model.

   b. Break the symmetry of the degenerate case very slightly by considering the inertia matrix

$$\mathbf{T} = \begin{pmatrix} 2 & 0.0001 \\ 0.0001 & 2 \end{pmatrix}$$

   Intuitively, you would expect that this should modify the eigenquantities by only a very slight amount, and yet the directly computed eigenvectors change dramatically.

Figure 5.21

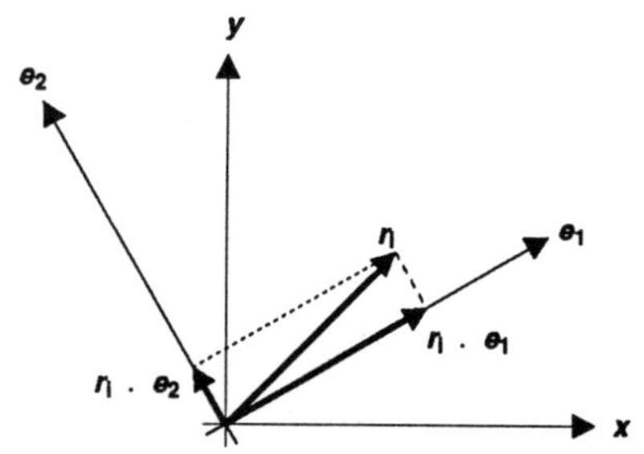

A position vector r_i that is defined in terms of the x-, y-coordinate system can be defined in terms of the e_1-, e_2-coordinate system by using Equation 5.99.

Figure 5.22

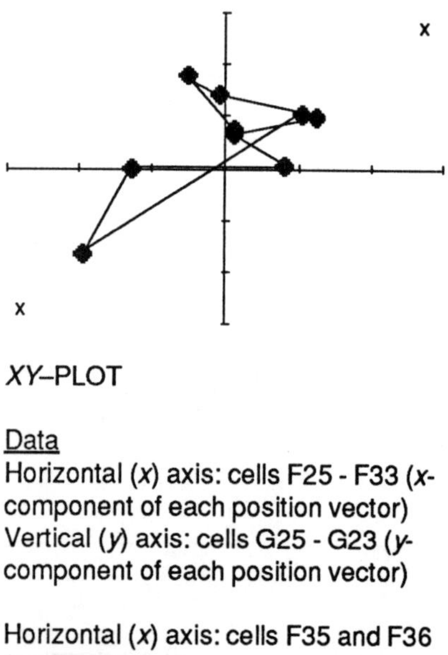

XY–PLOT

Data
Horizontal (*x*) axis: cells F25 - F33 (*x*-component of each position vector)
Vertical (*y*) axis: cells G25 - G23 (*y*-component of each position vector)

Horizontal (*x*) axis: cells F35 and F36 (scaling device)
Vertical (*y*) axis: cells G35 and G36 (scaling device)

Although they probably won't have physical meaning, connecting each plotting symbol with lines yields a pattern that it easy to recognize as it rotates.

Verify that, in fact, the characteristics of the matrix with the symmetry broken are essentially the same as those of the degenerate matrix by examing the directions of $\boldsymbol{v}$ and $\mathbf{T} \cdot \boldsymbol{v}$ for different values of $\theta(\boldsymbol{v})$.

3. It is sometimes said that an inertia matrix can be *diagonalized.* A diagonalized matrix has been calculated using a set of coordinate axes such that the symmetric off-diagonal components of the matrix become zero. The distribution of mass described by a diagonalized inertia matrix is the same as that in the initial matrix, but the mass is now described in a reference frame determined by the principal axes of the body.

Let's see how this works for the simple case of a two-dimensional body. Suppose you have calculated an inertia matrix of such a body and have found its eigenvectors $\boldsymbol{e}_1$ and $\boldsymbol{e}_2$. As you know, these eigenvectors are orthogonal to each other and so could be used to define a new coordinate system. What are the "*x*"- and "*y*"-components of the radius vector of each mass in the body in such a coordinate system? The answer is easy!

If $\boldsymbol{r}_i$ is the radius vector to mass *i* in the body, which you used in calculating the inertia matrix, then you can express it using the orthogonal vectors $\boldsymbol{e}_1$ and $\boldsymbol{e}_2$ as

$$\boldsymbol{r}_i = (\boldsymbol{r}_i \cdot \boldsymbol{e}_1)\boldsymbol{e}_1 + (\boldsymbol{r}_i \cdot \boldsymbol{e}_2)\mathbf{e}_2 \tag{5.99}$$

Equation 5.99 states that the two dot products are the *x*- and *y*-components of the vector $\boldsymbol{r}_i$ in a coordinate system adapted to the matrix (Figure 5.21). The inertia matrix of the body expressed in this *adapted* coordinate system will be diagonal.

Modify Model 5.6 to calculate the inertia tensor from a collection of point masses and then *recalculate* it in the adapted coordinate system (Model 5.7). If you build and verify directly from Model 5.7 take the time to explore its structure until you fully understand how it works. The model is designed so that you can easily add or remove masses below row 24 so long as you insert the masses *between* row 24 and the row of hyphens.

Notice that the new inertia tensor $\mathbf{T}_{\text{diag}}$ is diagonal and that the eigenvalues of $\mathbf{T}$ appear on the diagonal of $\mathbf{T}_{\text{diag}}$ in the order λ_1 and λ_2.

a. Plot Y vs X (Figure 5.22) and $\boldsymbol{e}_2 \cdot \boldsymbol{r}$ vs $\boldsymbol{e}_1 \cdot \boldsymbol{r}$ (Figure 5.23) including the scale setting points in cells F35 - G36 and K35 - L36. If you plot using both lines and symbols, you may find it easier to recognize the figure, but the manner in which the lines connect may have no physical meaning (depending on the plotting routines of your spreadsheet).

Explore both plots and verify that you see the same figure in both plots but that the one created by $\boldsymbol{e}_1 \cdot \boldsymbol{r}$ and $\boldsymbol{e}_2 \cdot \boldsymbol{r}$ is rotated about a normal to the plane by the angle $-\theta_1$. (Some spreadsheet programs permit you to display both plots at once; otherwise flip quickly between them.)

b. Prepare other mass distributions and observe how the eigenvectors of the inertia matrix provide the basis for an adapted coordinate system. You may wish to use your spreadsheet's random number generator to produce a variety of random mass

SPREADSHEET TECHNIQUE

Random Number Generators

It is often the case, in testing models for example, that you would like to just choose a lot of random numbers as input to the model to check that it works properly. Although it may seem trivial to make up a lot of random numbers for this purpose, it is actually harder that you might think because we tend to think in patterns (1,2,3 ... 2,4,6 ... your phone number, etc). Your spreadsheet program provides a function named RAND(), RANDOM () or something similar to solve this problem for you.

Everytime a model is recalculated, each occurrence of the function RAND evaluates to a new "random" value in the range 0 to 1. The word *random* is placed in quotes because the sequence of values that RAND produces in not random in the strict sense. For example, everytime you restart your model by reading it from the disk, the sequence of random values is the same (unless your spreadsheet is unusually clever and takes a seed from your onboard clock).

If you wish to generate random numbers in the range from a to b instead of 0 to 1, the simple formula

$$a + (b - a)\ \text{RANDOM}$$

produces them for you. For example, all masses in the initial data for Model 5.7 were produced with the formula ROUND(10*RAND,2), which first produces numbers in the range 0 to 10 and then rounds each to two decimal places. Similarly, the x- and y-coordinates were generated with the formula ROUND(-1+2*RAND,3).

As you begin to use your computer for independent investigations, you will find many applications of this simple technique.

Figure 5.23

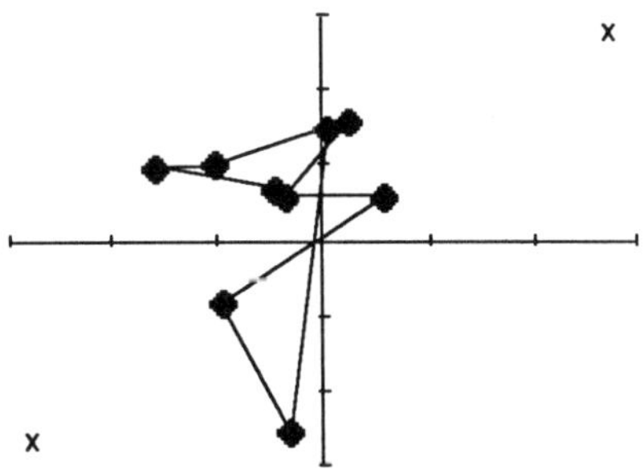

XY–PLOT

Data
Horizontal (x) axis: cells K25 - K33 (x-component of each position vector)
Vertical (y) axis: cells L25 - L23 (y-component of each position vector)

Horizontal (x) axis: cells K35 - K36 (scaling device)
Vertical (y) axis: cells L35 - L36 (scaling device)

Although they probably won't have physical meaning, connecting each plotting symbol with lines yields a pattern that it easy to recognize as it rotates.

distributions. See Spreadsheet Technique "Random Number Generators."

4. Build and verify Model 5.8. Model 5.8 uses Equations 5.81 - 5.90 to calculate the eigenvalues ($\lambda_1, \lambda_2, \lambda_3$) and eigenvectors ($\boldsymbol{e}_1, \boldsymbol{e}_2$, and $\boldsymbol{e}_3$) of a 3 x 3 symmetric matrix.

   Explore the structure of the model until you understand how it works. Cells B13 - B21 contain quantities, each of which should be exactly zero were it not for round-off errors. The sum of these nine numbers (cell B23) serves as a master check because it also should be zero. The nine model verifying quantities are the three dot products between pairs of eigenvectors (if these are zero, it means the eigenvectors are orthogonal); the difference in length of each of the eigenvectors from 1 (if these are zero, each eigenvector has been normalized to unit length); the differences between the three invariants of the original matrix and a diagonal matrix constructed from the eigenvalues of the original.

   a. Can the initial matrix of the model be an inertia matrix?

   b. Consider a matrix **M** whose value is twice the unit matrix:

   $$\mathbf{M} = \begin{pmatrix} 2 & 0 & 0 \\ 0 & 2 & 0 \\ 0 & 0 & 2 \end{pmatrix}$$

   Before inserting this matrix into your model, do you think that such a matrix could be the inertia matrix of a real body (a sphere, for example)? What do you know about eigenvalues of **M**? About the eigenvectors of **M**?

   Insert components of **M** into your model.

   You may be disappointed that your model fails for the simple matrix of a perfectly symmetric body. There are algorithms that correct for this problem, but we chose to keep the mathematics of the model somewhat more intuitive. You can still use your model by simply breaking the symmetry of your object that, except for objects that are molecular or smaller in size, is probably artificial anyway.

   First break the symmetry by multiplying M_{zz} by 1.0001. Examination of your check-sums demonstrates that you still have a problem.

Next multiply M_{yy} by 1.0002. Still no good. There is still too much symmetry for the simple formulas of the model.

Finally add $M_{xx}/10000$ to both M_{xy} and M_{xz}. Now you've got it!

Compare the final eigenvalues with what you know they should be.

The symmetry breaking we suggested selected a single orthogonal set of eigenvectors out of the infinite set of possible eigenvectors. Unless there was some physical reason to break the symmetry as described, the particular eigenvectors in the model are sufficient but hardly necessary.

c. Modify Model 5.8 slightly so that it automatically breaks the symmetry of highly symmetric matrices. MOVE (or CUT AND PASTE) cells A3 - D5 to A7 - D9 (formulas referring to these locations should automatically update) and now use cells A3 - D5 to input your "raw" data. If you now modify cells B3 - D5 to break the symmetry of your input tensor, your model should now work for symmetric objects. Study examples of all tensors you used earlier in this lesson (i.e., those in which M_{zx} and M_{zy} are both zero). Compare your results with your earlier results.

d. Finally, investigate several truly three-dimensional nondegenerate matrices and identify those that can be inertia matrices.

Figure 5.24

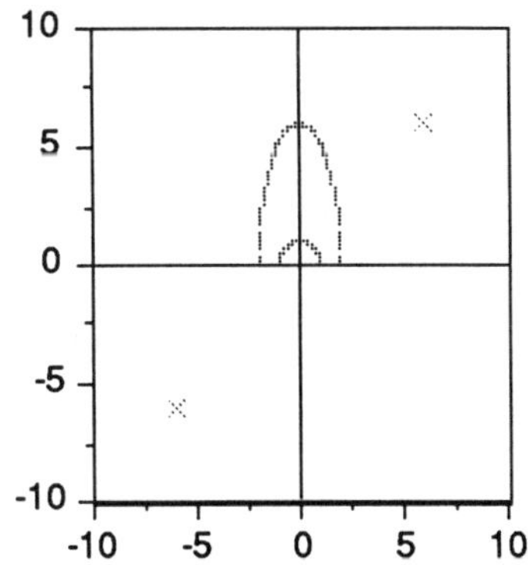

XY–PLOT

<u>Data</u>
Horizontal (*x*) axis: cells C17 - C107 (*x*-component of ***v***)
Vertical (*y*) axis: cells D17 - D107 (*y*-component of ***v***)

Horizontal (*x*) axis: cells E17 - E107 (*x*-component of **T** · ***v***)
Vertical (*y*) axis: cells F17 - F107 (*y*-component of **T** · ***v***)

Horizontal (*x*) axis: cells E109 and E110 (scaling device)
Vertical (*y*) axis: cells F109 and F110 (scaling device)

To make the plots in some spreadsheets (such as Lotus products), you must move some of the data in the table. MOVE C17.C107 to E112 and below. Then MOVE D17.D107 to I112 and below (this allows the end of vector ***v*** to be displayed in a different color). Finally, use E17.E202 as an *X* range; and with symbols only, plot the ranges F17.F107 and I17.I202.

PROBLEMS

1. Recall Model 5.5 to explore further the behavior of tensors as "operators" on vectors. Modify the model by adding the following formulas to the indicated cells

Cell	**Formula**
E126	=MAX(MAX(E34:F124),–MIN(E34:F124))
F126	=E126
E128	=–E126
F128	=–F126

Build Figure 5.24 to display the position of both ***v*** and **T** · ***v*** on the same screen. The formulas above help Figure 5.24 achieve the correct scaling. You will also use Figure 5.20 in this problem.

a. Consider first the unit matrix

$$\mathbf{T} = \begin{pmatrix} 1 & 0 \\ 0 & 1 \end{pmatrix}$$

Examine your plot of the positions of $\boldsymbol{v}$ and $\mathbf{T}\cdot\boldsymbol{v}$ (Figure 5.24) and verify that **T** has no effect on the vector $\boldsymbol{v}$.

Use Figure 5.20 to determine how many different eigenvectors **T** can have.

b. Now consider a matrix with eigenvalues twice those of the unit matrix

$$\mathbf{T} = \begin{pmatrix} 2 & 0 \\ 0 & 2 \end{pmatrix}$$

How does **T** "operate" on $\boldsymbol{v}$ (i.e., how is $\mathbf{T}\cdot\boldsymbol{v}$ different from $\boldsymbol{v}$)? How many different eigenvectors can **T** have?

c. Break the symmetry of Problem 1b by setting **T** such that

$$\mathbf{T} = \begin{pmatrix} 2 & 0 \\ 0 & 4 \end{pmatrix}$$

Now you should see that the "operation" that **T** performs on $\boldsymbol{v}$ is more complex. How many eigenvectors does **T** have now?

d. Now examine an inertia tensor with off-diagonal components:

$$\mathbf{T} = \begin{pmatrix} 2 & 1 \\ 1 & 4 \end{pmatrix}$$

Does the number of possible eigenvectors change from Problem 1c?

Pause for a moment and consider what you've seen. Perfectly symmetric matrices such as in Problem 1a or 1b yield an infinite number of eigenvectors (which correspond to an infinite number of principal axes for the perfectly symmetric bodies such inertia tensors represent). Once symmetry is broken, there are always two possible eigenvectors (or principal axes). Thus, although an inertia tensor can act on the angular velocity vector to yield very different patterns of angular momentum vectors (as Figure 5.24 plots show), there are always a minimum of two orthogonal principal axes about which the (two-dimensional) body can steadily rotate without external torque.

You might find it worthwhile to explore the behavior of tensors that are not symmetric and cannot represent the inertia tensor of any conceivable distribution of mass. Non-symmetric matrices arise in other parts of science and technology.

e. Restore the unit matrix of Problem 1a to your model and examine your plots (Figures 5.20 and 5.24).

Now add a single, nonzero off-diagonal component:

$$\mathbf{T} = \begin{pmatrix} 1 & 2 \\ 0 & 1 \end{pmatrix}$$

Describe qualitatively what the matrix does to a vector.

How many eigenvectors and eigenvalues does this tensor have? What are they?

Vary the nonzero off-diagonal element over a range of values from 0 to 4. What happens to the eigenquantities?

Notice the sharp contrast with symmetric matrices. Symmetric matrices always have at least two eigenvectors; if both off-diagonal components are zero and the matrix is degenerate, it has one eigenvalue and an infinite number of eigenvectors.

f. Next investigate

$$\mathbf{T} = \begin{pmatrix} 1 & 2 \\ 0 & 2 \end{pmatrix}$$

Describe qualitatively what the matrix does to a vector. How many eigenvectors and eigenvalues does this matrix have? What are they?

Notice the angle between the eigenvectors. Could this happen with an inertia tensor?

g. Investigate

$$\mathbf{T} = \begin{pmatrix} 1 & 1 \\ -1 & 3 \end{pmatrix}$$

How many eigenvectors and eigenvalues does this tensor have? Again notice the contrast to the case of degenerate symmetric tensors.

h. The tensor in Problem 1g is a borderline case. The border to what, you might ask? Perturb this tensor slightly and consider first the tensor

$$\mathsf{T} = \begin{pmatrix} 1 & .9 \\ -.9 & 3 \end{pmatrix}$$

and then

$$\mathsf{T} = \begin{pmatrix} 1 & 1.1 \\ -1.1 & 1.3 \end{pmatrix}$$

Discuss what happens in these cases.

i. Summarize your explorations of 2 x 2 matrices for real-valued solutions (complex eigenquantities are often of no physical interest and will not be considered here). There are four cases:

- Symmetric and degenerate
- Symmetric and nondegenerate
- Nonsymmetric and degenerate
- Nonsymmetric and nondegenerate

2. In this exercise, explicit formulas were given for the eigenvalues and eigenvectors of symmetric two-dimensional matrices. These formulas were used to build Model 5.6. Generalize these formulas to nonsymmetric matrices and modify Model 5.6 to include these cases. (What formulas might you build into your model to verify your derivation?)

3. Use Model 5.1 to find the inertia tensor of each of the following objects. Then diagonalize the inertia tensor and determine the directions of the principal axes on a drawing of each object.

   a. A carpenter's square (an L-shaped object)

   b. A doughnut

   c. A wedge

4. Distribute seven point masses (not necessarily of the same mass) in a plane. Guess at the location of the center of mass and the orientation of the principal axes and then use the models in this and

the previous exercise to test your predictions. Sketch the object and its principal axes.

5. Consider a real three-dimensional object with no axes of symmetry (say, a drinking cup with a handle). Construct the three-dimensional inertia tensor (or a reasonal approximation to it), diagonalize it, and sketch the object and its principal axes.

6. Build a model analogous to Model 5.7 but in three dimensions. Explore the adaption of a three-dimensional body to a coordinate system in which its inertia tensor is diagonal.

7. Model 5.4 in Exercise 5.3 can now be examined in the light of your new knowledge about tensors.

   a. Modify Model 5.4 in Exercise 5.3 to include the calculation of the three invariants defined in Equations 5.81 - 5.83 for the 3 x 3 inertia matrix in that model. Complete the following table (the time step is 1/10 of a revolution time):

t	**Trace**	**2nd Invariant**	**Determinant**
0	______	______	______
0.001666	______	______	______
0.003333	______	______	______
0.005	______	______	______
0.006666	______	______	______
0.008333	______	______	______
0.01	______	______	______
0.011666	______	______	______
0.013333	______	______	______
0.015	______	______	______
0.016666	______	______	______

   Do you see why the invariants are so called? What is the invariant of a vector?

   b. Modify Model 5.4 further by including calculations for the three eigenvalues of the inertia tensor (Equations 5.88 - 5.90). Use your model to complete the following table:

t	λ_1	λ_2	λ_3
0	______	______	______
0.001666	______	______	______
0.003333	______	______	______
0.005	______	______	______
0.006666	______	______	______
0.008333	______	______	______
0.01	______	______	______
0.011666	______	______	______
0.013333	______	______	______
0.015	______	______	______
0.016666	______	______	______

Your table may contain some ERR cells because of round-off errors in the calculation, but what is your conclusion about the behavior of the principal moments of a body? Are they a property of the mass distribution only, or do they also depend on the coordinate system used to describe the mass distribution? Can you find a reasonable way to avoid the ERRs in your table?

c. Use Equations 5.91 - 5.93 to try to calculate the eigenvector associated with λ_1. What happens? (Remember that a number of order 10^{-20} should be interpreted as zero when the order of size of numbers in the problem is unity.)

Instead, use these equations to calculate $\boldsymbol{e}_3$ (setting $i = 3$ in the equation). Describe the direction of this eigenvector with respect to the simple two-mass body. Can you find two other eigenvectors? (Remember they must be orthogonal to $\boldsymbol{e}_3$ and to each other.)

Notice that in this problem the body is highly symmetric. This is signaled by the fact that two of the eigenvalues are equal. Furthermore, the zero eigenvalue is due to the fact that all of the mass in the problem is on a line.

d. Insert one or more other masses into your planar rigid body so that it is not as symmetric as the simple dumbbell. In this case, Equations 5.91 - 93 can calculate the eigenvectors (i.e., the principal axes of the body). For simplicity you should arrange the model so that that the axle passes through the center of mass (although later you may wish to investigate what happens when it does not).

What is the direction of the eigenvector associated with the largest eigenvalue? How do the eigenvectors move as the body rotates?

MODELS

FORMULAS

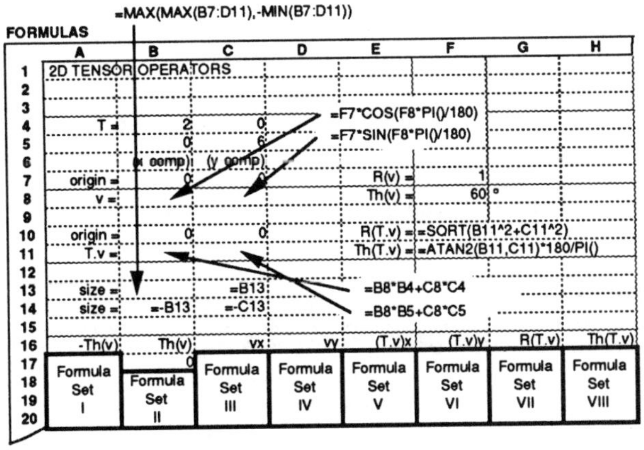

VALUES

	A	B	C	D	E	F	G	H
1	2D TENSOR OPERATORS							
2								
3								
4	T =	2	0					
5		0	6					
6		(x-comp)	(y-comp)					
7	origin =	0	0		R(v) =	1		
8	v =	0.5	0.866025		Th(v) =	60	°	
9								
10	origin =	0	0		R(T.v) =	5.291503		
11	T.v =	1	5.196152		Th(T.v) =	79.10661	°	
12								
13	size =	5.196152	5.196152					
14	size =	-5.19615	-5.19615					
15								
16	-Th(v)	Th(v)	vx	vy	(T.v)x	(T.v)y	R(T.v)	Th(T.v)
17	-180	0	1	0	2	0	2	0
18	-178	2	0.999391	0.034899	1.998782	0.209397	2.00972	5.980622
19	-176	4	0.997564	0.069756	1.995128	0.418539	2.038556	11.84773
20	-174	6	0.994522	0.104528	1.989044	0.627171	2.085579	17.50073

Formula Set I	Prototype Cell is A17	=B17-180
Formula Set II	Prototype Cell is B18	=B17+2
Formula Set III	Prototype Cell is C17	=**F7** *COS(B17*PI()/180)
Formula Set IV	Prototype Cell is D17	=**F7** *SIN(B17*PI()/180)
Formula Set V	Prototype Cell is E17	=C17* **B4** +D17* **C4**
Formula Set VI	Prototype Cell is F17	=C17* **B5** +D17* **C5**
Formula Set VII	Prototype Cell is G17	=SQRT(E17^2+F17^2)
Formula Set VIII	Prototype Cell is H17	=ATAN2(E17,F17)*180/PI()

Model 5.5

BUILD and verify Model 5.5. If you are using Lotus products, take special note of NOTICE below.

ENTER data using cells B4 - C5 (the tensor) and F7 and F8 (the magnitude and angle of vector ***v***).

READ output from the model at cells F8 and F11 (the angles with respect to the *x*-axis of ***v*** and **T** · ***v***). When these angles are equal they represent an eigenvector of **T**, and the magnitude of **T** · ***v*** (cell F10)is an eigenvalue. Cells B7 - C8 and B10 - C11 may be used to graphically represent **T** and **T** · ***v***. When they are pointed in the same direction the eigenquantities of **T** can be read from the model as already noted.

READ output from the table beginning in row 17 if you wish to simplify the process of finding eigenquantities. The table should be used to make the on-screen graphic shown in Figure 5.20.

NOTICE that some Lotus products require modification if you wish to plot the two vectors in different colors, and the caption to Figure 5.19 should be modified appropriately. The contents of cells C10 and C11 should be MOVED to cells D10 and D11. Be sure to build the model as shown before MOVING, or the formulas in the remainder of the model will point to incorrect locations!

The two rows labeled "origin" are arranged so that vectors can be drawn from the origin. The two rows labeled "size" cause the setting of the graphic scale to be the same in both *x* and *y*. If possible, arrange your screen layout so that the picture is geometrically correct.

Model 5.6

MODIFY Model 5.5 and verify it to create Model 5.6. Delete the table beginning in row 16 and enter the cell contents shown in the figure.

ENTER data using cells B4 - C5 (the tensor) and F7 and F8 (the magnitude and angle of vector *v*).

READ output from the model at cells F8 and F11 (the angles with respect to the *x*-axis of *v* and **T** · *v*). When these angles are equal, they represent an eigenvector of **T** and the magnitude of **T** · *v* is an eigenvalue. Cells B7 - C8 and B10 - C11 can be used to graphically represent **T** and **T** · *v*. When they are pointed in the same direction, the eigenquantities of **T** may be read from the model as already noted. This model also calculates eigenquantities directly in cells F13 - F21. The eigenvalues are located in cells F13 and F14, while the two components of each eigenvector are located in cells F18 - G19. The angle that each eigenvector makes with the *x*-axis is displayed in cells F20 and F21.

NOTICE that some Lotus products require modification if you wish to plot the two vectors in different colors, and the caption to Figure 5.19 should be modified appropriately. The contents of cells C10 and C11 should be MOVED to cells D10 and D11. Be sure to build the model as shown before MOVING, or the formulas in the remainder of the model will point to incorrect locations!

FORMULAS (to be added to Model 5.5)

	E	F	G
13	lam-1 =	=(B4+C5)/2+SQRT(((B4-C5)/2)^2+B5^2)	
14	lam-2 =	=(B4*C5-C4^2)/F13	
15	a =	=F13-C5	
16	N =	=SQRT(F15^2+C4^2)	
17		(x-comp)	(y-comp)
18	e1 =	=IF(C4<>0,+F15/F16,IF(B4>C5,1,0))	=IF(C4<>0,+C4/F16,IF(B4>C5,0,1))
19	e2 =	=-G18	=F18
20	Th1 =	=ATAN2(F18,G18)*180/PI()	°
21	Th2 =	=ATAN2(F19,G19)*180/PI()	°

VALUES

	A	B	C	D	E	F	G
1	EIGENPAIRS OF 2D TENSOR OPERATORS						
2							
3							
4	T =	2	1				
5		1	2				
6		(x-comp)	(y-comp)				
7	origin =	0	0		R(v) =	1	
8	v =	0.5	0.866025		Th(v) =	60	°
9							
10	origin =	0	0		R(T.v) =	2.909313	
11	T.v =	1.866025	2.232051		Th(T.v) =	50.10391	°
12							
13	size =	2.232051	2.232051		lam-1 =	3	
14	size =	-2.23205	-2.23205		lam-2 =	1	
15					a =	1	
16					N =	1.414214	
17						(x-comp)	(y-comp)
18					e1 =	0.707107	0.707107
19					e2 =	-0.70711	0.707107
20					Th1 =	45	°
21					Th2 =	135	°

FORMULAS to be added to Model 5.6

=SUM(H25:H34) (into B4)   =SUM(M25:M34) (into F4)

	A	B	C	D	E	F	G
4	T =		=SUM(J25:J34)		Tdiag =		=SUM(O25:O34)
5		=C4	=SUM(I25:I34)			=G4	=SUM(N25:N34)

	A	B	C
16	M =	=SUM(C25:C34)	
17	rcm =	=SUM(A25:A34)/B16	=SUM(B25:B34)/B16

	A	B	C	D	E	F	G	H
23				r		R		
24	mx	my	m	x	y	X	Y	iXX
25	Formula Set I	Formula Set II	6.25	0.51	0.55	Formula Set III	Formula Set IV	Formula Set V
26			9.97	-0.98	-0.82			
27			3.05	-0.65	0.03			
28			9.83	0.4	0.05			
29			2.42	0.05	0.35			
30			9.86	0.62	0.53			
31			4.34	-0.05	0.73			
32			6.93	-0.27	0.91			
33			9.49	0.05	0.41			
34	----------	----------	----------	----------	----------	----------	----------	----------
35							=F35	
36						=-F35	=-G35	

=MAX(MAX(F25:G34,K25:L34),-MIN(F25:G34,K25:L34)) (into F35)

	I	J	K	L	M	N	O
24	iYY	iXY	e1.R	e2.R	iXX	iYY	iXY
25	Formula Set VI	Formula Set VII	Formula Set VIII	Formula Set IX	Formula Set X	Formula Set XI	Formula Set XII
26							
27							
28							
29							
30							
31							
32							
33							
34	----------	----------			----------	----------	----------
35			=G35	=K35			
36			=-K35	=-L35			

Formula Set	Prototype Cell	Formula
Formula Set I	Prototype Cell is A25	=C25*D25
Formula Set II	Prototype Cell is B25	=D25*E25
Formula Set III	Prototype Cell is F25	=D25-**B17**
Formula Set IV	Prototype Cell is G25	=E25-**C17**
Formula Set V	Prototype Cell is H25	=C25*G25^2
Formula Set VI	Prototype Cell is I25	=C25*F25^2
Formula Set VII	Prototype Cell is J25	=-C25*F25*G25
Formula Set VIII	Prototype Cell is K25	=F25***F18** +G25***G18**
Formula Set IX	Prototype Cell is L25	=F25***F19** +G25***G19**
Formula Set X	Prototype Cell is M25	=C25*L25^2
Formula Set XI	Prototype Cell is N25	=C25*K25^2
Formula Set XII	Prototype Cell is O25	=-C25*K25*L25

Model 5.7

MODIFY and verify Model 5.6 to create Model 5.7. Change the contents of cells B4 - C5 to correspond to the figure and then add the remaining additional cell contents.

ENTER data using columns C - E below row 24 (the masses attached to a hypothetical rigid body and the *x*-, and *y*-components of their position vectors). Additional rows may be added below row 24 and above the row of hyphens to accommodate additional masses. Current rows may also be deleted. Formulas in the remainder of the model will automatically adjust.

READ output from the model at cells B4 - C5 (the inertia tensor of the rigid body specified below row 24) and F4 - G5 (the diagonalized inertia tensor). The remainder of the model yields the same information described for Model 5.6.

Model 5.7 (continued)

VALUES

	A	B	C	D	E	F	G	H
1	ADAPTING A COORDINATE SYSTEM IN A BODY TO							
2	DIAGONALIZE THE INERTIA TENSOR							
3								
4	T =	20.78165	-11.5211		Tdiag =	31.17253	5.55E-16	
5		-11.5211	18.39824			5.55E-16	8.007356	
6		(x-comp)	(y-comp)					
7	origin =	0	0		R(v) =	1		
8	v =	-0.70711	0.707107		Th(v) =	135	deg	
9								
10	origin =	0	0		R(T.v) =	31.13388		
11	T.v =	-22.8415	21.15618		Th(T.v) =	137.1936	deg	
12								
13	size =	22.8415	22.8415		lam-1 =	31.17253		
14	size =	-22.8415	-22.8415		lam-2 =	8.007356		
15					a =	12.77429		
16	M =	62.14			N =	17.20229		
17	rcm =	-0.00021	0.018812			(x-comp)	(y-comp)	
18					e1 =	0.742593	-0.66974	
19					e2 =	0.669743	0.742593	
20					Th1 =	-42.0473		
21					Th2 =	47.95273		
22								
23				r		R		
24	mx	my	m	x	y	X	Y	iXX
25	3.1875	0.2805	6.25	0.51	0.55	0.510209	0.531188	1.763502
26	-9.7706	0.8036	9.97	-0.98	-0.82	-0.97979	-0.83881	7.014954
27	-1.9825	-0.0195	3.05	-0.65	0.03	-0.64979	0.011188	0.000382
28	3.932	0.02	9.83	0.4	0.05	0.400209	0.031188	0.009561
29	0.121	0.0175	2.42	0.05	0.35	0.050209	0.331188	0.265438
30	6.1132	0.3286	9.86	0.62	0.53	0.620209	0.511188	2.576544
31	-0.217	-0.0365	4.34	-0.05	0.73	-0.04979	0.711188	2.195119
32	-1.8711	-0.2457	6.93	-0.27	0.91	-0.26979	0.891188	5.503913
33	0.4745	0.0205	9.49	0.05	0.41	0.050209	0.391188	1.452234
34	----------	----------	----------	----------	----------			----------
35						1.279104	1.279104	
36						-1.2791	-1.2791	

VALUES (Contd.)

	I	J	K	L	M	N	O
24	iYY	iXY	e1.R	e2.R	iXX	iYY	iXY
25	1.626959	-1.69386	0.023118	0.736165	3.387121	0.00334	-0.10637
26	9.5711	-8.19395	-0.1658	-1.2791	16.31199	0.274059	-2.11435
27	1.287796	0.022172	-0.49002	-0.42689	0.555805	0.732373	-0.63801
28	1.574446	-0.12269	0.276305	0.291197	0.833543	0.750464	-0.79091
29	0.006101	-0.04024	-0.18453	0.279565	0.189139	0.0824	0.12484
30	3.792742	-3.12605	0.118198	0.794985	6.231534	0.137752	-0.9265
31	0.010759	0.153682	-0.51329	0.494776	1.062445	1.143434	1.102196
32	0.504414	1.666209	-0.79721	0.481099	1.60399	4.404337	2.657915
33	0.023924	-0.1864	-0.22471	0.32412	0.996962	0.479195	0.691187
34	----------	----------			----------	----------	----------
35			1.279104	1.279104			
36			-1.2791	-1.2791			

FORMULAS

	A	B	C	D	E	F	G	H	I
1	DIAGONALIZING THREE DIMENSIONAL SYMMETRIC TENSORS								
2									
3		4	3	2		lam1 =	=G13		
4	M =	SYM	5	1		lam2 =	=I13		
5		SYM	SYM	6		lam3 =	=E13/3+E20*COS(E19+2*PI()/3)		
6									
7							x-comp	y-comp	z-comp
8						e1 =	=G18	=G19	=G20
9						e2 =	=I18	=I19	=I20
10						e3 =			
11						=H8*I9-I8*H9	=I8*G9-G8*I9		=G8*H9-H8*G9
12	CHECK AREA			SCRATCH AREA					
13	e1.e2 =	See Below		T =	See Below	lam1_ =	See Below	lam2_ =	See Below
14	e2.e3 =			S =		a11 =		a21 =	
15	e1.e3 =			D =		a12 =		a22 =	
16	e1^2-1 =					a13 =		a23 =	
17	e2^2-1 =			P =		N1 =		N2 =	
18	e3^2-1 =			Q =		alf11 =		alf21 =	
19	ΔT =			phi =		alf12 =		alf22 =	
20	ΔS =			R =		alf13 =		alf23 =	
21	ΔD =								
22									
23	SUM =		<-SHOULD BE ZERO						

VALUES

	A	B	C	D	E	F	G	H	I
1	DIAGONALIZING THREE DIMENSIONAL SYMMETRIC TENSORS								
2									
3		4	3	2		lam1 =	9		
4	M =	SYM	5	1		lam2 =	4.732051		
5		SYM	SYM	6		lam3 =	1.267949		
6									
7							x-comp	y-comp	z-comp
8						e1 =	0.57735	0.57735	0.57735
9						e2 =	0.211325	0.57735	-0.78868
10						e3 =	-0.78868	0.57735	0.211325
11									
12	CHECK AREA			SCRATCH AREA					
13	e1.e2 =	2.22E-16		T =	15	lam1_ =	9	lam2_ =	4.732051
14	e2.e3 =	2.78E-17		S =	60	a11 =	11	a21 =	2.464102
15	e1.e3 =	4.16E-17		D =	54	a12 =	11	a22 =	6.732051
16	e1^2-1 =	0				a13 =	11	a23 =	-9.19615
17	e2^2-1 =	2.22E-16		P =	5	N1 =	19.05256	N2 =	11.66025
18	e3^2-1 =	0		Q =	2	alf11 =	0.57735	alf21 =	0.211325
19	ΔT =	1.78E-15		phi =	0.463648	alf12 =	0.57735	alf22 =	0.57735
20	ΔS =	1.42E-14		R =	4.472136	alf13 =	0.57735	alf23 =	-0.78868
21	ΔD =	2.13E-14							
22									
23	SUM =	3.78E-14	<-SHOULD BE ZERO						

B13 contains=ABS(+G8*G9+H8*H9+I8*I9)
B14 contains =ABS(+G9*G10+H9*H10+I9*I10)
B15 contains =ABS(+G8*G10+H8*H10+I8*I10)
B16 contains =ABS(+G8^2+H8^2+I8^2-1)
B17 contains =ABS(+G9^2+H9^2+I9^2-1)
B18 contains =ABS(+G10^2+H10^2+I10^2-1)
B19 contains =ABS(+G3+G4+G5-E13)
B20 contains =ABS(+G3*G4+G3*G5+G4*G5-E14)
B21 contains=ABS(+G3*G4*G5-E15)

B23 contains =SUM(B13:B21)

E13 contains =B3+C4+D5
E14 contains =(B3*C4-C3^2)+(B3*D5-D3^2)+(C4*D5-D4^2)
E15 contains =B3*(C4*D5-D4^2)-C3*(C3*D5-D4*D3)+D3*(C3*D4-C4*D3)

E17 contains =E13^2/9-E14/3
E18 contains =E15/2-E14*E13/6+(E13/3)^3
E19 contains =ACOS(E18/SQRT(E17^3))/3
E20 contains =2*SQRT(E17)

G13 contains =E13/3+E20*COS(E19)
G14 contains =D3*(G13-C4)+D4*C3
G15 contains =D3*C3+D4*(G13-B3)
G16 contains =(G13-C4)*(G13-B3)-C3^2
G17 contains =SQRT(G14^2+G15^2+G16^2)
G18 contains =G14/G17
G19 contains =G15/G17
G20 contains =G16/G17

I13 contains =E13/3+E20*COS(E19-2*PI()/3)
I 14 contains =D3*(G4-C4)+D4*C3
I15 contains =D3*C3+D4*(G4-B3)
I16 contains =(G4-C4)*(G4-B3)-C3^2
I17 contains =SQRT(I14^2+I15^2+I16^2)
I18 contains =I14/I17
I19 contains =I15/I17
I20 contains =I16/I17

Model 5.8

BUILD and verify Model 5.8.

ENTER data using cells B3 - D3, C4 - D4, and D5 (the components of a three-dimensional, symmetric matrix).

READ output from the model at cells G3 - G5 (the eigenvalues of the matrix), G8 - I10 (the components of the eigenvectors of the matrix), and B13 - B23 (check-sums, all of which should be near zero for a properly functioning model).

NOTICE that not all of the components of the 3 x 3 matrix need be entered because the matrix must be symmetric. The model fails for highly symmetric objects but can still yield correct data if the symmetry of the entered object is slightly broken as described in the text.

EXERCISE **5.5**

FIXED-AXIS ROTATION WITH VARIABLE ANGULAR VELOCITY

TO GET THE MOST OUT OF THIS EXERCISE, YOU SHOULD ALREADY KNOW

The definitions of the terms:
angular position, velocity, and acceleration
angular momentum
torque
principal axes
principal moments of inertia

How to find the principal axes and principal moments of inertia.
How to solve differential equations using Runge-Kutta methods.

Earlier exercises in this chapter have explored the rotary motion of a rigid body in which the angular velocity **ω** is *constant* in both magnitude and direction. In this exercise, you study rigid bodies rotating with *variable* angular velocity about a fixed axle parallel to a principal axis.

Consider a circular wheel mounted on an axle that extends through the center of the wheel perpendicular to the plane that it defines — automobile wheels, computer disks, flywheels, etc. If the angular velocity of the wheel is constant, the bearings that support the axle produce no torque to maintain the simple motion. If the angular velocity of the wheel is not constant, however, a torque along the axle is mandatory. In this exercise you explore the behavior of the pulley in a block and tackle — Atwood's machine in the scientific literature (Figure 5.25).

From the symmetry of the circular pulley, you know that the axle passes through its center of mass and is parallel to one of its three principal axes. In this case then, the basic vector equation for angular momentum ($\boldsymbol{L} = \mathbf{I} \cdot \boldsymbol{\omega}$) reduces to an equation in magnitudes (Equation 5.100):

$$L = I\omega \tag{5.100}$$

where I is the principal moment of inertia about the axis of the axle and L and ω are the magnitudes of the angular momentum and angular velocity (the vectors of which are parallel to the axle). Similarly, the basic vector equation for torque ($\Gamma = dL/dt$) reduces to

$$\Gamma = \frac{dL}{dt} = \frac{d(I\omega)}{dt} \tag{5.101}$$

where the torque Γ is also parallel to the axle.

Exercise 5.3 (Lesson 4) shows that I is constant in time under these circumstances, and therefore it can be brought outside the derivative (Equation 5.102):

$$\Gamma = I\frac{d\omega}{dt} \tag{5.102}$$

To check this equation refer to Model 5.4; you will see that this component of Γ is always zero when the ω is constant, in accord with equation (5.102).

The angular velocity ω of a wheel rotating like this is simply the time rate of change on the angle θ (Figure 5.26, Equation 5.103):

$$\omega = \frac{d\theta}{dt} \tag{5.103}$$

Therefore the motion of the wheel is described by

$$\Gamma = I\left(\frac{d^2\theta}{dt^2}\right) = I\alpha \tag{5.104}$$

where α is the angular acceleration (Equation 5.105):

$$\alpha = \frac{d\omega}{dt} = \frac{d^2\theta}{dt^2} \tag{5.105}$$

With all these special but important conditions, the equation of motion has exactly the same form as that for one-dimensional translational motion. Rotational motion is not always that simple, as you have already seen, but for many technical applications, these simple equations are quite sufficient.

The following examples will illustrate how to use these equations.

LESSONS

Atwood's machine is a simple device consisting of a pulley of mass M_p and radius R that is suspended from a rigid support (Figure 5.25). A single rope hangs over the pulley and never slips with respect to it. This simple machine provides an excellent device with which to illustrate

Figure 5.25

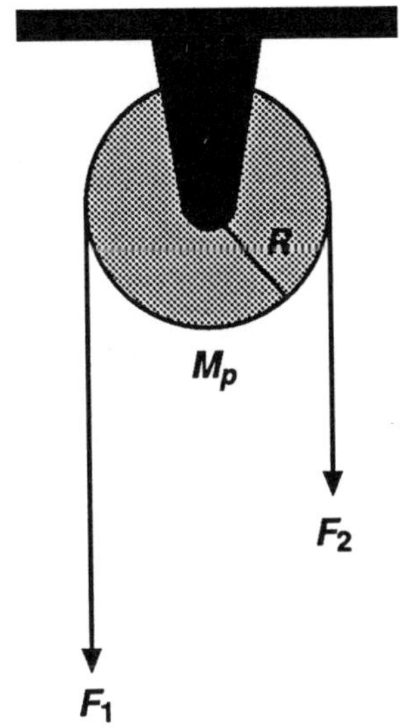

Atwood's machine is a simple device consisting of a massive pulley hanging from a rigid support. A massless rope hangs over the pulley with force F_1 pulling on the left side of the rope and F_2 pulling on the right side. The pulley itself has a mass of M_p uniformly distributed over a circle of radius R. Its moment of inertia about its axle is $M_pR^2/2$.

Figure 5.26

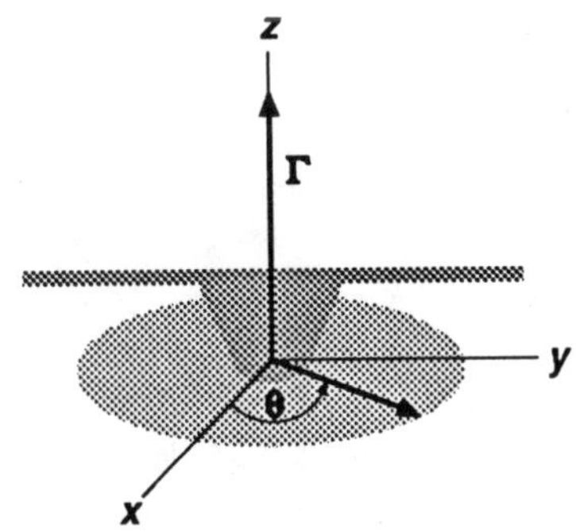

In these exercises, a right-handed coordinate system is used, and positive rotation is in the counterclockwise direction. Notice the direction of the angular momentum vector Γ when these conventions are adopted. θ increases with counter-clockwise rotation and is the angle made by the position vector of some chosen point on the pulley and the x-axis.

Figure 5.27

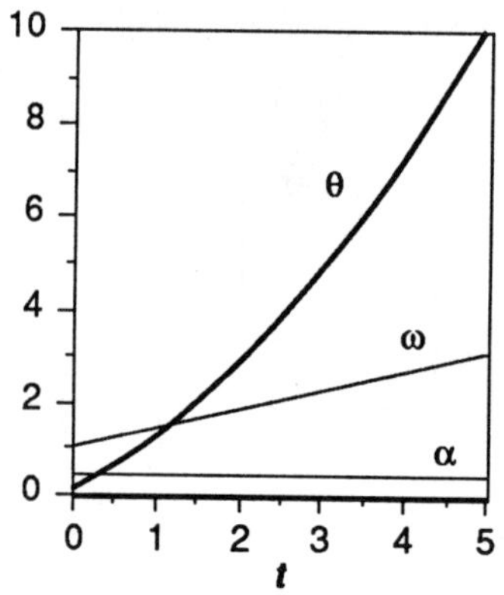

XY-PLOT

Data
Horizontal (*x*) axis: cells D11 - D60 (*t*)
Vertical (*y*) axis: cells C11 - C60 (α), E11 - E60 (θ), and F11 - F60 (ω)

Figure 5.28

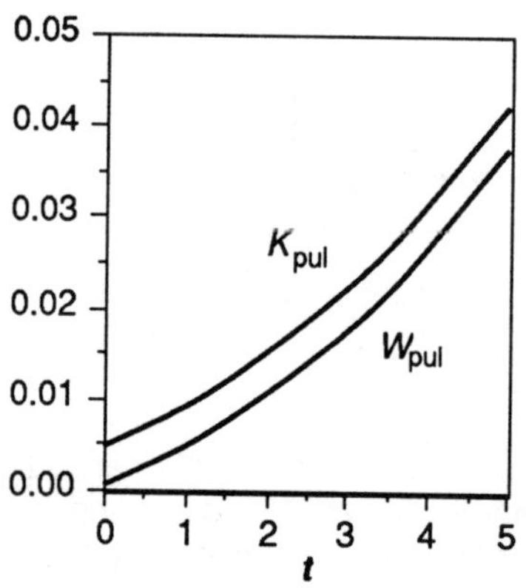

XY-PLOT

Data
Horizontal (*x*) axis: cells D11 - D60 (*t*)
Vertical (*y*) axis: cells A11 - A60 (K_{pul}), and B11 - B60 (W_{pul})

rotational dynamics and in the lessons and problems that follow you will build several versions of the machine of gradually increasing complexity.

As usual, a right-handed coordinate system is always assumed and positive rotation is in the counterclockwise direction (Figure 5.26).

1. A naive Atwood's machine consists of a frictionless pulley in which the forces $\boldsymbol{F}_1$ and $\boldsymbol{F}_2$ at the left and right ends of a massless rope are specified explicitly. From Equation 5.104 and a consideration of the forces acting on the pulley, it follows that

$$\Gamma = I_p\left(\frac{d^2\theta}{dt^2}\right) = F_1R - F_2R \tag{5.106}$$

where $I_p = \dfrac{M_pR^2}{2}$ (the pulley's moment of inertia about its axle).

and where Γ is the component of torque acting on the rotating pulley directed along the pulley's axle (and pointed out of the page in Figure 5.25), F_1 and F_2 are the magnitudes of the forces $\boldsymbol{F}_1$ and $\boldsymbol{F}_2$, R = radius of the pulley, and θ is the angular position of the pulley.

The equation of motion for the pulley of a naive Atwood's machine is therefore

$$\frac{d^2\theta}{dt^2} = \frac{F_1R - F_2R}{I_p} \tag{5.107}$$

The work done on the pulley W_{pul} is given by Equation 5.108 and the kinetic energy K_{pul} of the pulley by Equation 5.109:

$$W_{pul} = (F_1 - F_2)R\theta \tag{5.108}$$

$$K_{pul} = \frac{I_p\omega^2}{2} \tag{5.109}$$

where $\omega = \dfrac{d\theta}{dt}$ (the angular velocity).

a. Build and verify Model 5.9. Also build on-screen graphics that plot θ, ω, α vs t (Figure 5.27) and W_{pul}, K_{pul} vs t (Figure 5.28).

b. Model 5.9 describes the behavior of an object rotating on a fixed axis and subject to constant angular acceleration. What parallels can you draw between the motion of this object and an object subject to constant linear acceleration?

c. Study the dependence of the angular velocity ω at $t = 1$ s as the mass of the pulley and its radius are changed by completing the following two tables and plotting $\omega\,(t = 1\text{ s})$ vs the mass (for fixed radius) and $\omega\ (t = 1\text{ s})$ vs radius (for fixed mass). What is the functional dependence on these parameters? Notice that ω is more sensitive to changes in radius than to mass.

[NOTE: The list of mass values was generated by taking the reciprocals of a list of numbers going from 10 to 0.5 in increments of 0.5. The table of radii were similarly produced.]

M_p	$\omega(t = 1\text{ s})$	R	$\omega(t = 1\text{ s})$
0.100	5.839	0.010	6.000
0.105	5.608	0.011	5.545
0.111	______	0.012	______
0.118	______	0.014	______
0.125	______	0.016	______
0.133	______	0.018	______
0.143	______	0.020	______
0.154	______	0.024	______
0.167	______	0.028	______
0.182	______	0.033	______
0.200	______	0.040	______
0.222	______	0.049	______
0.250	______	0.063	______
0.286	______	0.082	______
0.333	______	0.111	______
0.400	______	0.160	______
0.500	______	0.250	______
0.667	______	0.444	______
1.000	______	1.000	______
2.000	______	4.000	______

d. Explore the relationship between work done and the kinetic energy of the pulley for different initial angular velocities ω_0. Explain your results.

2. Your naive Atwood's machine can be modified to include the frictional torque Γ_{frict} acting in the bearings that support the pulley's axle. Often the magnitude of the friction force F_{frict} can be approximated as a constant independent of the angular velocity, and we will make this assumption (Figure 5.29, Equation 5.110):

$$F_{frict} = \mu(F_1 + F_2 + M_p g) \tag{5.110}$$

where μ is the coefficient of kinetic friction in the bearings. This force is tangential to the axle and so produces a torque parallel to the

Figure 5.29

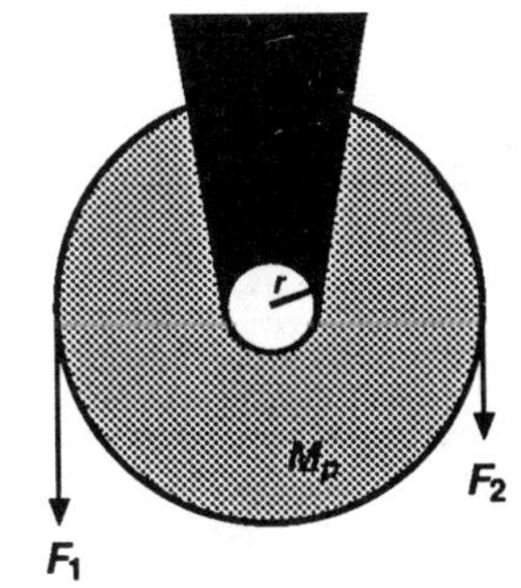

$F_{frict} = \mu\,(F_1 + F_2 + M_p g)$

One way to model the friction force at the interface between a pulley and a pulley shaft is to add up all the forces that cause the pulley to press against the shaft ($F_1 + F_2 + M_p g$) and multiply this value by a coefficient of kinetic friction μ. Such a force is used in Models 5.10 and 5.11.

Figure 5.30

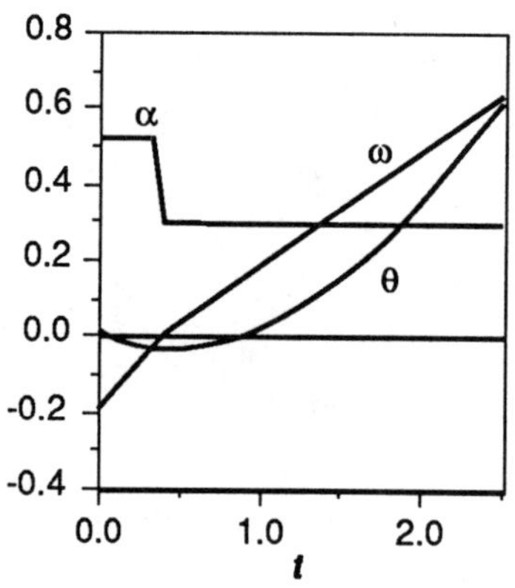

XY-PLOT

Data
Horizontal (*x*) axis: cells F11 - F60 (*t*)
Vertical (*y*) axis: cells E11 - E60 (α), G11 - G60 (θ), and H11 - H60 (ω)

Figure 5.31

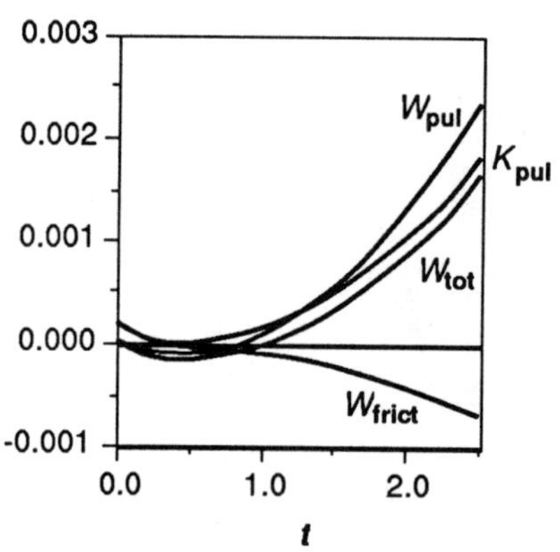

XY-PLOT

Data
Horizontal (*x*) axis: cells F11 - F60 (*t*)
Vertical (*y*) axis: cells A11 - A60 (K_{pul}), B11 - B60 (W_{frict}), C11 - C60 (W_{pul}), and D11 - D60 (W_{net})

axle. The magnitude of this torque is $F_{frict}r$ (where r = radius of the pulley's axle where it goes through the bearings).

As always, friction acts in the direction that opposes the motion, and so the friction torque itself requires a function for ω similar to that in Exercise 4.3:

$$\Gamma_{frict} = -F_{frict} r\, \varepsilon(\omega) \tag{5.111}$$

$$\text{where } \varepsilon(\omega) = \begin{cases} +1 \text{ when} & \omega > 0 \\ 0 \text{ when} & \omega = 0 \\ -1 \text{ when} & \omega < 0 \end{cases}$$

Now the pulley has the equation of motion

$$\frac{d^2\theta}{dt^2} = \frac{F_1R - F_2R - F_{frict}\, r\, \varepsilon(\omega)}{I_p} \tag{5.112}$$

Because static friction effects have not been taken into account, you must use this equation with some care to avoid obvious physical nonsense. A more detailed treatment along the lines of Exercise 4.3 (Problem 1) could be made to describe the physics more accurately.

W_{pul} and K_{pul} remain unchanged, but the work of friction must now be added to the system (Equation 5.113):

$$W_{frict} = \int_0^t \Gamma_{frict}\, \omega\, dt \tag{5.113}$$

You cannot evaluate this integral without knowing the motion because Γ_{frict} is a function of ω.

The net work W_{net} on the pulley is $W_{pul} + W_{frict}$.

a. Build and verify Model 5.10. Also build on-screen graphics that plot θ, ω, α vs t (Figure 5.30) and K_{pul}, W_{pul}, W_{frict}, W_{tot} vs *t* (Figure 5.31).

b. Examine the shapes of the plots of θ, ω, and α vs *t* for various values of ω_0 (cell E6) between –1 and +1. Under what conditions is there a step in the angular acceleration? Why?

How is a stepwise change in angular acceleration reflected in the angular velocity?

c. How are K_{pul} and W_{tot} related? Why?

Explain why the work done by the friction force becomes more negative with time, whereas the net work done by the two forces F_1 and F_2 eventually become positive.

d. Explore the effect of other factors (e.g., $F_1, F_2, M_p, r, R, \mu, \theta_0, \omega_0$) on the behavior of this version of Atwood's machine.

3. Your naive Atwood's machine (with friction) also can be modified by letting the forces F_1 and F_2 be produced by the weights of two masses M_1 and M_2 attached to the ends of a rope that hangs over the pulley (Figure 5.32). We assume for simplicity that the rope is massless and inextensible.

Now the pulley has the equation of motion

$$\frac{d^2\theta}{dt^2} = \frac{T_1R - T_2R - F_{frict}\, r\, \varepsilon(\omega)}{I_p} \qquad (5.114)$$

where T_1 and T_2 are the tensions in the left and right parts of the rope that hangs over the pulley. Notice that these tensions produce downward forces on the rim of the pulley and upward forces on the two hanging masses. The friction force of Equation 5.110 is also modified to be

$$F_{frict} = \mu(T_1 + T_2 + M_p g) \qquad (5.115)$$

Newton's second law of motion provides a relationship between T_1, T_2 and M_1, M_2 (Equation 5.116):

$$M_1 g - T_1 = M_1 a \quad \text{and} \quad M_2 g - T_2 = -M_2 a \qquad (5.116)$$

where a is the downward acceleration of M_1. Because the rope is inextensible, both the velocity and acceleration of M_2 must be equal in magnitude but opposite in direction to those of M_1.

The magnitudes of linear acceleration and angular acceleration are related by $a = R(d^2\theta/dt^2)$, and so it follows that

$$T_1 = M_1 g - M_1 R\left(\frac{d^2\theta}{dt^2}\right) \quad \text{and} \quad T_2 = M_2 g + M_2 R\left(\frac{d^2\theta}{dt^2}\right) \qquad (5.117)$$

Substituting Equations 5.117 into Equation 5.114 and solving for $d^2\theta/dt^2$ yields the equation of motion for the pulley (Equation 5.118):

Figure 5.32

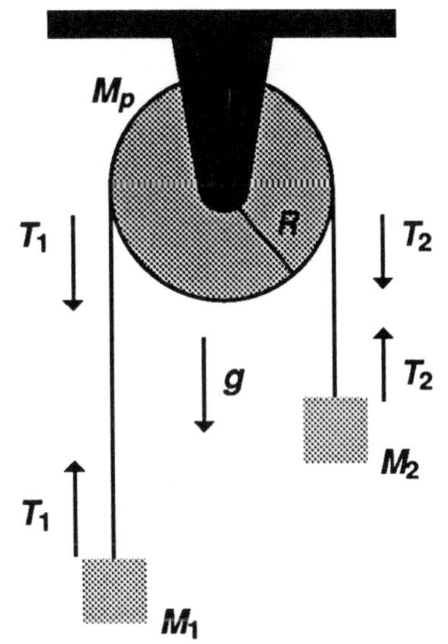

The forces responsible for motion in Atwood's machine can be generated by two masses attached to either end of the massless rope. Model 5.7 represents such a machine.

$$\frac{d^2\theta}{dt^2} = \left[\frac{(M_1 - M_2)R - \mu(M_1 + M_2 + M_p)r\ \varepsilon(\omega)}{I_p + (M_1 + M_2)R^2 - \mu(M_1 - M_2)Rr\ \varepsilon(\omega)}\right]g \quad (5.118)$$

The kinetic energy of the system now has two components — the kinetic energy of the pulley K_{pul} and the kinetic energy of the attached masses. The total kinetic energy of the system K is now

$$K = \frac{I_p\omega^2}{2} + \frac{(M_1 + M_2)(R\omega)^2}{2} \quad (5.119)$$

The work done by gravity to raise and lower the masses is

$$W_{grav} = (M_1 - M_2)gR\theta = K - \int_0^t \Gamma_{frict}\,\omega\, dt \quad (5.120)$$

a. Build and verify Model 5.11. Build appropriate on-screen graphics including the time dependencies of the tensions T_1 and T_2. Temporarily eliminate the friction by setting $\mu = 0$ and examine the plots. Explain their behavior and especially why the tensions are not equal to the weights of the two masses.

   Reset the coefficient of dynamic friction μ to 0.02 and discuss the changes that you observe in the plots.

b. Despite its simplicity, the dynamics of Atwood's machine is now rather complex, but acceleration remains piecewise constant.[1] Be sure you understand why.

   Your model does not include the effects of static friction. What happens as μ increases?

c. Dissect the forces, works, and energies associated with this simple version of Atwood's machine. You might consider the following issues:

   The kinetic energy of the pulley and net work (as defined by the model) are no longer equal when ω_0 equals zero. Why? What changes should you make to the model to clarify the accounting?

   Holding $M_1 - M_2$ constant, how do the masses affect the angular velocity of the pulley?

   What happens when the pulley mass is zero?

   Why are the rope tensions generally larger when the friction force is large?

PROBLEMS

1. Suppose that Atwood's machine in Lesson 3 is built with M_1 and M_2 representing bags of sand and that M_2 leaks sand at a constant rate. In this circumstance, T_2 changes with elapsed time according to

$$T_2 - M_2(t)\, g = \frac{d[M_2(t)v]}{dt} = v\,\frac{dM_2(t)}{dt} + M_2(t)a \qquad (5.121)$$

where v and a are the upward velocity and acceleration of M_2 and

$$M_2(t) = \begin{cases} M_f + \dfrac{(M_e - M_f)t}{T} & \text{for } 0 \le t \le T \\ M_e & \text{for } t > T \end{cases} \qquad (5.122)$$

M_f is the mass of the bag when it is full of sand, M_e is its mass when empty, and T is the time that it takes for a full bag of sand to empty.

Suppose that M_1 is a constant equal to M_f and that M_1 is initially 5 m above the ground.

Develop a formula for the upward acceleration of the leaking bag and then modify Model 5.11 to describe this system.

a. Assume that the sand has an average specific gravity of 1.8, a full bag holds 1 ft^3, and the bag itself weights 1 lb. If the bag has a fairly small leak such that it takes 10 hours to empty (and if the system starts balanced with zero velocity), plot the position, velocity, acceleration, and the tensions in the rope as a function of time.

b. How long will it take for the full bag to hit the ground. How fast will it be moving when it hits?

Compare this velocity with the velocity of the bag had it simply fallen unimpeded from a height of 5 m. Why the large difference?

c. How do your answers to Problems 1a and 1b change if the bag takes 100 hours to empty?

Why is there so little change in your answers?

d. How do your answers to Problems 1a and 1b change if the bag develops a catastrophic leak such that it empties in 0.5 s?

Figure 5.33

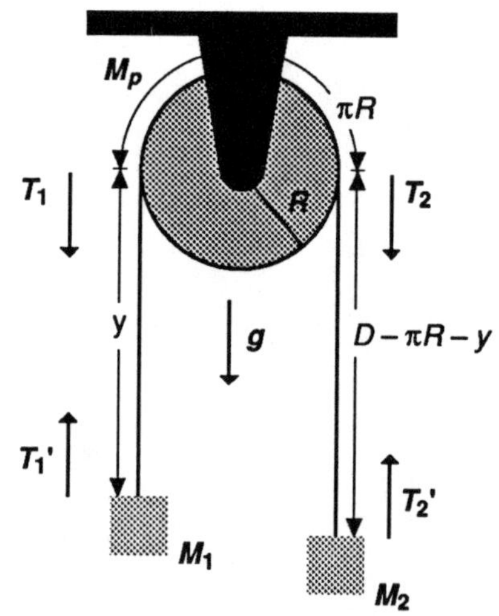

If the rope in Atwood's machine is not massless, the motion can no longer be described by piecewise constant acceleration, and the mathematics becomes somewhat more complex. If the rope is divided into segments, as shown in the figure, then Equations 5.123 - 5.125 can be used to build a description of the motion.

The forces T_1 and T_2 are shown as they act on the pulley. They act in the opposite direction on the mass below the pulley, as indicated in Equations 5.123 - 5.125.

Notice that the rope tension at the rim of the pulley differs from that acting on the mass. The difference provides the force necessary to accelerate the massive rope between these two places. Try writing the equations for T_1' and T_2'.

[NOTE: If you wish to include friction effects in this problem, you will find it essential to include the effect of static friction for any reasonable value of the coefficient of dynamic friction.]

2. Suppose that Atwood's machine in Lesson 3 is built with a rope of length D and uniform mass density λ. The rope hangs over the pulley with the masses attached at either end (Figure 5.33). Ignore friction in this problem.

Build, verify, and explore a model of Atwood's machine with these characteristics. Equations 5.123 - 5.125 give the equations of motion of the system.

$$(M_1 + \lambda y)g - T_1 = (M_1 + \lambda y)a \tag{5.123}$$

$$[M_2 + \lambda(D - \pi R - y)]g - T_2 = -[M_2 + \lambda(D - \pi R - y)]a \tag{5.124}$$

$$\Gamma = [I_p + (\pi R\lambda)R^2]\alpha = (T_1 - T_2)R \tag{5.125}$$

where a is the downward acceleration M_1; y is the length of rope extending from M_1 to where it just touches the pulley; T_1, T_2 are the tensions in the rope where it just leaves the rim of the pulley.

Using the relationship $a = R\alpha$ you can solve Equations 5.123 - 5.125 for the vertical acceleration a.

$$a = \left[\frac{2\lambda y + (M_1 - M_2) - \lambda(D - \pi R)}{\lambda D + M_1 + M_2 + \frac{M_p}{2}}\right]g \tag{5.126}$$

Notice that both the pulley and the rope increase the total mass of the system in motion but that the rope's weight gets transferred from one side of the pulley to the other, producing the dynamical effect.

a. Build and verify a model of the Atwood's machine described above. Build on-screen graphics.

The kinetic energy K for this system is

$$K = \frac{(M_1 + M_2 + \lambda D)\left(\frac{dy}{dt}\right)^2}{2} + \frac{I\omega^2}{2} \tag{5.127}$$

The potential energy U is

$$U = -M_1gy - M_2g(D - \pi R - y) - \lambda yg\left(\frac{y}{2}\right)$$
$$- \lambda g(D - \pi R - y)\left(\frac{D - \pi R - y}{2}\right) \quad (5.128)$$

b. Suppose $M_1 = M_2 = 100$ g and $D = 100$ cm, $R = 2$ cm, and the total mass of the rope is 20 g. Find settings that render the machine motionless.

c. Increase y slightly. What happens and why?

d. Suppose you place a penny on M_1 (a penny has a mass of about 1.9 g). Now what value of y renders the machine motionless?

f. What difference in weight between M_1 and M_2 makes it impossible for the system to remain motionless?

g. Verify that the kinetic and potential energies of the system are accurately given by Equations 5.127 and 5.128.

h. This model is a rather nice example of a real machine. You probably have numerous intuitions about how such a machine should behave. Test them.

3. Another interesting problem in rotational motion involves the behavior of a rigid body swinging with its center of mass a distance R from the pivot (Figure 5.34). This simple system is usually called a physical pendulum. The problem involves both the rotation of the body about its center of mass and the motion of the center of mass itself. Until now, you have only seen problems in which the center of mass is stationary in an inertial system, and so we have only talked about the angular momentum about the center of mass (sometimes called its *spin*). If the center of mass is moving, it also contributes to the angular momentum, and so the total angular momentum is

$$\mathbf{L} = \mathbf{I} \cdot \omega + \mathbf{R} \times \mathbf{P} \quad (5.129)$$

where $\mathbf{L}$ is the *full* angular momentum of the body, $\mathbf{I}$ is its inertia tensor about its center of mass, ω is its angular velocity of rotation about the center of mass, $\mathbf{R}$ is the radius vector from a point in an inertial frame to the center of mass, and $\mathbf{P}$ is the total momentum of the body. If the total torque on the body is Γ, then its equation of motion is

Figure 5.34

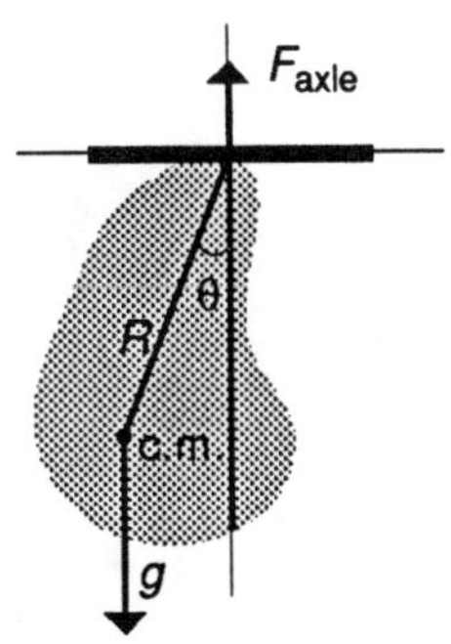

A physical pendulum swings with its center of mass a distance R from its pivot (Problem 3). The supporting axle of the pendulum is parallel to a principal axis and the principal moment of inertia about the center of mass (and associated with this direction is I_{cm}). Try to derive a formula for F_{axle} that you can use after you know the motion.

$$\Gamma = \frac{dL}{dt} \tag{5.130}$$

Now back to the physical pendulum. Assume that it is planar and that it swings on an axle perpendicular to its plane, then the axle is parallel to one of the body's principal moments and so no torque acting perpendicular to the axle is required. However, there is a torque acting parallel to the axle due to the external forces acting on the pendulum.

These forces are gravity (acting downward) and an upward force acting at the axle (ignore friction effects at the axle). All of the downward force can be taken to act at the center of mass (also called the *center of gravity* because of this fact). So if you shrewdly take the origin of the inertial coordinate system at the center of the axle, the unknown force on the axle makes no contribution to the torque, and it is all due to the known gravity force acting at the center of mass. The equation describing the motion of the pendulum is then

$$\Gamma = -RMg \sin\theta = (I_{cm} + MR^2)\left(\frac{d^2\theta}{dt^2}\right) \tag{5.131}$$

where Γ is the torque acting about the origin and directed into the page in Figure 5.34, M is the mass of the pendulum, and I_{cm} is its principal moment of inertia about an axis perpendicular to its plane and through its center of mass. The term involving I_{cm} is due to the body's rotation about its center of mass, and that involving MR^2 is the contribution due to the motion of the center of mass.

Introductory physics textbooks typically use the small-angle approximation for $\sin\theta$ and solve this problem as an example of simple harmonic motion. Your computer, however, allows you to study the motion without this approximation and to observe interesting deviations from simple harmonic oscillation.

Build, verify, and explore the behavior of a physical pendulum using a one-dimensional Runge-Kutta model. Suppose the body is a circular wheel of radius $R = 10$ cm, mass 1 kg, and that its axle is off-center by an amount r. Study what happens as r changes, including $r = 0$, as you give the wheel larger and larger shoves to cause it to move about its axle.

Figure 5.35

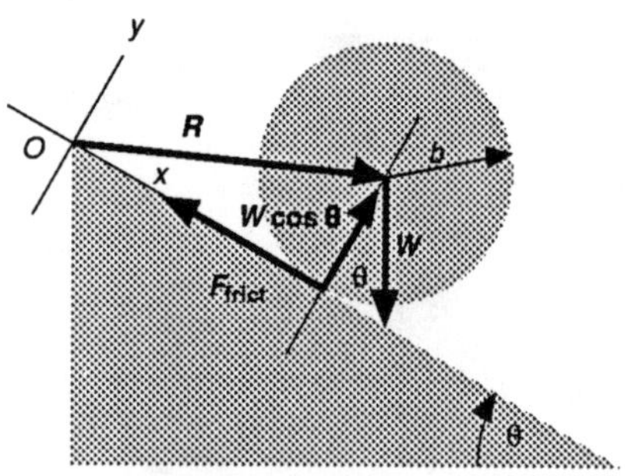

A uniform disk of known mass M and radius b rolls down an incline of angle θ. The z-direction is normal to the plane of the page and directed out of it. The weight W is Mg. The only forces acting on the disk are the three shown (W due to gravity and $W\cos\theta$ and F_{frict} due to the surface of the incline.

4. Consider a uniform disk of radius b and mass M rolling (without slipping) down an plane inclined by an angle θ to the horizontal (Figure 5.35). The distance x is the distance from the origin to the

point of contact between the ramp and the wheel. This distance is just $b\phi$ where ϕ is the angle through which the wheel has turned.

The net torque Γ about O is given by

$$\Gamma_z = bW \sin\theta \tag{5.132}$$

and the angular momentum L_z about A is

$$\begin{aligned} L_z &= -I\omega + (\mathbf{R} \times M\mathbf{V})_z \\ &= -\frac{Mb^2\omega}{2} - Mb^2\omega \\ &= -\frac{3Mb^2\omega}{2} \end{aligned} \tag{5.133}$$

where $\mathbf{V} = d\mathbf{R}/dt$, the velocity of the disk's center of mass.

Notice that the z-axis is parallel to a principal axis of the wheel. Also, be sure to realize that it was not *necessary* to locate the origin on the surface of the plane, but it was certainly a shrewd problem solving strategy because this choice eliminates any contribution to the torque of the unknown force F_{frict}. In general, a clever choice of the location of the origin can simplify the algebraic work in problems such as this.

Because $\Gamma_z = d(L_z)/dt$, it follows that

$$\alpha = \frac{2g \sin\theta}{3b} \tag{5.134}$$

where α is the angular acceleration of the disk.

It is now easy to use the equation of motion for the center of mass itself to calculate the friction force (Equation 5.135):

$$F_x = W \sin\theta - F_{frict} = M\left(\frac{d^2x}{dt^2}\right) = Mb\alpha \tag{5.135}$$

where $d^2x/dt^2 = d(b\omega)/dt = b\alpha$ has been used.

$$F_{frict} = \frac{W \sin\theta}{3} \tag{5.136}$$

The kinetic energy of the wheel has one part from its rotation about the center of mass and a second part from the translational motion of the center of mass. Thus,

$$K = \frac{I\omega^2}{2} + \frac{Mv_x^2}{2} = \frac{3Mb^2\omega^2}{4} \qquad (5.137)$$

The gravitational potential energy of the wheel relative to the the origin O is

$$U = -Mgx \sin \phi \qquad (5.138)$$

Recall that ϕ is the angle through which the wheel has turned to reach the distance x from the origin to the point of contact between the ramp and the wheel.

a. Build, verify, and explore a model of this system calculating F_{frict}, K, U, and $K + U$ (the total mechanical energy). Because this problem includes a friction force, you might not expect the sum of the kinetic energy and the gravitational potential energy to be conserved. Is it? You will probably find this a puzzling result. Read on for insight!

b. The x-component of the velocity of the point on the rim of the wheel that touches the ramp at $t = 0$ is

$$v_x = v_{cm} - b\omega \cos(\omega t) \qquad (5.139)$$

where v_{cm} and ω are functions of time. Add a column to your model that calculates this quantity.

Plot this quantity vs time and notice the velocity of this point every time it is in contact with the ramp. Now you should see that the friction force never pushes against anything *that is moving!* Thus, F_{frict} does no work and energy is conserved, even though the problem contains a friction force.

c. Suppose the disk were slipping as it moved down the ramp so that the relationship $x = b\phi$ (where ϕ is the angle through which the wheel has turned) is violated. Try to formulate equations of motion for this situation and build a model to solve them.

d. Suppose the ramp is not a plane but rather a gently flattening slope. Assume the ramp's shape can be modeled by the equation

$$y = ae^{-x/b} \qquad (5.140)$$

where gravity acts along the negative y-axis. Derive the equation of motion for a wheel rolling without slipping down a ramp shaped like this. Assume that the wheel has a radius of curvature smaller than that of the ramp so that it touches the ramp at just one point. Use the total turn angle of the wheel, ϕ, as your variable.

Build and verify a model capable of exploring energy, final angular and linear velocities, and the friction force necessary to prevent slipping.

5. In the lessons and problems of this exercise, the important effect of static friction was not included. Modify your models to incorporate this effect, using the methods of Exercise 4.3.

MODELS

FORMULAS

	A	B	C	D	E	F	G	H	I	J	K	L	M
1				Atwood's machine (naive)				F1 =	4.03	N			
2				DD(th)= A				F2 =	4	N			
3								Mp =	1.2	kg	Ip =	=(0.5*I3*I4*I4)	
4				Δt =	0.1	sec		R =	0.124	m	A =	=(I1-I2)*I4/L3	
5				th0 =	0	m							
6				w0 =	1	m/s							
7													
8				0					1				
9	K(pul)	W(pul)	a	t	th	w	k(th0)	k(w0)	t	th	w	k(th1)	k(w1)
10	J	J	rad/s^2	s	rad	rad/s			s	rad	rad/s		
11	Form	Form	Form	0	=E5	=E6	Form	Form	Form	Form	Form	Form	Form
12	Set	Set	Set	Form	Form	Form	Set	Set	Set	Set	Set	Set	Set
13	I	II	III	Set IV	Set V	Set VI	VII	VIII	IX	X	XI	XII	XIII

FORMULAS (Contd.)

	N	O	P	Q	R	S	T	U	V	W
8	2					3				
9	t	th	w	k(th2)	k(w2)	t	th	w	k(th3)	k(w3)
10	s	rad	rad/s			s	rad	rad/s		
11	Form	Form	Form	Form	Form	Form	Form	Form	Form	Form
12	Set	Set	Set	Set	Set	Set	Set	Set	Set	Set
13	XIV	XV	XVI	XVII	XVIII	XIX	XX	XXI	XXII	XXIII

VALUES

	A	B	C	D	E	F	G	H	I	J	K	L	M
1				Atwood's machine (naive)				F1 =	4.03	N			
2				DD(th)= A				F2 =	4	N			
3								Mp =	1.2	kg	Ip =	0.0092	kg m^2
4				Δt =	0.1	sec		R =	0.124	m	A =	0.4032	
5				th0 =	0	m							
6				w0 =	1	m/s							
7													
8				0					1				
9	K(pul)	W(pul)	a	t	th	w	k(th0)	k(w0)	t	th	w	k(th1)	k(w1)
10	J	J	rad/s^2	s	rad	rad/s			s	rad	rad/s		
11	0.0046	0	0.4032	0	0	1	0.1	0.0403	0.05	0.05	1.0202	0.102	0.0403
12	0.005	0.0004	0.4032	0.1	0.102	1.0403	0.104	0.0403	0.15	0.154	1.0605	0.106	0.0403
13	0.0054	0.0008	0.4032	0.2	0.2081	1.0806	0.1081	0.0403	0.25	0.2621	1.1008	0.1101	0.0403

VALUES (Contd.)

	N	O	P	Q	R	S	T	U	V	W
8	2					3				
9	t	th	w	k(th2)	k(w2)	t	th	w	k(th3)	k(w3)
10	s	rad	rad/s			s	rad	rad/s		
11	0.05	0.051	1.0202	0.102	0.0403	0.1	0.102	1.0403	0.104	0.0403
12	0.15	0.155	1.0605	0.106	0.0403	0.2	0.2081	1.0806	0.1081	0.0403
13	0.25	0.2631	1.1008	0.1101	0.0403	0.3	0.3181	1.121	0.1121	0.0403

Model 5.9

BUILD and verify Model 5.9.

ENTER data using cells E4 - E6 (Δt, θ_0, ω_0) and I1 - I4 (forces, pulley mass, pulley radius).

READ output from the model from columns A - F (K_{pul}, W_{pul}, α, t, θ, ω).

NOTICE that the model uses a one-dimensional Runge-Kutta method to solve the equation, but that, in this instance, Simpson's rule might also be used. Cell L4 (A) can contain the entire expression for acceleration because acceleration is not a function of position or velocity.

Model 5.9 (continued)

Formula Set I	Prototype Cell is A11	=**L3** *F11*F11/2
Formula Set II	Prototype Cell is B11	=(**I1** -**I2**)***I4** *E11
Formula Set III	Prototype Cell is C11	=**L4**
Formula Set IV	Prototype Cell is D12	=D11+**E4**
Formula Set V	Prototype Cell is E12	=E11+(G11+2*(L11+Q11)+V11)/6
Formula Set VI	Prototype Cell is F12	=F11+(H11+2*(M11+R11)+W11)/6
Formula Set VII	Prototype Cell is G11	=**E4** *F11
Formula Set VIII	Prototype Cell is H11	=**E4** ***L4**
Formula Set IX	Prototype Cell is I11	=D11+**E4** /2
Formula Set X	Prototype Cell is J11	=E11+G11/2
Formula Set XI	Prototype Cell is K11	=F11+H11/2
Formula Set XII	Prototype Cell is L11	=**E4** *K11
Formula Set XIII	Prototype Cell is M11	=**E4** ***L4**
Formula Set XIV	Prototype Cell is N11	=D11+**E4** /2
Formula Set XV	Prototype Cell is O11	=E11+L11/2
Formula Set XVI	Prototype Cell is P11	=F11+M11/2
Formula Set XVII	Prototype Cell is Q11	=**E4** *P11
Formula Set XVIII	Prototype Cell is R11	=**E4** ***L4**
Formula Set XIX	Prototype Cell is S11	=D11+**E4**
Formula Set XX	Prototype Cell is T11	=E11+Q11
Formula Set XXI	Prototype Cell is U11	=F11+R11
Formula Set XXII	Prototype Cell is V11	=**E4** *U11
Formula Set XXIII	Prototype Cell is W11	=**E4** ***L4**

Model 5.10

BUILD and verify Model 5.10.

ENTER data using cells E4 - E6 (Δt, θ_0, ω_0) and I1 - I7 (acceleration of gravity, forces, pulley mass, pulley radius, shaft radius, coefficient of kinetic friction).

READ output from the model from columns A - H (K_{pul}, W_{frict}, W_{pul}, W_{net}, α, t, θ, ω).

FORMULAS

	A	B	C	D	E	F	G	H	I	J	K	L	M
1	Atwood's machine (with frict)							g:	9.8	m/s^2			
2	DD(th)=A-Afr*epsilon							F1 =	4.03	N			
3								F2 =	4	N	Ip =	=I4*I5*I5/2	
4				Δt =	0.05	sec		Mp =	1.2	kg	Ffr =	=I7*(I2+I3+I4*I1)	
5				th0 =	0	rad		R =	0.124	m	A =	=(I2-I3)*I5/L3	
6				w0 =	-0.2	rad/s		r =	0.01	m	Afr =	=L4*I6/L3	
7								μ =	0.005				
8	K	Work	Work	Work					0				
9	(pul)	(fr)	(pul)	(net)	a	t	th	w	k(th0)	k(w0)	t	th	w
10	(J)	(J)	(J)	(J)	(r/s^2)	(s)	(rad)	(rad/s)					
11		0				0	=E5	=E6					
12	Form	Form	Form	Form	Form	Form	Form	Form	Form	Form	Form	Form	Form
13	Set	Set	Set	Set	Set	Set	Set	Set	Set	Set	Set	Set	Set
14	I	II	III	IV	V	VI	VII	VIII	IX	X	XI	XII	XIII

FORMULAS (Contd.)

	N	O	P	Q	R	S	T	U	V	W	X	Y
8	1					2					3	
9	kt	kw	t	th	w	kt	kw	t	th	w	kt	kw
10												
11	Form	Form	Form	Form	Form	Form	Form	Form	Form	Form	Form	Form
12	Set	Set	Set	Set	Set	Set	Set	Set	Set	Set	Set	Set
13	XIV	XV	XVI	XVII	XVIII	XIX	XX	XXI	XXII	XXIII	XXIV	XXV
14												

VALUES

	A	B	C	D	E	F	G	H	I	J	K	L	M
1	Atwood's machine (with frict)							g:	9.8	m/s^2			
2	DD(th)=A-Afr*epsilon							F1 =	4.03	N			
3								F2 =	4	N	Ip =	0.0092	kg m^2
4				Δt =	0.05	sec		Mp =	1.2	kg	Ffr =	0.099	N
5				th0 =	0	rad		R =	0.124	m	A =	0.4032	rad/s^2
6				w0 =	-0.2	rad/s		r =	0.01	m	Afr =	0.1073	rad/s^2
7								μ =	0.005				
8	K	Work	Work	Work					0				
9	(pul)	(fr)	(pul)	(net)	a	t	th	w	k(th0)	k(w0)	t	th	w
10	(J)	(J)	(J)	(J)	(r/s^2)	(s)	(rad)	(rad/s)					
11	0.0002	0	0	0	0.5105	0	0	-0.2	-0.01	0.0255	0.025	-0.005	-0.187
12	0.0001	-9E-06	-3E-05	-4E-05	0.5105	0.05	-0.009	-0.174	-0.009	0.0255	0.075	-0.014	-0.162
13	0.0001	-2E-05	-6E-05	-8E-05	0.5105	0.1	-0.017	-0.149	-0.007	0.0255	0.125	-0.021	-0.136
14	7E-05	-2E-05	-9E-05	-1E-04	0.5105	0.15	-0.024	-0.123	-0.006	0.0255	0.175	-0.027	-0.111

VALUES (Contd.)

	N	O	P	Q	R	S	T	U	V	W	X	Y
8	1					2					3	
9	kt	kw	t	th	w	kt	kw	t	th	w	kt	kw
10												
11	-0.009	0.0255	0.025	-0.005	-0.187	-0.009	0.0255	0.05	-0.009	-0.174	-0.009	0.0255
12	-0.008	0.0255	0.075	-0.013	-0.162	-0.008	0.0255	0.1	-0.017	-0.149	-0.007	0.0255
13	-0.007	0.0255	0.125	-0.021	-0.136	-0.007	0.0255	0.15	-0.024	-0.123	-0.006	0.0255
14	-0.006	0.0255	0.175	-0.027	-0.111	-0.006	0.0255	0.2	-0.03	-0.098	-0.005	0.0255

Model 5.10 (continued)

Formula Set I	Prototype Cell is A11	=**L3** *H11^2/2
Formula Set II	Prototype Cell is B12	=B11 -**L4** ***I6** *H12*IF(H12>0,1,IF(H12=0,0,-1))* **E4**
Formula Set III	Prototype Cell is C11	=(**I2-I3**)***I5** *G11
Formula Set IV	Prototype Cell is D11	=B11+C11
Formula Set V	Prototype Cell is E11	=J11/**E4**
Formula Set VI	Prototype Cell is F12	=F11+**E4**
Formula Set VII	Prototype Cell is G12	=G11+(I11+2*(N11+S11)+X11)/6
Formula Set VIII	Prototype Cell is H12	=H11+(J11+2*(O11+T11)+Y11)/6
Formula Set IX	Prototype Cell is I11	=**E4** *H11
Formula Set X	Prototype Cell is J11	=**E4** *(**L5-L6** *IF(H11>0,1,IF(H11=0,0,-1)))
Formula Set XI	Prototype Cell is K11	=F11+**E4** /2
Formula Set XII	Prototype Cell is L11	=G11+I11/2
Formula Set XIII	Prototype Cell is M11	=H11+J11/2
Formula Set XIV	Prototype Cell is N11	=**E4** *M11
Formula Set XV	Prototype Cell is O11	=**E4** *(**L5-L6** *IF(M11>0,1,IF(M11=0,0,-1)))
Formula Set XVI	Prototype Cell is P11	=F11+**E4** /2
Formula Set XVII	Prototype Cell is Q11	=G11+N11/2
Formula Set XVIII	Prototype Cell is R11	=H11+O11/2
Formula Set XIX	Prototype Cell is S11	=**E4** *R11
Formula Set XX	Prototype Cell is T11	=**E4** *(**L5-L6** *IF(R11>0,1,IF(R11=0,0,-1)))
Formula Set XXI	Prototype Cell is U11	=F11+**E4**
Formula Set XXII	Prototype Cell is V11	=G11+S11
Formula Set XXIII	Prototype Cell is W11	=H11+T11
Formula Set XXIV	Prototype Cell is X11	=**E4** *W11
Formula Set XXV	Prototype Cell is Y11	=**E4** *(**L5-L6** *IF(W11>0,1,IF(W11=0,0,-1)))

FORMULAS

	A	B	C	D	E	F	G	H	I	J	K	L	M	N
1	Atwood's machine (weights & frict)							g =	9.8	m/s^2				
2	DD(th)=Num(omega)/Denom(omega)							M1 =	1.015	kg	M1g =	=I2*I1	N	
3								M2 =	1	kg	M2g =	=I3*I1	N	
4	Denoma =	=(I4/2+I2+I3)*I5*I5		Δt =	0.05	sec		Mp =	1.2	kg				
5	Denomb =	=(I2-I3)*I5*I6*I7		th0 =	0	rad		Rpul =	0.124	m				
6	Numa =	=(I2-I3)*I1*I5		w0 =	-0.2	rad/s		r =	0.01	m				
7	Numb =	=(I2+I3+I4)*I1*I6*I7						μ =	0.02					
8											0			
9	K(pul)	W(fr)	W(grav)	W(net)	T1	T2	a	t	th	w	kth	kw	t	th
10	(J)	(J)	(J)	(J)	(N)	(N)	(r/s^2)	(s)	(rad)	(rad/s)				
11		0						0	=E5	=E6				
12	Form	Form	Form	Form	Form	Form	Form	Form	Form	Form	Form	Form	Form	Form
13	Set	Set	Set	Set	Set	Set	Set	Set	Set	Set	Set	Set	Set	Set
14	I	II	III	IV	V	VI	VII	VIII	IX	X	XI	XII	XIII	XIV

FORMULAS (Contd.)

	O	P	Q	R	S	T	U	V	W	X	Y	Z	AA
8		1					2					3	
9	w	kt	kw	t	th	w	kt	kw	t	th	w	kt	kw
10													
11	Form	Form	Form	Form	Form	Form	Form	Form	Form	Form	Form	Form	Form
12	Set	Set	Set	Set	Set	Set	Set	Set	Set	Set	Set	Set	Set
13	XV	XVI	XVII	XVIII	XIX	XX	XXI	XXII	XXIII	XXIV	XXV	XXVI	XXVII
14													

VALUES

	A	B	C	D	E	F	G	H	I	J	K	L	M	N
1	Atwood's machine (weights & frict)							g =	9.8	m/s^2				
2	DD(th)=Num(omega)/Denom(omega)							M1 =	1.015	kg	M1g =	9.947	N	
3								M2 =	1	kg	M2g =	9.8	N	
4	Denoma =	0.0402	kg m^2	Δt =	0.05	sec		Mp =	1.2	kg				
5	Denomb =	4E-07	kg m^2	th0 =	0	rad		Rpul =	0.124	m				
6	Numa =	0.0182	N	w0 =	-0.2	rad/s		r =	0.01	m				
7	Numb =	0.0063	N					μ =	0.02					
8											0			
9	K(pul)	W(fr)	W(grav)	W(net)	T1	T2	a	t	th	w	kth	kw	t	th
10	(J)	(J)	(J)	(J)	(N)	(N)	(r/s^2)	(s)	(rad)	(rad/s)				
11	0.00080416	0	0	0	9.8702	9.8756	0.6101	0	0	-0.2	-0.01	0.0305	0.025	-0.005
12	0.00057758	-5E-05	-2E-04	-2E-04	9.8702	9.8756	0.6101	0.05	-0.009	-0.169	-0.008	0.0305	0.075	-0.013
13	0.0003884	-1E-04	-3E-04	-4E-04	9.8702	9.8756	0.6101	0.1	-0.017	-0.139	-0.007	0.0305	0.125	-0.02
14	0.00023664	-1E-04	-4E-04	-6E-04	9.8702	9.8756	0.6101	0.15	-0.023	-0.108	-0.005	0.0305	0.175	-0.026

VALUES (Contd.)

	O	P	Q	R	S	T	U	V	W	X	Y	Z	AA
8		1					2					3	
9	w	kt	kw	t	th	w	kt	kw	t	th	w	kt	kw
10													
11	-0.185	-0.009	0.0305	0.025	-0.005	-0.185	-0.009	0.0305	0.05	-0.009	-0.169	-0.008	0.0305
12	-0.154	-0.008	0.0305	0.075	-0.013	-0.154	-0.008	0.0305	0.1	-0.017	-0.139	-0.007	0.0305
13	-0.124	-0.006	0.0305	0.125	-0.02	-0.124	-0.006	0.0305	0.15	-0.023	-0.108	-0.005	0.0305
14	-0.093	-0.005	0.0305	0.175	-0.025	-0.093	-0.005	0.0305	0.2	-0.028	-0.078	-0.004	0.0305

Model 5.11

BUILD and verify Model 5.11.

ENTER data using cells E4 - E6 (Δt, θ_0, ω_0) and I1 - I7 (acceleration of gravity, masses attached to rope, pulley mass, pulley radius, shaft radius, coefficient of kinetic friction).

READ output from the model from columns A - H (K_{pul}, W_{frict}, W_{pul}, W_{net}, α, t, θ, ω).

Model 5.11 (continued)

Formula Set I Prototype Cell is A11
=**B4** *J11^2/2

Formula Set II Prototype Cell is B12
=B11 - (E12+F12+ **I4** ***I1**)***I6** ***I7** *IF(J12>0,J12,IF(J12=0,0, -J12))***E4**

Formula Set III Prototype Cell is C11
=(**L2-L3**)***I5** *I11

Formula Set IV Prototype Cell is D11
=B11+C11

Formula Set V Prototype Cell is E11
=**I2** *(**I1-I5** *G11)

Formula Set VI Prototype Cell is F11
=**I3** *(**I1** +**I5** *G11)

Formula Set VII Prototype Cell is G11
=L11/ **E4**

Formula Set VIII Prototype Cell is H12
=H11+**E4**

Formula Set IX Prototype Cell is I12
=I11+(K11+2*(P11+U11)+Z11)/6

Formula Set X Prototype Cell is J12
=J11+(L11+2*(Q11+V11)+AA11)/6

Formula Set XI Prototype Cell is K11
=**E4** *J11

Formula Set XII Prototype Cell is L11
=**E4** *((**B6-B7** *IF(J11>0,1,IF(J11=0,0,-1)))/(**B4** - **B5** *IF(J11>0,1,IF(J11=0,0,-1))))

Formula Set XIII Prototype Cell is M11
=H11+ **E4** /2

Formula Set XIV Prototype Cell is N11
=I11+K11/2

Formula Set XV Prototype Cell is O11
=J11+L11/2

Formula Set XVI Prototype Cell is P11
=**E4** *O11

Formula Set XVII Prototype Cell is Q11
=**E4** *((**B6-B7** *IF(O11>0,1,IF(O11=0,0,-1)))/(**B4** - **B5** *IF(O11>0,1,IF(O11=0,0,-1))))

Formula Set XVIII Prototype Cell is R11
=H11+ **E4** /2

Formula Set XIX Prototype Cell is S11
=I11+P11/2

Formula Set XX Prototype Cell is T11
=J11+Q11/2

Formula Set XXI Prototype Cell is U11
=**E4** *T11

Formula Set XXII Prototype Cell is V11
=**E4** *((**B6-B7** *IF(T11>0,1,IF(T11=0,0,-1)))/(**B4** - **B5** *IF(T11>0,1,IF(T11=0,0,-1))))

Formula Set XXIII Prototype Cell is W11
=H11+ **E4**

Formula Set XXIV Prototype Cell is X11
=I11+U11

Formula Set XXV Prototype Cell is Y11
=J11+V11

Formula Set XXVI Prototype Cell is Z11
=**E4** *Y11

Formula Set XXVII Prototype Cell is AA11
=**E4** *((**B6-B7** *IF(Y11>0,1,IF(Y11=0,0,-1)))/(**B4** - **B5** *IF(Y11>0,1,IF(Y11=0,0,-1))))

GENERAL ROTATIONAL MOTION

Earlier exercises in this chapter explored the motions of rigid bodies rotating about an axle fixed in an inertial reference frame. This last exercise removes this constraint and asks you to explore the general rotational motion of a rigid body — the dynamics of a gyrocompass, for example, or the motion of a tennis racket thrown up into the air.

A tennis racket thrown into the air exhibits two types of motion — the translational motion of its center of mass and the rotational motion about that center of mass.

- The translational motion of an object's center of mass is governed by the equation

  $$\boldsymbol{F} = M\frac{d^2\boldsymbol{R}_{cm}}{dt^2} \tag{5.141}$$

  or, more generally

  $$\boldsymbol{F} = \frac{d\boldsymbol{P}}{dt} \tag{5.142}$$

  where $\boldsymbol{F}$ is the net external force acting on the body, M is the body's mass, $\boldsymbol{R}_{cm}$ is the position of the center of mass, and $\boldsymbol{P} = d(M\boldsymbol{R}_{cm})/dt$ is the linear momentum of the body.

TO GET THE MOST OUT OF THIS EXERCISE, YOU SHOULD ALREADY KNOW

The definitions of the terms:
angular velocity and its relationship to linear velocity
angular momentum
torque
principal axes
principal moments of inertia

How to find the principal axes and principal moments of inertia
How to solve differential equations using Runge-Kutta methods

Figure 5.36

When you toss a tennis racket up into the air, its center of mass moves in the usual parabolic orbit of a ball or rock or other "point" particle. It also, however, tumbles and turns in its flight. To see this tumbling and turning motion, you can watch the time dependence of the three unit vectors e_1, e_2, and e_3 as seen from the inertial frame (i, j, k). These three unit vectors are fixed in the body and lie along the principal axes of the racket such that e_1 is parallel to the symmetry (long) axis of the racket, e_3 is perpendicular to the plane formed by the racket's face, and e_2 is perpendicular to the other two and directed such that the system is right-handed.

The angular velocity ω gives the instantaneous axis of rotation of the racket so that for any point r in the body at time t, its position at $t + dt$ is $r + \omega \times r + V_{cm} dt$. The components of ω can be given as either ω_1, ω_2, ω_3 (the components in body coordinates) or as ω_x, ω_y, ω_z (the components in the inertial frame). Euler's equations are expressed in terms of the components of ω in body coordinates.

The equations of motion of such a body are usually written in a coordinate system that has its origin at the center of mass and its axes along the principal axes of the body. These equations are relatively simple because the inertia tensor is both diagonal and constant in time in this system.

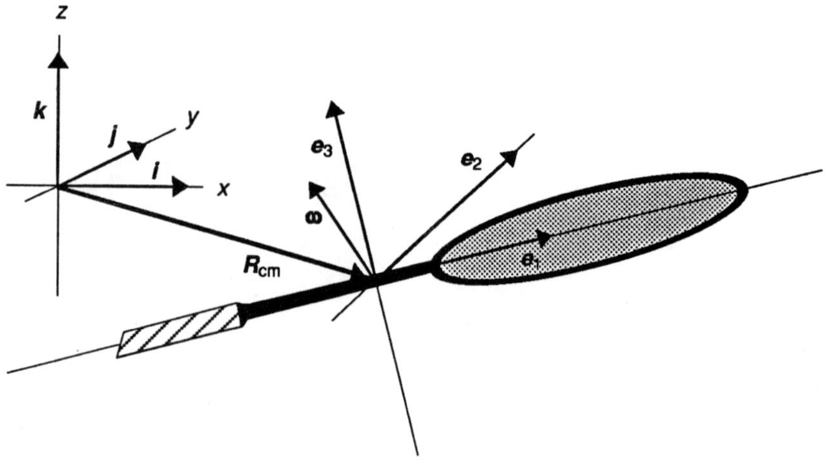

- The rotational motion about an object's center of mass is governed by the equation

$$\frac{dL}{dt} = \Gamma \qquad (5.143)$$

where $L = \mathbf{I} \cdot \omega$ is the angular momentum about the center of mass and Γ is the torque about the center of mass due to external forces. $\mathbf{I}$ is the inertia tensor of the rigid body, and ω is its angular velocity. Equation 5.143 contains all the dynamics necessary to fully describe this aspect of the motion.

This exercise is about Equation 5.143 in its most general form of expression.

If necessary, return to Exercise 5.3 and recall that the components of the inertia tensor $\mathbf{I}$ depend on the coordinate system in which the problem is posed. Furthermore, if you choose an inertial frame, these components are complicated functions of time because the coordinates of each element of mass changes as the body tumbles and turns. Until now you have used only inertial frames of reference to set up and explore equations of motion. This is about to change — the equations of motion for a freely rotating body are greatly simplified when referred to a coordinate system fixed in the rigid body and rotating with it (Figure 5.36). Of course, in the end, you want to describe the body's motion in an inertial frame. For convenience,

- The inertial reference frame will be called "the laboratory system."

- The body-fixed frame will be called "the body system." The body system is a noninertial reference frame.

One reason that the body system yields greatly simplified equations of motion is that, within it, the components of **I** are invariant with time. Also, if the body system has its origin at the body's center of mass and its coordinate axes along the principal axes of the body, the inertia tensor is diagonal and the expression for angular momentum is particularly simple. If $\boldsymbol{e}_1$, $\boldsymbol{e}_2$, and $\boldsymbol{e}_3$ are three unit vectors along the body's principal axes, and if I_1, I_2, and I_3 are the corresponding eigenvalues, then the angular momentum is

$$\boldsymbol{L} = I_1\omega_1\boldsymbol{e}_1 + I_2\omega_2\boldsymbol{e}_2 + I_3\omega_3\boldsymbol{e}_3 \tag{5.144}$$

You should review these matters in Exercise 5.4, if necessary.

Equation 5.144 is more conveniently written as

$$\boldsymbol{L} = \sum_{i=1}^{3} I_i\omega_i\boldsymbol{e}_i \tag{5.145}$$

To find dL/dt (and hence Γ) you must realize that while I_1, I_2, and I_3 are time-independent, ω_1, ω_2, ω_3, $\boldsymbol{e}_1$, $\boldsymbol{e}_2$, $\boldsymbol{e}_3$ are not. Therefore, the time derivative of the angular momentum is

$$\boldsymbol{\Gamma} = \frac{d\boldsymbol{L}}{dt} = \sum_{i=1}^{3} I_i\left(\frac{d\omega_i}{dt}\right)\boldsymbol{e}_i + \sum_{i=1}^{3} I_i\omega_i\left(\frac{d\boldsymbol{e}_i}{dt}\right) \tag{5.146}$$

A final transformation of Equation 5.146 depends on the fact that a vector of constant length (e.g., the three unit vectors $\boldsymbol{e}_i$) has a time derivative that can be neatly expressed as a cross product with the angular velocity. Thus,

$$\frac{d\boldsymbol{e}_i}{dt} = \boldsymbol{\omega} \times \boldsymbol{e}_i \tag{5.147}$$

With this, Equation 5.146 becomes

$$\boldsymbol{\Gamma} = \frac{d\boldsymbol{L}}{dt} = \left[\sum_{i=1}^{3} I_i\left(\frac{d\omega_i}{dt}\right)\boldsymbol{e}_i\right] + \sum_{i=1}^{3} I_i\omega_i\left(\frac{d\boldsymbol{e}_i}{dt}\right) \tag{5.148}$$

This is the generalization of Equation 5.53 in Exercise 5.3.

You express Equation 5.148 in components along the three unit vectors $\boldsymbol{e}_i$ to obtain Euler's equations, named after Leonhard Euler who first derived them in the 1700s (Equations 5.149a - 5.149c):

$$I_1\frac{d\omega_1}{dt} + \omega_2\omega_3(I_3 - I_2) = \Gamma_1 \tag{5.149a}$$

$$I_2\frac{d\omega_2}{dt} + \omega_3\omega_1(I_1 - I_3) = \Gamma_2 \tag{5.149b}$$

$$I_3\frac{d\omega_3}{dt} + \omega_1\omega_2(I_2 - I_1) = \Gamma_3 \tag{5.149c}$$

When the torque components Γ_1, Γ_2, and Γ_3 can be expressed without knowing the orientation of the body in the lab, these equations can be immediately solved using the Runge-Kutta method to find the three components ω_i as functions of time. You might get slightly dizzy trying to interpret these functions, however, because they describe how the body is turning and tumbling in a coordinate system fixed in the very body that is turning and tumbling — and they are functions of time at that! Without more work you still have little insight into how the body is moving. In other words, just as in any mechanics problem, you must further integrate the velocities to get the positions.

For rigid bodies, Equation 5.147 is exactly what you need. As usual, we use $\boldsymbol{i}, \boldsymbol{j}$, and $\boldsymbol{k}$ as the unit vectors along the axes of the laboratory system and so you can write

$$e_i = e_{ix}\,\boldsymbol{i} + e_{iy}\boldsymbol{j} + e_{iz}\boldsymbol{k}\,, \quad i = 1, 2, 3 \tag{5.150}$$

Knowing these nine components of $\boldsymbol{e}_1$, $\boldsymbol{e}_2$, and $\boldsymbol{e}_3$, along the laboratory axes as functions of time, fully solves the problem. With these three unit vectors (and the position of the center of mass), you can calculate the time history of any point in the body. Putting Equations 5.147 and 5.150 together, you get

$$\frac{de_{1x}}{dt} = \omega_3 e_{2x} - \omega_2 e_{3x} \tag{5.151a}$$

$$\frac{de_{2x}}{dt} = \omega_1 e_{3x} - \omega_3 e_{1x} \tag{5.151b}$$

$$\frac{de_{3x}}{dt} = \omega_2 e_{1x} - \omega_1 e_{2x} \tag{5.151c}$$

and the three similar equations for the y-components, and three more for the z-components.

It appears that you must simultaneously solve twelve first-order differential equations to fully understand the motion, but this is really unnecessary. For example, if you know $\boldsymbol{e}_1$ and $\boldsymbol{e}_2$, then, remembering that the three principal axes are mutually orthogonal, you can directly calculate

$$e_3 = e_1 \times e_2 \tag{5.152}$$

Three of the differential equations are thereby eliminated. In principle, you can eliminate three others of them by using the facts that both e_1 and e_2 are of unit length and orthogonal to each other, but it is rather difficult to use these equations in numeric work because of sign ambiguities and the possibility of dividing by zero. For these reasons, only Equation 5.152 is used and nine equations are solved using the Runge-Kutta method.

In the lessons that follow, you explore only the case in which a body is torque-free (other cases are saved for the problems). In the torque-free case Euler's equations do not depend on the lab components of e_i, and so they can be solved all by themselves (but the results are hard to interpret unless you are living on the moving body). On the other hand, if there is a torque in the problem and you directly know its components in the lab, then Euler's equations depend on the directions of the principal axes. In particular, suppose that Γ_x, Γ_y, and Γ_z are known. Then, taking the dot product of

$$\Gamma = \Gamma_x i + \Gamma_y j + \Gamma_z k \tag{5.153}$$

with e_i, you get

$$\Gamma_i = \Gamma_x e_{ix} + \Gamma_y e_{iy} + \Gamma_z e_{iz} \quad i = 1, 2, 3 \tag{5.154}$$

In this case, you see that the nine equations of motion are completely coupled and Euler's equations cannot be solved independently.

LESSONS

1. Build and verify Model 5.12. This model uses the Runge-Kutta method to solve for the torque-free rotary motion of a rigid body. The model is quite large and would be even larger if the "condensed" format of the Runge-Kutta method were not used. Depending on your computer, it may take a very long time to calculate.

   a. Extend your model to include at least 6 s of time. If you run into memory restrictions, use fewer rows and a larger Δt. If you have the memory and processor power to include more than 61 rows, you may wish to decrease Δt and obtain smoother solutions.

   b. Prepare plots showing ω_1, ω_2, and ω_3 vs t (Figure 5.37); ω_x, ω_y, and ω_z vs t (Figure 5.38); e_{1x}, e_{1y}, and e_{1z} vs t (Figure 5.39); and similarly for e_2 and e_3. You might also enjoy making orbit plots

Figure 5.37

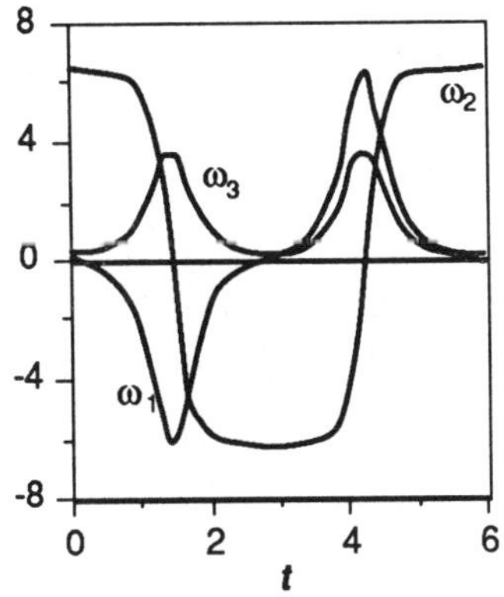

XY–PLOT

Data
Horizontal (*x*) axis: cells O21 - O80 (*t*)
Vertical (*y*) axis: cells P21 - P80 (ω_1), Q21 - Q80 (ω_2), and R21 - R80 (ω_3)

Note
If you plot with symbols and the lines you will also get a sense of the time development from the point density.

Figure 5.38

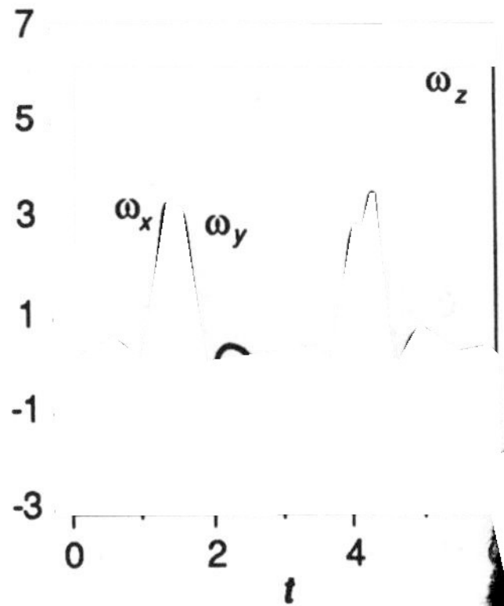

XY–PLOT

Data
Horizontal (*x*) axis: cells O21 - C
Vertical (*y*) axis: cells H21 - HE
I21 - I80 (ω_y), and J21 - J80 (

Note
If you plot with symbols
you will also get a sens
development from th

Figure 5.39

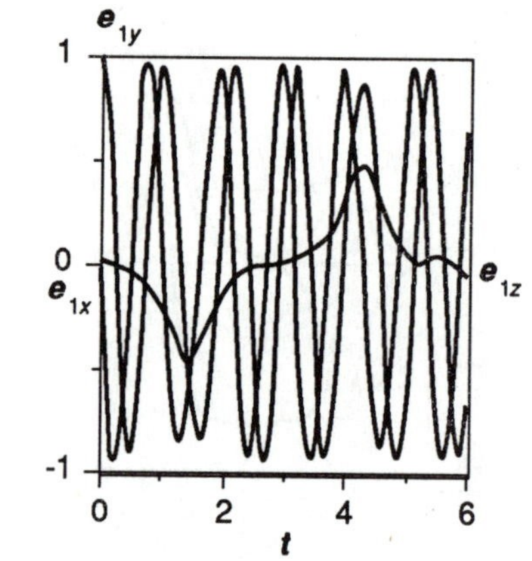

XY–PLOT

Data
Horizontal (x) axis: cells O21 - O80 (t)
Vertical (y) axis: cells S21 - S80 (e_{1x}), T21 - T80 (e_{1y}), and U21 - U80 (e_{1z})

Note
If you plot with symbols and the lines, you will also get a sense of the time development from the point density.

Figure 5.40

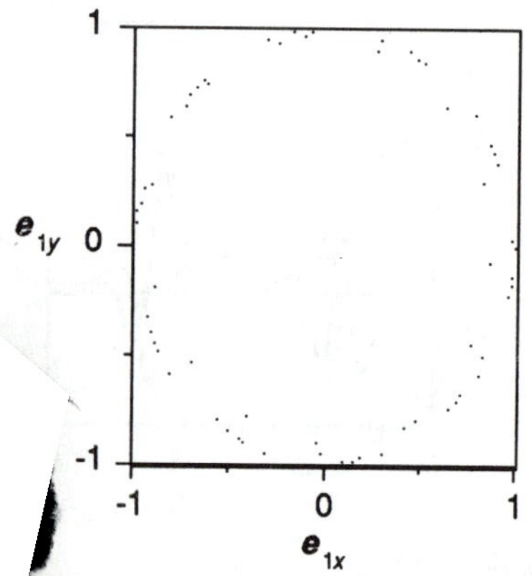

of e_{1y} vs e_{1x} (Figure 5.40) and similarly for $\boldsymbol{e}_2$ and $\boldsymbol{e}_2$; however, such plots will not be explicitly requested in the lessons that follow. Plots such as Figure 5.40 show the motion of the tip of the principal axis vectors as seen by an observer looking down the inertial z-axis. You may find it interesting to look down one or both of the other axes and also look down the laboratory z-axis at the tip of the angular velocity vector.

Notice that the input data for the model does not include the initial values of the three unit vectors along the principal axes, $\boldsymbol{e}_{10}$, $\boldsymbol{e}_{20}$, and $\boldsymbol{e}_{30}$. Instead, these are calculated using formulas that force the initial angular momentum of the body to be along the laboratory z-axis. Arbitrarily, the vector $\boldsymbol{e}_{30}$ is initially taken in the laboratory x-, y-plane.

Because there are no external forces or torques acting on the body, kinetic energy and angular momentum are conserved (in the laboratory system). Therefore, you can use columns E - G of your model to monitor numeric method accuracy.[1] The model also includes calculations that permit you to verify the unit lengths and orthogonality relationships of the three unit vectors $\boldsymbol{e}_1$, $\boldsymbol{e}_2$, and $\boldsymbol{e}_3$. You can save some calculation time by deleting these calculations from your spreadsheet.

2. A body in which all three principal moments of inertia are the same ($I_1 = I_2 = I_3$) is called a *dynamical sphere.* A ball is a dynamical sphere, but so is a cube. Toss such an object into the air and observe its motion. Notice that it just "turns"; it doesn't "tumble."

   a. Picture in your mind how the axes $\boldsymbol{e}_1$, $\boldsymbol{e}_2$, and $\boldsymbol{e}_3$ (that are fixed to the body and rotate with it) move with respect to the laboratory system.

   (One of the principal axes constitutes the axis about which the body rotates. The other two axes orthogonal to the angular velocity rotate at a constant rate about it. Of course, for a dynamical sphere, *any* line through the center of mass is a principal axis.)

   b. Simulate this motion on your computer by using Model 5.12 and setting $I_1 = I_2 = I_3 = 2$ kg m². Also set an initial angular velocity of magnitude π rad s^{-1} (i.e., 0.5 rev s^{-1}) in any direction (e.g., $\omega_{10} = \pi$, $\omega_{20} = \omega_{30} = 0$).

How do the principal axes move? Take the time to relate your plots to images of the actual motion. Which axis is motionless with respect to the laboratory system? Which axes move in a sinusoidal motion? What happens if you change the direction of the initial angular velocity (e.g., $\omega_{20} = \pi$, $\omega_{10} = \omega_{30} = 0$)? Discuss.

Look at the time development of the angular velocity in both the laboratory and the body systems. Describe the motion qualitatively. What is the angle between the angular momentum and the angular velocity and does it depend on time? (Remember that the angular momentum is along the *z*-axis.)

3. Now set your model such that $I_1 = I_2 = 1$ kg m² and $I_3 = 2$ kg m². These settings approximate the inertia tensor for a flat object such as a shallow square box or a roll of masking tape. Picture in your mind the location of such an object's principal axes.

   a. Simulate this object's motion when the initial angular velocity is π rad s^{-1} and fully along one or the other of the principal axes:

Case	ω_{10}	ω_{20}	ω_{30}
1	π	0	0
2	0	π	0
3	0	0	π

   Study your plots and picture the resulting motions.

   b. Qualitatively describe the motion in each of the above three cases. What do you conclude about the motion when the angular velocity is initially along one of the principal axes? What is the angle between the angular momentum and the angular velocity in each of these cases, and does it depend upon time?

   Verify your conclusions by establishing these initial conditions in a suitable body as you toss it up in the air.

   c. You can put the initial angular velocity into the symmetry plane of the body (but keep its magnitude at π rad s^{-1}) by setting $\omega_1 = \omega_2 = \pi/2^{0.5}$ rad s^{-1} and $\omega_3 = 0$. Is this motion qualitatively different from the motions in Lesson 3a?

   d. Now let $\omega_1 = 0$ and $\omega_2 = \omega_3 = \pi/2^{0.5}$ rad s^{-1}. The motion is more complex than in any of the earlier cases — you might say that the body wobbles as it spins (look at $\boldsymbol{e}_1$ and $\boldsymbol{e}_2$ from down the z-axis).

Notice that the tip of $\boldsymbol{e}_3$ (which is the body's axis of symmetry and which people call the body's *figure axis*) moves in a circle centered on the angular momentum. The figure axis is said to *precess* about the angular momentum.

What is the precession frequency? Is the magnitude of the angular velocity constant in time? What about the angle between the angular momentum and the angular velocity? Do all of the components of $\boldsymbol{e}_1$ and $\boldsymbol{e}_2$ execute sinusoidal motion?

4. The motion you saw in Lesson 3d is characteristic of a body with a symmetry axis and that is flatter than it is thick (e.g., the bodies in Exercise 5.2). Suppose the flat disk (or box) in Lesson 3 is stretched such that it becomes a pencil (or a tall box). This kind of symmetric body has a moment of inertia about its symmetry axis that is smaller than the one about its symmetry plane (a football is another good example).

   Let $I_1 = I_2 = 1$ kg m² and $I_3 = 0.5$ kg m² and study the character of the motion. How does this motion differ from that in Lesson 3?

5. Finally, let $I_1 = 1$, $I_2 = 2$, and $I_3 = 3$ kg m². This is the inertia tensor for a body that does not have rotational symmetry about any axis—a tennis racket, for example. Return to Figure 5.36 (or Exercise 5.2) and determine which principal axis has the largest moment of inertia and which the smallest.

   a. Examine the motion of a tennis racket thrown into the air when the initial angular velocity is π rad s^{-1} along each of the principal axes in succession:

Case	ω_{10}	ω_{20}	ω_{30}
1	π	0	0
2	0	π	0
3	0	0	π

   Is the motion qualitatively different from what you saw in Lessons 3 or 4?

   b. In the real world, of course, it is impossible to give a body an initial spin that is *exactly* along a principal axis. It is more realistic to consider cases in which slight errors exist in your attempt to put all of the spin along the axes.

Case	ω_{10}	ω_{20}	ω_{30}
4	π	0.02	0.09
5	0.02	π	0.09
6	0.02	0.09	π

Compare Cases 1 and 4, 2 and 5, 3 and 6. You may be surprised to discover that Cases 2 and 5 differ dramatically in their qualitative behavior. You will want to verify this behavior with a real tennis racket. It is a nice parlor trick to toss the handle under and over the strings (with the face of the racket held horizontally) and force it to *not* flip over before catching it again. It can be done, but it's difficult.

c. To see the full periodicity of the motion, change ω_2 to 1 rev s^{-1} $= 2\pi$ rad s^{-1}. You see more of the motion, but the plots get a bit ragged because of the relatively large time steps. Despite the violent fluctuations of the angular velocity in the body frame, notice that ω_z behaves quite modestly. What feature of the body's motion converts the near constancy of ω_z into the flip-flop behavior of ω_2?

d. Notice also that the magnitude of the angular velocity increases momentarily during the flip, whereas the angular momentum stays rock-solid constant. During the twisting and turning of its flip, the body presents a reduced moment of inertia about the direction of its angular momentum. Just as a dancer spins faster as he pulls his arms in, reducing his moment of inertia, so the tennis racket spins faster during its flip.

e. Study the "flipping" frequency as a function of the component of the angular velocity normal to $\boldsymbol{e}_2$. For example, set the initial length of the angular velocity to π rad s^{-1} and let it lie in the $\boldsymbol{e}_2 - \boldsymbol{e}_1$ plane, making an angle θ with $\boldsymbol{e}_2$. Look at all of your plots for θ values from 10° to 90° in steps of 20° (although the plots will appear rather rough with $\Delta t = 0.1$ s, you may wish to let the magnitude of the angular velocity be 2π rad s^{-1} to see the periodicity of the motion more clearly).

f. Design and toss other objects into the air. You can use your models from previous exercises to calculate and diagonalize the inertia tensors for these objects. (See how it all finally comes together!)

PROBLEMS

1. To a good approximation, the Earth is a symmetric body with $I_1 = I_2 < I_3$. The value of the ratio $(I_3 - I_1)/I_1$ is 0.00327, and the Earth's angular velocity of 1 rev day^{-1} makes an angle of 0.2 seconds of arc from the principal axis $\boldsymbol{e}_3$. Find the period of precession of the angular velocity vector about the figure axis. Describe what astronomical observations you might make to verify this calculation. Remember that you are living *on* the rotating rigid body!

   [NOTE: Although it is not necessary, you will probably want to calculate the motion for a much longer time than six days. To do this problem you do not need the vectors $\boldsymbol{e}_1$, $\boldsymbol{e}_2$, and $\boldsymbol{e}_3$. If you simply delete all columns from Model 5.12 that refer to these vectors, you will be able to greatly increase the time range that you can view.]

2. While studying the free motion of a rigid body, you verified that the energy and angular momentum are conserved. These two facts can be used to reduce the number of differential equations that must be considered, but only by increasing the complexity of the equations. Recall that you can often calculate motion from the energy integral, but with the introduction of a sign ambiguity that must be carefully handled (Exercise 4.4). On the other hand, the angular momentum is linear in the angular velocities and so does not suffer from this problem. In this problem you will reduce the number of differential equations to be solved by explicitly using the conservation of angular momentum.

   Let the three lab components of the angular momentum be L_x, L_y, and L_z. These are constants and can be written as

   $$I_1\omega_1 e_{1x} + I_2\omega_2 e_{2x} + I_3\omega_3 e_{3x} = L_x \tag{5.155a}$$

   $$I_1\omega_1 e_{1y} + I_2\omega_2 e_{2y} + I_3\omega_3 e_{3y} = L_y \tag{5.155b}$$

   $$I_1\omega_1 e_{1z} + I_2\omega_2 e_{2z} + I_3\omega_3 e_{3z} = L_z \tag{5.155c}$$

   You can look upon these three equations as three simultaneous equations in ω_1, ω_2, and ω_3, and they can be solved by Cramer's method. This gives the three body components of $\boldsymbol{\omega}$ as functions of the lab components of the the three principal axis vectors. You can then numerically solve the six equations of which Equations 5.151a

- 5.151c are the first three. Build a model using this idea to find the motion of a free rigid body.

3. Model 5.12 simulates the torque-free rotational motion of an object tossed into the air. By adding torques to the equations used in Model 5.12, you can use it to study the behavior of a top or gyroscope spinning in the constant gravitational field of the Earth and having one point fixed in the laboratory system (Figure 5.41). In these circumstances, Euler's equations are even sufficient to describe the motion of the center of mass.

   The gravity force on the body can be taken as acting at the top's center of mass and the laboratory supplies another force at the support point. You are not given the value of this support force, and in fact one of the goals of this sort of problem is to find it. If you cleverly use the support point (rather than the center of mass) as the origin of the body system, the unknown support force does not enter in the equations of rotational motion. This approach differs from your usual procedure only in that the inertia tensor needs to be calculated and diagonalized about the support point rather than about the center of mass.

   In general, the center of mass will not be on a principal axis, but it may be. Consider only the special case in which the center of mass is at

$$R_{cm} = De_3 \tag{5.156}$$

   where the origin is at the point of support, e_3 is a unit vector along one of the principal axes (calculated with the origin at the support point, remember), and D is the length of the top's support rod.

   If the body is rotationally symmetric about e_3 (like a top or a gyroscope) and the point of support is on the figure axis, then Equation 5.156 is guaranteed.

   By reference to the figure, you can see that the torque about the support point is

$$\Gamma = De_3 \times (-Mgk) \tag{5.157}$$

   Taking the dot product of this with e_1, e_2, and e_3, you get

$$\Gamma_1 = MgDe_{2z} \tag{5.158}$$

Figure 5.41

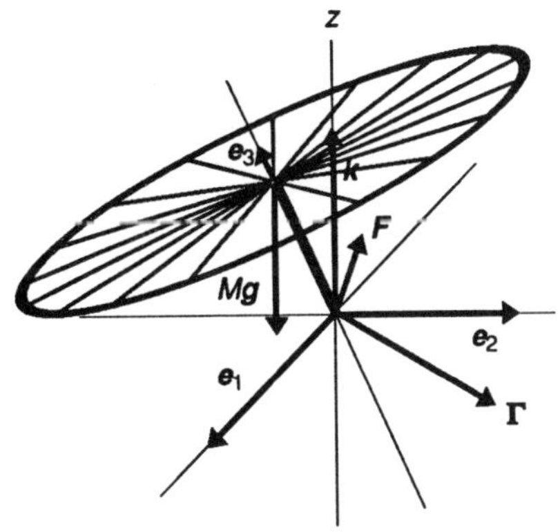

The spinning, symmetric heavy top (or gyroscope) with one point fixed in the lab is an excellent example of a rigid body moving under the influence of torques. Taking torques about the fixed (or support) point is a clever ploy to eliminate the support force ***F*** (supplied by the lab) from the equations of motion. We assume that the center of mass is located on one of the principal axes of the inertia tensor (evaluated with the support point as center.

$$\Gamma_2 = -MgDe_{1z} \qquad (5.159)$$

$$\Gamma_3 = 0 \qquad (5.160)$$

To study gyroscopic and related motions, you will use a large, heavy, donut-shaped object like a bicycle wheel (Figure 5.41). The wheel's major diameter (1.5 m), its minor diameter (.04 m), and its density (8.1 g cm^{-3}, iron) will be specified. As with a bicycle wheel, light spokes connect the rim to the center of the wheel, and the center is connected to a fixed support point by a light rod of length D (0.4 m). This body is symmetric, but you will make it unsymmetric later.

This wheel may seem exceptionally large, but it is chosen so that interesting effects can be observed with small values of angular velocity. Most tops or gyroscopes are much smaller and lighter and achieve large angular momenta by spinning very rapidly. Otherwise, they behave essentially as you will see here.

a. Modify Model 5.12 to include the above torques and dynamical parameters of the rigid body described above. Include suitable checks for the accuracy of the numeric method for different values of Δt. Also calculate the ratio of the support force supplied by the lab to the weight of the body by using Equation 5.141. [HINT: To calculate the support force, add the vector $M g \boldsymbol{k}$ to $Md^2(D\boldsymbol{e}_3)/dt^2$. Notice that $d^2(D\boldsymbol{e}_3)/dt^2$ is the acceleration of the center of mass. It can be calculated from ingredients already in the Runge-Kutta table.] Model 5.13 is our solution of this problem.

   In Model 5.13, you must specify the three body components of the angular velocity and the angle θ between the z-axis and the unit vector $\boldsymbol{e}_3$ [arccos(e_{3z})]. The vector $\boldsymbol{e}_3$ is taken in the x-, z-plane, and the principal axis $\boldsymbol{e}_2$ is taken along the laboratory system's y-axis. The model should be extended to at least 4.8 s.

   The most informative plots are

   - The three components of $\boldsymbol{e}_3$ vs time
   - Views of this vector along the z-axis and the y-axis
   - The body system and laboratory system angular velocity components vs time

- The body system and laboratory system angular momentum components vs time

- The three components of the support force vs time

b. It should be clear that if you set $D = 0$, the body is supported at its center of mass and moves as a free rigid body. Verify this. What force must the support supply? [NOTE: Your plots will look different from those in Model 5.12 because the z-axis is not along the lab angular momentum in Model 5.13.]

c. Observe the motion with the following parameters: $D = 0.4$ m, $\omega_1 = \omega_3 = 0$, $\omega_2 = 1$ rad s^{-1}, $\theta = 180°$, and $g = 0.001$ m s^{-2} (essentially free space. Zero is not used because the support force is normalized to the weight). You should recognize free-body motion about a principal axis, but why must the support point supply a force?

Now turn on the Earth's gravity by setting $g = 9.8$ m s^{-2} and look at the motion. This should be a very familiar situation to you! Describe it qualitatively.

d. Set $\theta = 30°$ and drop the pendulum from rest. How does the motion differ qualitatively from the small-angle oscillations in Problem 3c? Try other initial values of θ and angular velocities along the y-axis. Describe your results qualitatively.

e. Look at the motion with $\theta = 90$ and $\omega = 0$.

Now change the system from a pendulum to a gyroscope or top by putting some spin into the body. Set $\omega_3 = 5$ rad s^{-1}.

You will notice a dramatic change in the character of the motion. The body starts to fall as a pendulum, but promptly begins to move about the lab z-axis. The fall is quickly arrested, and the figure axis *rises* back up to its original level (90°). The general drift about the vertical is called *precession*, and the falling and rising part of the motion is called *nutation*. Clearly, adding a significant amount of angular momentum along the figure axis has induced the type of motion you have seen executed by a top or a gyroscope.

Investigate the motion for various other initial conditions. For example, is the precession always in the same direction or can there be periods of retrograde motion (a temporary change in the direction of the precession)? Can you adjust the initial conditions to prevent nutation?

Prepare a short report on your observations. Explain as much of the physics of your observations as you can. For example, why are L_z and L_3 conserved? What is the character of the force that the support must supply? How do the precession and nutation frequencies depend on the parameters of the problem (at least qualitatively)? How does the motion depend on the density?

Use simple English to describe the physics and understand the motion.

f. In the real world, nominally symmetric bodies are always asymmetric to some extent. Observe such effects by making your body quite unsymmetric.

Add two diametrically opposing masses of value $fM_{rim}/2$ to the rim of the wheel (f is an adjustable parameter). The center of mass stays at the center of the wheel, but the mass is increased (which increases I_1 and I_2 by $fM_{rim}D^2$). Furthermore, I_2 and I_3 are additionally increased by $fM_{rim}(d/2)^2$, where d is the diameter of the wheel.

Verify these assertions and study the resulting motions for at least $f = 1$. Compare the motion to that of the torqued symmetric body and the asymmetric free body. (Notice that L_3 is not conserved for an asymmetric body, so this check is automatically removed from the check table in Model 5.13.)

4. A sleeping top is one that stably spins upright without falling down. Mathematically, any top that spins about only the $\boldsymbol{e}_3$ axis and has $\theta = 0$ will stay that way forever. In the real world, however, the angle is never exactly $\theta = 0$ but some small angle, say 1°. Use Model 5.13 to find the smallest value of ω_3 (with $\omega_1 = \omega_2 = 0$) for which the motion stays near $\theta = 0$ if it starts with $\theta = 1°$. What happens when the body is asymmetric? [HINT: Start with $\boldsymbol{\omega} = 0$ and watch the pendulum fall. Then increase ω_{30} in steps of 1 rad s^{-1} and observe the motion.]

5. Recall in Problem 1 that the value of $(I_3 - I_1)/I_1$ is 0.00327 for the Earth. Introduce some extra wobble by making I_3 time-dependent via a small extra contribution from, for example, snowfall on the tundra. Try either a sinusoidal term with about 0.1 percent effect on I_3 or allow some short-term perturbation. Describe the effect on the precession of the angular velocity vector about the symmetry axis.

[HINT: If the moments of inertia are changing in time, then terms such as $\omega_1(dI_1/dt)$ must be added to Euler's equations. Also notice that you need not include the equations describing the motion of the principal axes to solve this problem.]

MODELS

Model 5.12

BUILD and verify Model 5.12.

ENTER data using cells B3 - B5 (eigenvalues of the inertia tensor), B8 - B10 (components of the initial angular velocity in the body system), and B16 (Δt).

READ output from the model in columns A - AA. Of particular interest are columns H - J (components of angular velocity in the laboratory system), P - R (components of angular velocity in the body system), and S - AA (components of the unit vectors $\boldsymbol{e}_1$, $\boldsymbol{e}_2$, $\boldsymbol{e}_3$ in the laboratory system).

NOTICE that this model determines the rotational motion of a torque-free rigid body by solving nine first-order differential equations using the Runge-Kutta method. The other three differential equations are eliminated by noting that two orthogonal vectors determine the third. Initial directions of the body axes $\boldsymbol{e}_1$, $\boldsymbol{e}_2$, $\boldsymbol{e}_3$ are adjusted so that the angular momentum is along the z-axis.

FORMULAS

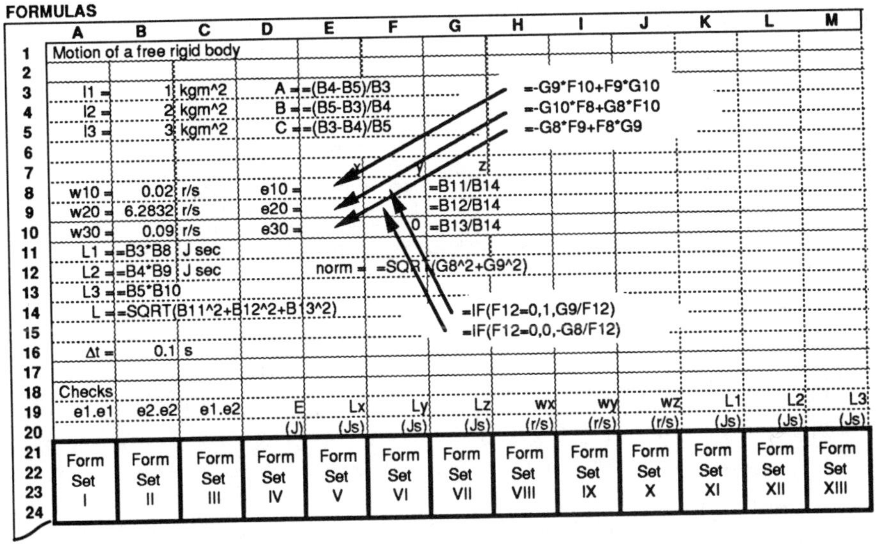

FORMULAS (Contd.)

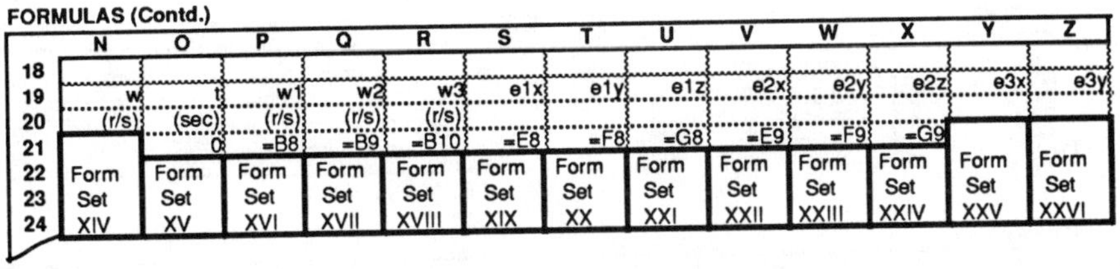

FORMULAS (Contd.)

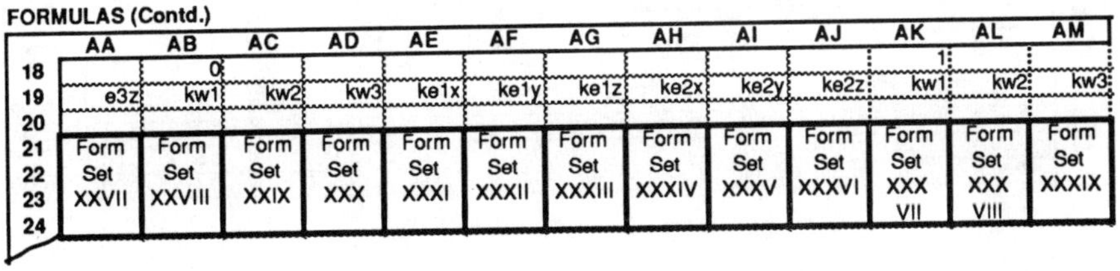

FORMULAS (Contd.)

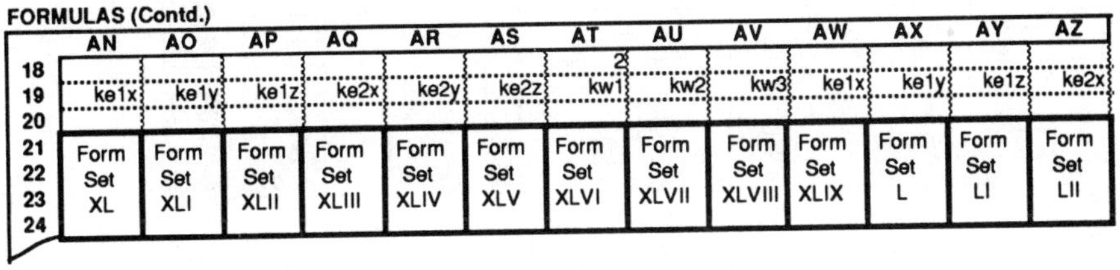

FORMULAS (Contd.)

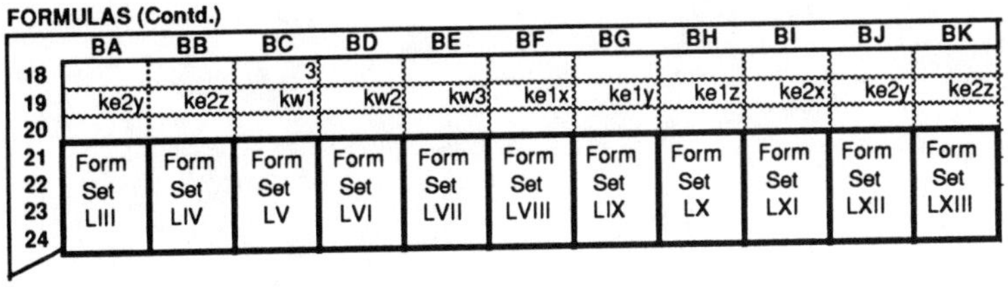

Model 5.12 (continued)

VALUES

	A	B	C	D	E	F	G	H	I	J	K	L	M
1	Motion of a free rigid body												
2													
3	I1 =	1	kgm^2	A =	-1								
4	I2 =	2	kgm^2	B =	1								
5	I3 =	3	kgm^2	C =	-0.333								
6													
7					x	y	z						
8	w10 =	0.02	r/s	e10 =	-3E-05	1	0.0016						
9	w20 =	6.2832	r/s	e20 =	-0.021	-0.002	0.9998						
10	w30 =	0.09	r/s	e30 =	0.9998	0	0.0215						
11	L1 =	0.02	J sec										
12	L2 =	12.566	J sec		norm =	0.9998							
13	L3 =	0.27	J sec										
14	L =	12.569	J sec										
15													
16	Δt =	0.1	s										
17													
18	Checks												
19	e1.e1	e2.e2	e1.e2	E	Lx	Ly	Lz	wx	wy	wz	L1	L2	L3
20				(J)	(Js)	(Js)	(Js)	(r/s)	(r/s)	(r/s)	(Js)	(Js)	(Js)
21	1	1	-2E-19	39.491	0	-3E-18	12.569	-0.045	0.01	6.2837	0.02	12.566	0.27
22	0.9992	1	5E-05	39.491	3E-07	0.0001	12.569	-0.026	-0.042	6.2837	-0.036	12.566	0.2751
23	0.9984	1	0.0001	39.491	-2E-05	0.0003	12.569	0.0301	-0.065	6.2837	-0.098	12.565	0.3168
24	0.9976	1	0.0002	39.491	-1E-04	0.0005	12.569	0.1022	-0.037	6.2837	-0.172	12.562	0.4007

VALUES (Contd.)

	N	O	P	Q	R	S	T	U	V	W	X	Y	Z
18													
19	w	t	w1	w2	w3	e1x	e1y	e1z	e2x	e2y	e2z	e3x	e3y
20	(r/s)	(sec)	(r/s)	(r/s)	(r/s)								
21	6.2839	0	0.02	6.2832	0.09	-3E-05	1	0.0016	-0.021	-0.002	0.9998	0.9998	7E-21
22	6.2839	0.1	-0.036	6.2831	0.0917	-0.587	0.809	-0.003	-0.019	-0.01	0.9998	0.8088	0.587
23	6.2841	0.2	-0.098	6.2825	0.1056	-0.95	0.31	-0.008	-0.015	-0.022	0.9997	0.3097	0.9496
24	6.2846	0.3	-0.172	6.2809	0.1336	-0.95	-0.307	-0.013	-0.003	-0.034	0.9994	-0.307	0.9499

VALUES (Contd.)

	AA	AB	AC	AD	AE	AF	AG	AH	AI	AJ	AK	AL	AM
18		0									1		
19	e3z	kw1	kw2	kw3	ke1x	ke1y	ke1z	ke2x	ke2y	ke2z	kw1	kw2	kw3
20													
21	0.0215	-0.057	0.0002	-0.004	-0.628	-1E-05	-0.004	0.002	-0.009	3E-05	-0.055	-7E-05	0.0017
22	0.0219	-0.058	-3E-04	0.0076	-0.508	-0.369	-0.005	0.0024	-0.01	-5E-05	-0.06	-6E-04	0.0137
23	0.0251	-0.066	-0.001	0.0205	-0.195	-0.597	-0.005	0.007	-0.013	-2E-04	-0.073	-0.002	0.0274
24	0.0318	-0.084	-0.002	0.036	0.1928	-0.597	-0.007	0.018	-0.012	-4E-04	-0.095	-0.003	0.0448

VALUES (Contd.)

	AN	AO	AP	AQ	AR	AS	AT	AU	AV	AW	AX	AY	AZ
18							2						
19	ke1x	ke1y	ke1z	ke2x	ke2y	ke2z	kw1	kw2	kw3	ke1x	ke1y	ke1z	ke2x
20													
21	-0.628	-0.197	-0.005	0.0019	-0.009	-1E-05	-0.057	-7E-05	0.0016	-0.566	-0.197	-0.004	0.0022
22	-0.392	-0.529	-0.006	0.004	-0.011	-1E-04	-0.062	-7E-04	0.0139	-0.342	-0.492	-0.004	0.0041
23	-0.007	-0.658	-0.007	0.012	-0.014	-3E-04	-0.075	-0.002	0.0281	0.012	-0.599	-0.005	0.0116
24	0.3803	-0.536	-0.009	0.0259	-0.009	-6E-04	-0.098	-0.003	0.046	0.3613	-0.477	-0.007	0.0245

VALUES (Contd.)

	BA	BB	BC	BD	BE	BF	BG	BH	BI	BJ	BK
18			3								
19	ke2y	ke2z	kw1	kw2	kw3	ke1x	ke1y	ke1z	ke2x	ke2y	ke2z
20											
21	-0.008	-6E-06	-0.058	-3E-04	0.0078	-0.504	-0.356	-0.004	0.0022	-0.009	-6E-05
22	-0.011	-9E-05	-0.066	-0.001	0.0206	-0.199	-0.584	-0.005	0.0067	-0.012	-2E-04
23	-0.013	-2E-04	-0.084	-0.002	0.0362	0.1817	-0.589	-0.006	0.0175	-0.012	-4E-04
24	-0.008	-5E-04	-0.113	-0.005	0.0565	0.4926	-0.37	-0.008	0.0317	-0.002	-8E-04

Model 5.12 (continued)

Formula Set I
Prototype Cell is A21

=S21^2+T21^2+U21^2

Formula Set II
Prototype Cell is B21

=V21^2+W21^2+X21^2

Formula Set III
Prototype Cell is C21

=S21*V21+T21*W21+U21*X21

Formula Set IV
Prototype Cell is D21

=0.5*(**B3***P21^2+**B4***Q21^2+**B5***R21^2)

Formula Set V
Prototype Cell is E21

=K21*S21+L21*V21+M21*Y21

Formula Set VI
Prototype Cell is F21

=K21*T21+L21*W21+M21*Z21

Formula Set VII
Prototype Cell is G21

=K21*U21+L21*X21+M21*AA21

Formula Set VIII
Prototype Cell is H21

=P21*S21+Q21*V21+R21*Y21

Formula Set IX
Prototype Cell is I21

=P21*T21+Q21*W21+R21*Z21

Formula Set X
Prototype Cell is J21

=P21*U21+Q21*X21+R21*AA21

Formula Set XI
Prototype Cell is K21

=**B3***P21

Formula Set XII
Prototype Cell is L21

=**B4***Q21

Formula Set XIII
Prototype Cell is M21

=**B5***R21

Formula Set XIV
Prototype Cell is N21

=SQRT(P21^2+Q21^2+R21^2)

Formula Set XV
Prototype Cell is O22

=O21+**B16**

Formula Set XVI
Prototype Cell is P22

=P21+(AB21+2*(AK21+AT21)+BC21)/6

Formula Set XVII
Prototype Cell is Q22

=Q21+(AC21+2*(AL21+AU21)+BD21)/6

Formula Set XVIII
Prototype Cell is R22

=R21+(AD21+2*(AM21+AV21)+BE21)/6

Formula Set XIX
Prototype Cell is S22

=S21+(AE21+2*(AN21+AW21)+BF21)/6

Formula Set XX
Prototype Cell is T22

=T21+(AF21+2*(AO21+AX21)+BG21)/6

Formula Set XXI
Prototype Cell is U22

=U21+(AG21+2*(AP21+AY21)+BH21)/6

Formula Set XXII
Prototype Cell is V22

=V21+(AH21+2*(AQ21+AZ21)+BI21)/6

Formula Set XXIII
Prototype Cell is W22

=W21+(AI21+2*(AR21+BA21)+BJ21)/6

Formula Set XXIV
Prototype Cell is X22

=X21+(AJ21+2*(AS21+BB21)+BK21)/6

Formula Set XXV
Prototype Cell is Y21

=T21*X21-W21*U21

Formula Set XXVI
Prototype Cell is Z21

=U21*V21-S21*X21

Formula Set XXVII
Prototype Cell is AA21

=S21*W21-T21*V21

Formula Set XXVIII
Prototype Cell is AB21

=**E3***Q21*R21***B16**

Formula Set XXIX
Prototype Cell is AC21

=**E4***P21*R21***B16**

Formula Set XXX
Prototype Cell is AD21

=**E5***P21*Q21***B16**

Model 5.12 (continued)

Formula Set XXXI	Prototype Cell is AE21	=-**B16** *(Q21*Y21-R21*V21)
Formula Set XXXII	Prototype Cell is AF21	=-**B16** *(Q21*Z21-R21*W21)
Formula Set XXXIII	Prototype Cell is AG21	=-**B16** *(Q21*AA21-R21*X21)
Formula Set XXXIV	Prototype Cell is AH21	=-**B16** *(R21*S21-P21*Y21)
Formula Set XXXV	Prototype Cell is AI21	=-**B16** *(R21*T21-P21*Z21)
Formula Set XXXVI	Prototype Cell is AJ21	=-**B16** *(R21*U21-P21*AA21)
Formula Set XXXVII	Prototype Cell is AK21	=**E3** *(Q21+AC21/2)*(R21+AD21/2)* **B16**
Formula Set XXXVIII	Prototype Cell is AL21	=**E4** *(P21+AB21/2)*(R21+AD21/2)* **B16**
Formula Set XXXIX	Prototype Cell is AM21	=**E5** *(P21+AB21/2)*(Q21+AC21/2)* **B16**
Formula Set XL	Prototype Cell is AN21	=-**B16** *((Q21+AC21/2)*((T21+AF21/2) *(X21+AJ21/2)-(W21+AI21/2)*(U21 +AG21/2))-(R21+AD21/2)*(V21+AH21/2))
Formula Set XLI	Prototype Cell is AO21	=-**B16** *((Q21+AC21/2)*((U21+AG21/2) *(V21+AH21/2)-(S21+AE21/2)*(X21 +AJ21/2))-(R21+AD21/2)*(W21+AI21/2))
Formula Set XLII	Prototype Cell is AP21	=-**B16** *((Q21+AC21/2)*((S21+AE21/2) *(W21+AI21/2)-(T21+AF21/2)*(V21 +AH21/2))-(R21+AD21/2)*(X21+AJ21/2))
Formula Set XLIII	Prototype Cell is AQ21	=-**B16** *((R21+AD21/2)*(S21+AE21/2) -(P21+AB21/2)*((T21+AF21/2)*(X21+AJ21 /2)-(W21+AI21/2)*(U21+AG21/2)))
Formula Set XLIV	Prototype Cell is AR21	=-**B16** *((R21+AD21/2)*(T21+AF21/2) -(P21+AB21/2)*((U21+AG21/2)*(V21+AH21 /2)-(S21+AE21/2)*(X21+AJ21/2)))
Formula Set XLV	Prototype Cell is AS21	=-**B16** *((R21+AD21/2)*(U21+AG21/2) -(P21+AB21/2)*((S21+AE21/2)*(W21+AI21 /2)-(T21+AF21/2)*(V21+AH21/2)))
Formula Set XLVI	Prototype Cell is AT21	=**E3** *(Q21+AL21/2)*(R21+AM21/2)* **B16**
Formula Set XLVII	Prototype Cell is AU21	=**E4** *(P21+AK21/2)*(R21+AM21/2)* **B16**
Formula Set XLVIII	Prototype Cell is AV21	=**E5** *(P21+AK21/2)*(Q21+AL21/2)* **B16**
Formula Set XLIX	Prototype Cell is AW21	=-**B16** *((Q21+AL21/2)*((T21+AO21/2) *(X21+AS21/2)-(W21+AR21/2)*(U21 +AP21/2))-(R21+AM21/2)*(V21+AQ21/2))
Formula Set L	Prototype Cell is AX21	=-**B16** *((Q21+AL21/2)*((U21+AP21/2) *(V21+AQ21/2)-(S21+AN21/2)*(X21 +AS21/2))-(R21+AM21/2)*(W21+AR21/2))
Formula Set LI	Prototype Cell is AY21	=-**B16** *((Q21+AL21/2)*((S21+AN21/2) *(W21+AR21/2)-(T21+AO21/2)*(V21 +AQ21/2))-(R21+AM21/2)*(X21+AS21/2))
Formula Set LII	Prototype Cell is AZ21	=-**B16** *((R21+AM21/2)*(S21+AN21/2) -(P21+AK21/2)*((T21+AO21/2)*(X21+AS21 /2)-(W21+AR21/2)*(U21+AP21/2)))
Formula Set LIII	Prototype Cell is BA21	=-**B16** *((R21+AM21/2)*(T21+AO21/2) -(P21+AK21/2)*((U21+AP21/2)*(V21+AQ21 /2)-(S21+AN21/2)*(X21+AS21/2)))
Formula Set LIV	Prototype Cell is BB21	=-**B16** *((R21+AM21/2)*(U21+AP21/2) -(P21+AK21/2)*((S21+AN21/2)*(W21+AR21 /2)-(T21+AO21/2)*(V21+AQ21/2)))
Formula Set LV	Prototype Cell is BC21	=**E3** *(Q21+AU21)*(R21+AV21)* **B16**

Model 5.12 (continued)

Formula Set LVI Prototype Cell is BD21
=**E4** *(P21+AT21)*(R21+AV21)* **B16**

Formula Set LVII Prototype Cell is BE21
=**E5** *(P21+AT21)*(Q21+AU21)* **B16**

Formula Set LVIII Prototype Cell is BF21
=-**B16** *((Q21+AU21)*((T21+AX21)*(X21+BB21)-(W21+BA21)*(U21+AY21))-(R21+AV21)*(V21+AZ21))

Formula Set LIX Prototype Cell is BG21
=-**B16** *((Q21+AU21)*((U21+AY21)*(V21+AZ21)-(S21+AW21)*(X21+BB21))-(R21+AV21)*(W21+BA21))

Formula Set LX Prototype Cell is BH21
=-**B16** *((Q21+AU21)*((S21+AW21)*(W21+BA21)-(T21+AX21)*(V21+AZ21))-(R21+AV21)*(X21+BB21))

Formula Set LXI Prototype Cell is BI21
=-**B16** *((R21+AV21)*(S21+AW21)-(P21+AT21)*((T21+AX21)*(X21+BB21)-(W21+BA21)*(U21+AY21)))

Formula Set LXII Prototype Cell is BJ21
=-**B16** *((R21+AV21)*(T21+AX21)-(P21+AT21)*((U21+AY21)*(V21+AZ21)-(S21+AW21)*(X21+BB21)))

Formula Set LXIII Prototype Cell is BK21
=-**B16** *((R21+AV21)*(U21+AY21)-(P21+AT21)*((S21+AW21)*(W21+BA21)-(T21+AX21)*(V21+AZ21)))

Model 5.13

BUILD and verify Model 5.13.

ENTER data using cells B5 - B8 (wheel parameters), B20 (a parameter defining the amount of asymmetric mass on the wheel rim; $f = 0$ generates a symmetric body), B24 - B27 (components of the initial angular velocity in the body system and the initial angle between the laboratory system z-axis and $\boldsymbol{e}_3$), and B29 (Δt).

READ output from the model in columns A - AA. Of particular interest are columns H - J (components of angular velocity in the laboratory system), P - R (components of angular velocity in the body system), and S - AA (components of the unit vectors $\boldsymbol{e}_1$, $\boldsymbol{e}_2$, $\boldsymbol{e}_3$ in the laboratory system).

NOTICE that this model determines the motion of a wheel in a uniform gravitational field with one point on its axle fixed in the laboratory system. The origin of both the lab and the body coordinate systems are taken at the fixed point. The center of mass is assumed to be at $D\boldsymbol{e}_3$. Initially, $\boldsymbol{e}_3$ is in the lab x-, z-plane, and $\boldsymbol{e}_2$ is along the lab y-axis.

FORMULAS

	A	B	C	D	E	F	G	H	I	J	K	L	M
1	Motion of a wheel in a uniform gravitational fidd												
2	with one point on its axle fixed in the lab.					=B16+B19*B6^2*B20							
3						=B17+B19*((B7/2)^2+B6^2)*B20							
4	Wheel parameters					=B18+B19*B7^2/4*B20							
5	density =	8.1	gm/cm^3(iron)			=B19*(1+B20)							
6	D =	0.4	m			=0.5*B18+B19*B6^2							
7	wheel dia =	1.5	m			=B16							
8	minor dia =	0.04	m			=B15*B7^2/4							
9						=PI()*(B8/2*100)^2*2*PI()*(B7*100/2)*B5/10							
10	Dynamical parameters			Mathematical parameters									
11	g =	9.8	m/s/s	A =		<-(I2-I3)/I1		=(B13-B14)/B12					
12	I1 =		kgm^2	B =		<-(I3-I1)/I2		=(B14-B12)/B13					
13	I2 =		kgm^2	C =		<-(I1-I2)/I3		=(B12-B13)/B14					
14	I3 =		kgm^2	D =		<-MgD		=B15*B11*B6					
15	M =		kg	E =		<-MgD/I1		=E14/B12					
16	rim I1 =		kgm^2	F =		<-MgD/I2		=E14/B13					
17	rim I2 =		kgm^2	G =		<-MD/dt		=B15*B6/B29					
18	rim I3 =		kgm^2	H =		<-Mg		=B15*B11					
19	rim M =		kg										
20	f =	0	<-defines assymetric mass on rim (f=0 gives zero extra mass)										
21													
22			=MAX(MAX(D31:F32),-MIN(D31:F32))										
23	Initial conditions				x	y	z						
24	w10 =	0	r/s	e10	=G26	0	=-E26						
25	w20 =	0	r/s	e20	0	1	0						
26	w30 =	5	r/s	e30			0	=COS(B27*PI()/180)					
27	theta =	90	°										
28						=SIN(B27*PI()/180)							
29	Δt =	0.12	s	Checks (all should be small compared to 1)									
30				e1.e1	e2.e2	e1.e2							
31	max chk			See Below			<-Geometry						
32							<-Physics						
33	Init L		Js	E	Lz	L3	(L3 conserved only for symmetric bodies)						
34	nit E	=D38	J										
35	Checks			=SQRT((B12*B24)^2+(B13*B25)^2+(B14*B26)^2)									
36	e1.e1	e2.e2	e1.e2	E	Lx	Ly	Lz	wx	wy	wz	L1	L2	L3
37				(J)	(Js)	(Js)	(Js)	(r/s)	(r/s)	(r/s)	(Js)	(Js)	(Js)
38-41	Form Set I	Form Set II	Form Set III	Form Set IV	Form Set V	Form Set VI	Form Set VII	Form Set VIII	Form Set IX	Form Set X	Form Set XI	Form Set XII	Form Set XIII

FORMULAS (Contd.)

	N	O	P	Q	R	S	T	U	V	W	X	Y	Z
35													
36	w	t	w1	w2	w3	e1x	e1y	e1z	e2x	e2y	e2z	e3x	e3y
37	(r/s)	(sec)	(r/s)	(r/s)	(r/s)								
38		0	=B24	=B25	=B26	=E24	=F24	=G24	=E25	=F25	=G25		
39-41	Form Set XIV	Form Set XV	Form Set XVI	Form Set XVII	Form Set XVIII	Form Set XIX	Form Set XX	Form Set XXI	Form Set XXII	Form Set XXIII	Form Set XXIV	Form Set XXV	Form Set XXVI

FORMULAS (Contd.)

	AA	AB	AC	AD	AE	AF	AG	AH	AI	AJ	AK	AL	AM
35		0									1		
36	e3z	kw1	kw2	kw3	ke1x	ke1y	ke1z	ke2x	ke2y	ke2z	kw1	kw2	kw3
37													
38-41	Form Set XXVII	Form Set XXVIII	Form Set XXIX	Form Set XXX	Form Set XXXI	Form Set XXXII	Form Set XXXIII	Form Set XXXIV	Form Set XXXV	Form Set XXXVI	Form Set XXXVII	Form Set XXXVIII	Form Set XXXIX

FORMULAS (Contd.)

	AN	AO	AP	AQ	AR	AS	AT	AU	AV	AW	AX	AY	AZ
35							2						
36	ke1x	ke1y	ke1z	ke2x	ke2y	ke2z	kw1	kw2	kw3	ke1x	ke1y	ke1z	ke2x
37													
38-41	Form Set XL	Form Set XLI	Form Set XLII	Form Set XLIII	Form Set XLIV	Form Set XLV	Form Set XLVI	Form Set XLVII	Form Set XLVIII	Form Set XLIX	Form Set L	Form Set LI	Form Set LII

FORMULAS (Contd.)

	BA	BB	BC	BD	BE	BF	BG	BH	BI	BJ	BK	BL	BM	BN
35			3											
36	ke2y	ke2z	kw1	kw2	kw3	ke1x	ke1y	ke1z	ke2x	ke2y	ke2z	Fx/Mg	Fy/Mg	Fz/Mg
37														
38-41	Form Set LIII	Form Set LIV	Form Set LV	Form Set LVI	Form Set LVII	Form Set LVIII	Form Set LIX	Form Set LX	Form Set LXI	Form Set LXII	Form Set LXIII	Form Set LXIV	Form Set LXV	Form Set LXVI

D31 contains =A79-A38
E31 contains =B79-B38
F31 contains =C79-C38

D32 contains =IF(B34=0,D38-D79,(+D38-D79)/B34)
E32 contains =IF(B33=0,G38-G79,(G38-G79)/B33)
F32 contains =IF(B20<>0,0,IF(B33=0,M38-M79,(M38-M79)/B33))

Model 5.13 (continued)

VALUES

	A	B	C	D	E	F	G	H	I	J	K	L	M
1	Motion of a wheel in a uniform gravitational field												
2	with one point on its axle fixed in the lab.												
3													
4	Wheel parameters												
5	density =	8.1	gm/cm^3(iron)										
6	D =	0.4	m										
7	wheel dia =	1.5	m										
8	minor dia =	0.04	m										
9													
10	Dynamical parameters			Mathematical parameters									
11	g =	9.8	m/s/s	A =	-0.275	<-(I2-I3)/I1							
12	I1 =	21.165	kgm^2	B =	0.2748	<-(I3-I1)/I2							
13	I2 =	21.165	kgm^2	C =	0	<-(I1-I2)/I3							
14	I3 =	26.981	kgm^2	D =	188.03	<-MgD							
15	M =	47.966	kg	E =	8.8839	<-MgD/I1							
16	rim I1 =	21.165	kgm^2	F =	8.8839	<-MgD/I2							
17	rim I2 =	21.165	kgm^2	G =	159.89	<-MD/dt							
18	rim I3 =	26.981	kgm^2	H =	470.07	<-Mg							
19	rim M =	47.966	kg										
20	f =	0	<-defines assymetric mass on rim (f=0 gives zero extra mass)										
21													
22													
23	Initial conditions				x	y	z						
24	w10 =	0	r/s	e10	6E-17	0	-1						
25	w20 =	0	r/s	e20	0	1	0						
26	w30 =	5	r/s	e30	1	0	6E-17						
27	theta =	90	°										
28													
29	Δt =	0.12	s	Checks (all should be small compared to 1)									
30				e1.e1	e2.e2	e1.e2							
31	max chk	0.0416		-0.039	-0.042	-0.007	<-Geometry						
32				-9E-04	0.0025	0	<-Physics						
33	Init L	134.91	Js	E	Lz	L3	(L3 conserved only for symmetric bodies)						
34	nit E	337.26	J										
35	Checks												
36	e1.e1	e2.e2	e1.e2	E	Lx	Ly	Lz	wx	wy	wz	L1	L2	L3
37				(J)	(Js)	(Js)	(Js)	(r/s)	(r/s)	(r/s)	(Js)	(Js)	(Js)
38	1	1	0	337.26	134.91	0	8E-15	5	0	3E-16	0	0	134.91
39	0.99926771	0.9993	0.0002	337.3	134.73	22.457	0.0164	4.9955	1.0401	0.084	4.7489	21.472	134.91
40	0.99900494	0.9979	0.0007	337.36	133.42	44.538	0.0455	4.9707	1.9495	0.2854	17.086	36.783	134.91
41	0.99938636	0.9953	0.0006	337.32	128.69	65.235	0.0013	4.8696	2.6442	0.4694	31.648	41.471	134.91

VALUES (Contd.)

	N	O	P	Q	R	S	T	U	V	W	X	Y	Z
35													
36	w	t	w1	w2	w3	e1x	e1y	e1z	e2x	e2y	e2z	e3x	e3y
37	(r/s)	(sec)	(r/s)	(r/s)	(r/s)								
38	5	0	0	0	5	6E-17	0	-1	0	1	0	1	0
39	5.1068	0.12	0.2244	1.0145	5	-0.059	0.5637	-0.823	0.0216	0.8255	0.5633	0.9973	0.0152
40	5.3546	0.24	0.8073	1.7379	5	-0.18	0.9221	-0.341	0.1517	0.3692	0.9158	0.9704	0.1127
41	5.5745	0.36	1.4953	1.9594	5	-0.226	0.9308	0.2861	0.4093	-0.174	0.8929	0.8811	0.3188

VALUES (Contd.)

	AA	AB	AC	AD	AE	AF	AG	AH	AI	AJ	AK	AL	AM
35		0									1		
36	e3z	kw1	kw2	kw3	ke1x	ke1y	ke1z	ke2x	ke2y	ke2z	kw1	kw2	kw3
37													
38	6E-17	0	1.0661	0	0	0.6	0	-4E-17	0	0.6	0.2319	1.0661	0
39	-0.061	0.4333	0.9148	0	-0.108	0.4935	0.3454	0.062	-0.338	0.4924	0.6204	0.7665	0
40	-0.206	0.6897	0.4969	0	-0.111	0.198	0.5924	0.2018	-0.542	0.1848	0.7472	0.2379	0
41	-0.342	0.6288	-0.059	0	0.0384	-0.18	0.6161	0.2936	-0.501	-0.233	0.5095	-0.335	0

VALUES (Contd.)

	AN	AO	AP	AQ	AR	AS	AT	AU	AV	AW	AX	AY	AZ
35							2						
36	ke1x	ke1y	ke1z	ke2x	ke2y	ke2z	kw1	kw2	kw3	ke1x	ke1y	ke1z	ke2x
37													
38	-0.07	0.6	0.18	-4E-17	-0.18	0.6	0.2319	0.9892	0	-0.059	0.5453	0.182	0.0337
39	-0.16	0.3839	0.5064	0.125	-0.483	0.3843	0.5749	0.6961	0	-0.101	0.3408	0.4775	0.1411
40	-0.095	0.005	0.6718	0.2841	-0.581	-0.012	0.6638	0.2003	0	-0.011	0.0014	0.6104	0.2555
41	0.1239	-0.369	0.5538	0.3207	-0.398	-0.439	0.4225	-0.312	0	0.1885	-0.317	0.4823	0.2481

VALUES (Contd.)

	BA	BB	BC	BD	BE	BF	BG	BH	BI	BJ	BK	BL	BM	BN
35			3											
36	ke2y	ke2z	kw1	kw2	kw3	ke1x	ke1y	ke1z	ke2x	ke2y	ke2z	Fx/Mg	Fy/Mg	Fz/Mg
37														
38	-0.18	0.5456	0.4185	0.9102	0	-0.095	0.4916	0.3352	0.0622	-0.327	0.4889	2E-17	0	0.6374
39	-0.45	0.3329	0.6734	0.5006	0	-0.095	0.2076	0.5803	0.1861	-0.534	0.1877	-0.064	0.2498	0.7423
40	-0.526	-0.038	0.6164	-0.044	0	0.0459	-0.158	0.6073	0.2651	-0.505	-0.223	-0.187	0.3352	1.027
41	-0.357	-0.414	0.2386	-0.503	0	0.2555	-0.423	0.3631	0.1841	-0.248	-0.549	-0.207	0.154	1.3324

Model 5.13 (continued)

Formula Set	Prototype Cell	Formula
I	A38	=S38^2+T38^2+U38^2
II	B38	=V38^2+W38^2+X38^2
III	C38	=S38*V38+T38*W38+U38*X38
IV	D38	=0.5*(**B12** *P38^2+**B13** *Q38^2+**B14** *R38^2)+**E14** *AA38
V	E38	=K38*S38+L38*V38+M38*Y38
VI	F38	=K38*T38+L38*W38+M38*Z38
VII	G38	=K38*U38+L38*X38+M38*AA38
VIII	H38	=P38*S38+Q38*V38+R38*Y38
IX	I38	=P38*T38+Q38*W38+R38*Z38
X	J38	=P38*U38+Q38*X38+R38*AA38
XI	K38	=**B12** *P38
XII	L38	=**B13** *Q38
XIII	M38	=**B14** *R38
XIV	N38	=SQRT(P38^2+Q38^2+R38^2)
XV	O39	=O38+**B29**
XVI	P39	=P38+(AB38+2*(AK38+AT38)+BC38)/6
XVII	Q39	=Q38+(AC38+2*(AL38+AU38)+BD38)/6
XVIII	R39	=R38+(AD38+2*(AM38+AV38)+BE38)/6
XIX	S39	=S38+(AE38+2*(AN38+AW38)+BF38)/6
XX	T39	=T38+(AF38+2*(AO38+AX38)+BG38)/6
XXI	U39	=U38+(AG38+2*(AP38+AY38)+BH38)/6
XXII	V39	=V38+(AH38+2*(AQ38+AZ38)+BI38)/6
XXIII	W39	=W38+(AI38+2*(AR38+BA38)+BJ38)/6
XXIV	X39	=X38+(AJ38+2*(AS38+BB38)+BK38)/6
XXV	Y38	=T38*X38-W38*U38
XXVI	Z38	=U38*V38-S38*X38
XXVII	AA38	=S38*W38-T38*V38
XXVIII	AB38	=**B29** *(**E11** *Q38*R38+ **E15** *X38)

Model 5.13 (continued)

Formula Set XXIX
Prototype Cell is AC38

=**B29** *(**E12** *P38*R38 -**E16** *U38)

Formula Set XXX
Prototype Cell is AD38

=**E13** *P38*Q38* **B29**

Formula Set XXXI
Prototype Cell is AE38

=-**B29** *(Q38*Y38-R38*V38)

Formula Set XXXII
Prototype Cell is AF38

=-**B29** *(Q38*Z38-R38*W38)

Formula Set XXXIII
Prototype Cell is AG38

=-**B29** *(Q38*AA38 -R38*X38)

Formula Set XXXIV
Prototype Cell is AH38

=-**B29** *(R38*S38-P38*Y38)

Formula Set XXXV
Prototype Cell is AI38

=-**B29** *(R38*T38-P38*Z38)

Formula Set XXXVI
Prototype Cell is AJ38

=-**B29** *(R38*U38-P38*AA38)

Formula Set XXXVII
Prototype Cell is AK38

=**B29** *(**E11** *(Q38+AC38/2)*(R38+AD38/2)
+**E15** *(X38+AJ38/2))

Formula Set XXXVIII
Prototype Cell is AL38

=**B29** *(**E12** *(P38+AB38/2)*(R38+AD38/2)
-**E16** *(U38+AG38/2))

Formula Set XXXIX
Prototype Cell is AM38

=**E13** *(P38+AB38/2)*(Q38+AC38/2)* **B29**

Formula Set XL
Prototype Cell is AN38

=-**B29** *((Q38+AC38/2)*((T38+AF38/2)
(X38+AJ38/2)-(W38+AI38/2)(U38
+AG38/2))-(R38+AD38/2)*(V38+AH38/2))

Formula Set XLI
Prototype Cell is AO38

=-**B29** *((Q38+AC38/2)*((U38+AG38/2)
(V38+AH38/2)-(S38+AE38/2)(X38
+AJ38/2))-(R38+AD38/2)*(W38+AI38/2))

Formula Set XLII
Prototype Cell is AP38

=-**B29** *((Q38+AC38/2)*((S38+AE38/2)
(W38+AI38/2)-(T38+AF38/2)(V38
+AH38/2))-(R38+AD38/2)*(X38+AJ38/2))

Formula Set XLIII
Prototype Cell is AQ38

=-**B29** *((R38+AD38/2)*(S38+AE38/2)
-(P38+AB38/2)*((T38+AF38/2)*(X38
+AJ38/2)-(W38+AI38/2)*(U38+AG38/2)))

Formula Set XLIV
Prototype Cell is AR38

=-**B29** *((R38+AD38/2)*(T38+AF38/2)
-(P38+AB38/2)*((U38+AG38/2)*(V38
+AH38/2)-(S38+AE38/2)*(X38+AJ38/2)))

Formula Set XLV
Prototype Cell is AS38

=-**B29** *((R38+AD38/2)*(U38+AG38/2)
-(P38+AB38/2)*((S38+AE38/2)*(W38
+AI38/2)-(T38+AF38/2)*(V38+AH38/2)))

Formula Set XLVI
Prototype Cell is AT38

=**B29** *(**E11** *(Q38+AL38/2)*(R38+AM38/2)
+**E15** *(X38+AS38/2))

Formula Set XLVII
Prototype Cell is AU38

=**B29** *(**E12** *(P38+AK38/2)*(R38+AM38/2)
-**E16** *(U38+AP38/2))

Formula Set XLVIII
Prototype Cell is AV38

=**E13** *(P38+AK38/2)*(Q38+AL38/2)* **B29**

Formula Set XLIX
Prototype Cell is AW38

=-**B29** *((Q38+AL38/2)*((T38+AO38/2)
(X38+AS38/2)-(W38+AR38/2)(U38
+AP38/2))-(R38+AM38/2)*(V38+AQ38/2))

Formula Set L
Prototype Cell is AX38

=-**B29** *((Q38+AL38/2)*((U38+AP38/2)
(V38+AQ38/2)-(S38+AN38/2)(X38
+AS38/2))-(R38+AM38/2)*(W38+AR38/2))

Formula Set LI
Prototype Cell is AY38

=-**B29** *((Q38+AL38/2)*((S38+AN38/2)
(W38+AR38/2)-(T38+AO38/2)(V38
+AQ38/2))-(R38+AM38/2)*(X38+AS38/2))

Formula Set LII
Prototype Cell is AZ38

=-**B29** *((R38+AM38/2)*(S38+AN38/2)
-(P38+AK38/2)*((T38+AO38/2)*(X38
+AS38/2)-(W38+AR38/2)*(U38+AP38/2)))

Model 5.13 (continued)

Formula Set LIII Prototype Cell is BA38
=-**B29** *((R38+AM38/2)*(T38+AO38/2)-(P38+AK38/2)*((U38+AP38/2)*(V38+AQ38/2)-(S38+AN38/2)*(X38+AS38/2)))

Formula Set LIV Prototype Cell is BB38
=-**B29** *((R38+AM38/2)*(U38+AP38/2)-(P38+AK38/2)*((S38+AN38/2)*(W38+AR38/2)-(T38+AO38/2)*(V38+AQ38/2)))

Formula Set LV Prototype Cell is BC38
=**B29** *(**E11** *(Q38+AU38)*(R38+AV38)+**E15** *(X38+BB38))

Formula Set LVI Prototype Cell is BD38
=**B29** *(**E12** *(P38+AT38)*(R38+AV38)-**E16** *(U38+AY38))

Formula Set LVII Prototype Cell is BE38
=**E13** *(P38+AT38)*(Q38+AU38)* **B29**

Formula Set LVIII Prototype Cell is BF38
=-**B29** *((Q38+AU38)*((T38+AX38)*(X38+BB38)-(W38+BA38)*(U38+AY38))-(R38+AV38)*(V38+AZ38))

Formula Set LIX Prototype Cell is BG38
=-**B29** *((Q38+AU38)*((U38+AY38)*(V38+AZ38)-(S38+AW38)*(X38+BB38))-(R38+AV38)*(W38+BA38))

Formula Set LX Prototype Cell is BH38
=-**B29** *((Q38+AU38)*((S38+AW38)*(W38+BA38)-(T38+AX38)*(V38+AZ38))-(R38+AV38)*(X38+BB38))

Formula Set LXI Prototype Cell is BI38
=-**B29** *((R38+AV38)*(S38+AW38)-(P38+AT38)*((T38+AX38)*(X38+BB38)-(W38+BA38)*(U38+AY38)))

Formula Set LXII Prototype Cell is BJ38
=-**B29** *((R38+AV38)*(T38+AX38)-(P38+AT38)*((U38+AY38)*(V38+AZ38)-(S38+AW38)*(X38+BB38)))

Formula Set LXIII Prototype Cell is BK38
=-**B29** *((R38+AV38)*(U38+AY38)-(P38+AT38)*((S38+AW38)*(W38+BA38)-(T38+AX38)*(V38+AZ38)))

Formula Set LXIV Prototype Cell is BL38
=**E17** *(AC38*S38+Q38*AE38-AB38*V38-P38*AH38)/ **E18**

Formula Set LXV Prototype Cell is BM38
=**E17** *(AC38*T38+Q38*AF38-AB38*W38-P38*AI38)/ **E18**

Formula Set LXVI Prototype Cell is BN38
=**E17** *(AC38*U38+Q38*AG38-AB38*X38-P38*AJ38)/ **E18** +1

CHAPTER 6

PERIODIC MOTIONS

Human beings are influenced by day-night rhythms. Everyone recongizes the "owls and the fowls" among us — those who are most active at night and those whose energies are stimulated by the rising of the sun.

Nature contains numerous types of physical objects that exhibit many different kinds of motions. Periodic motions, however, are so important that they have been singled out for special attention in this chapter. Periodic motion is a central feature of objects from the very small to the very large. Much of the atomic and subatomic structure of Nature as well as many aspects of her macroscopic structure can be understood in terms of such motions. You have already seen many examples of periodic motion in earlier exercises.

The vibration pattern of a hummingbird wing shows the general features of periodic motion:

- The wing motion is a general, periodic vibration exhibiting a fixed time interval between repetitions (Figure 6.1).

- The motion is not symmetric (the upstroke is of shorter time duration than the downstroke) because different external forces contribute to the motion in the two different directions.

  Most periodic motion exhibits some form of time asymmetry.

In general, periodic motions can be described by mathematical functions that are suitable sums of sines and cosines, and any such function is called *harmonic.* When a motion is described by a single sinusoid function, it is called *simple harmonic motion.* Simple harmonic motion can be used to approximate many periodic motions and also can be taken as the basis of more general periodic motions.

Simple harmonic motion arises in mechanics when a mass is subjected to a force that always acts toward a single place and whose magnitude is proportional to its distance from that place. Such a force is called a *linear restoring force.*

Figure 6.1

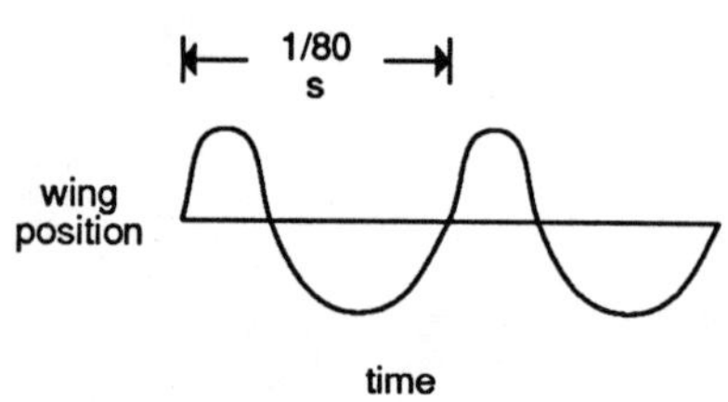

The wing movement of the hummingbird is an example of a general periodic vibration. Notice that the upstroke is shorter than the downstroke. This asymmetry is typical of periodic motion. The period of the motion in this example is 1/80 s.

In one dimension, the equation of motion for an object subject to a linear restoring force can be given as

$$m\frac{d^2x}{dt^2} + kx = 0 \qquad (6.1)$$

where k is a constant and where $x = 0$ is the place to which the mass is attracted. You have already seen numerous instances of this equation— the motion of a mass attached to a horizontal spring, for example (Equation 3.12 in Exercise 3.1).

Equation 6.1 is a differential equation in x that has an analytic solution (i.e., the solution can be expressed in terms of simple, well-known functions) (Equation 6.2):

$$x = A \sin(\omega t + \phi) \tag{6.2}$$

where $\omega = \sqrt{k/m}$.

The quantity A is called the *amplitude* of the function; x varies between $-A$ and $+A$. The angle ϕ is called the *phase* of the motion; it serves to locate the sinusoid relative to time $t = 0$. Notice that x is zero when $t = -\phi/\omega$ so that changing ϕ "slides" the sinusoid along the time axis. Finally, ω is called the *angular frequency* of the motion; it measures how frequently the motion assumes any particular value between $-A$ and $+A$ in unit time. The ordinary frequency of the motion is $f = \omega/2\pi$, and the period is $T = 1/f = 2\pi/\omega$. You might wish to build a spreadsheet model that plots x as a function of t and verify the significance of A, ϕ, and ω.

Solving Equation 6.1 to yield Equation 6.2 is not a trivial matter, but it is easy to work backward to verify that Equation 6.2 is a correct answer: Take the second derivative of Equation 6.2 (Equation 6.3) and recall that $x = A \sin(\omega t + \phi)$ (Equation 6.4):

$$\frac{d^2x}{dt^2} = -A\omega^2 \sin(\omega t + \phi) \tag{6.3}$$

$$\frac{d^2x}{dt^2} + \omega^2 x = 0 \tag{6.4}$$

Equation 6.4 is, of course, identical to Equation 6.1.

Besides the simplest example given by Equation 6.1, there are several other physical situations that can be described by related differential equations:

- The viscously damped, harmonic oscillator is a simple harmonic oscillator (Equation 6.1) subject to a dissipative force f that follows the rule

  $$f = -b\frac{dx}{dt} \tag{6.5}$$

  The equation of motion for a viscously damped, harmonic oscillator is

  $$m\frac{d^2x}{dt^2} + b\frac{dx}{dt} + kx = 0 \tag{6.6}$$

Because this rule for dissipation both is very common and is easily treated mathematically, the word *viscous* is often dropped from its name. However, as you will see, there are many other force laws that can arise to damp an oscillation (i.e., cause the amplitude to decrease with time).

- The driven harmonic oscillator is acted on by a further force $F(t)$ so that Equation 6.1 becomes

$$m\frac{d^2x}{dt^2} + kx = F(t) \tag{6.7}$$

- The damped driven harmonic oscillator is subject to both viscous dissipative forces (Equation 6.5) and another force $F(t)$. It has the equation of motion

$$m\frac{d^2x}{dt^2} + b\frac{dx}{dt} + kx = F(t) \tag{6.8}$$

Several linear oscillators can also be coupled so that they exchange energy. Examples of systems like this are bedsprings, automobile chasses with their complex supporting systems, and the atoms of a crystal. In general, the motion is very complex, and none of the oscillators execute simple harmonic motion. However, under special initial conditions, an undamped system develops a collective response called *normal mode behavior* in which each oscillator does execute simple harmonic motion. In this interesting collective motion, all oscillators in the system move with the same frequency, and each has its own characteristic amplitude and phase.

The equations of motion for all these oscillators have analytic solutions that belong to advanced courses in mechanics. Your personal computer and the Runge-Kutta method offer you the opportunity to explore these motions now and to gain insight into these important physical situations. Furthermore, even more advanced texts deal almost exclusively with linear restoring forces, damping forces that are linear in the velocity, and relatively simple external forces. Although these are the only cases you will study in the lessons, you will find examples of nonlinear restoring forces, realistic damping forces, and more general driving forces in the problems.

EXERCISE **6.1**

THE DRIVEN HARMONIC OSCILLATOR

Many systems in Nature display periodic motion about a fixed location called the *equilibrium position* when undisturbed. A system like this remains motionless at its equilibrium position, but it is subject to a force that acts to bring it back to its equilibrium position otherwise. Often this force increases approximately linearly with the distance of the system from its equilibrium position (a *linear restoring force*). If the force is exactly linear with this distance, the motion is called simple harmonic motion, and the system is called a *linear oscillator*.

Recall, for example, the behavior of a mass m attached to a horizontal spring of spring constant k (Figure 6.2). When the mass is displaced a distance x from its equilibrium position, it experiences a force of magnitude kx directed back toward the equilibrium position so that its equation of motion is given by

$$m\frac{d^2x}{dt^2} + kx = 0 \qquad (6.9)$$

This system executes simple harmonic motion with a natural frequency $\omega_0 = (k/m)^{1/2}$.

When dealing with linear oscillators in general, Equation 6.9 is often expressed as

TO GET THE MOST OUT OF THIS EXERCISE, YOU SHOULD ALREADY KNOW

The definitions of the terms:
oscillator
simple harmonic motion

How to solve equations of motion using the Runge-Kutta method.

Figure 6.2

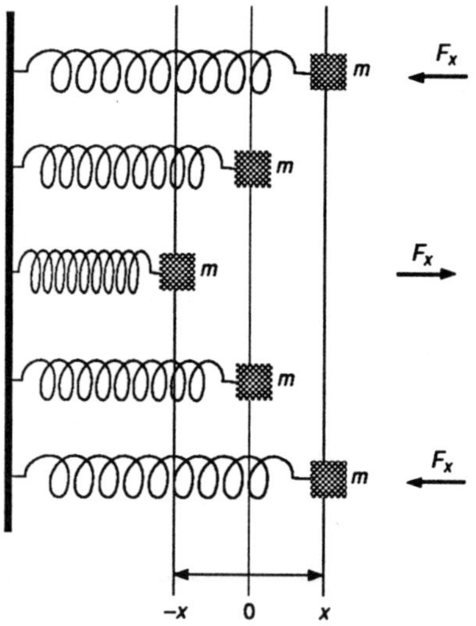

An ideal spring with a mass m attached to its free end oscillates, displacing the mass a distance, $x(t)$. The spring exerts a linear restoring force on the mass, and so the system executes simple harmonic motion. The equilibrium position is $x = 0$, where the spring is neither stretched nor compressed.

$$\frac{d^2x}{dt^2} + \omega_0^2 x = 0 \tag{6.10}$$

Equations 6.9 and 6.10 are examples of linear differential equations with constant coefficients. Advanced texts in mechanics devote considerable time to exploring the analytic solutions of Equation 6.10 and its various generalizations.[1]

Suppose the mass in Figure 6.2 is subject to a friction force F_{frict} where

$$F_{frict} = -b\frac{dx}{dt} \tag{6.11}$$

In macroscopic systems, this sort of friction force is usually caused by the viscosity of the material surrounding the mass or some other part of the system. It is therefore called a *viscous friction force.* In practical systems, such as the suspension of an automobile, viscous damping forces are explicitly built in with special devices like shock absorbers. As you will see, energy is removed from the oscillation by this friction force, and the amplitude of the motion dies away; people say that the motion is "damped."

The equation of motion for this damped oscillator is

$$m\frac{d^2x}{dt^2} + b\frac{dx}{dt} + kx = 0 \tag{6.12}$$

Or, dividing out the mass, you get the general form

$$\frac{d^2x}{dt^2} + \gamma\frac{dx}{dt} + \omega_0^2 x = 0 \tag{6.13}$$

where $\gamma = b/m$ is called the damping coefficient.

Finally, the damped oscillator might be driven by an applied force $F(t)$ so that Equation 6.13 becomes

$$\frac{d^2x}{dt^2} + \gamma\frac{dx}{dt} + \omega_0^2 x = \frac{F(t)}{m} \tag{6.14}$$

In this exercise, you will build and explore a damped oscillator driven sinusoidally and subject to the equation of motion

$$\frac{d^2x}{dt^2} + \gamma\frac{dx}{dt} + \omega_0^2 x = \frac{F_0}{m}\cos(\omega t) \tag{6.15}$$

where the applied force oscillates with angular frequency ω and amplitude F_0.

This driving force might be caused by an external agent acting directly on the mass or by something moving the spring's support. An example of the former is an electrically charged mass acted on by both a restoring force represented by the spring and a sinusoidally varying electric field. For the second case, you might think of the delicately supported mass of a seismograph during an earthquake or the suspension system of an automobile traveling on a rough road.

This model, of course, can also be used to explore the properties of the simple linear oscillator and the damped oscillator. It can also be used to explore other laws of damping and other driving forces.

LESSONS

Build and verify Model 6.1 for a driven damped harmonic oscillator. Verify that the model uses the Runge-Kutta method to solve Equation 6.15. Examine the columns that calculate kinetic energy K, potential energy U, and total mechanical energy E_0 and verify that these formulas are correct. Also build two on-screen graphics that plot the position and velocity of the oscillating mass with time (Figure 6.3) and the potential, kinetic, and total mechanical energy of the mass (Figure 6.4).

1. *A simple harmonic oscillator* can be represented by your model when the friction force and the driving force F_0 are both equal to zero.

   a. Using the initial values in Model 6.1 but setting $b = 0$ N m^{-1}s and $F_0 = 0$ N, verify that your model displays simple harmonic motion. You may wish to reduce calculation time in this part of the lesson by reducing the number of rows so that the model computes only to 5 s (when $\Delta t = 0.1$ s).

   b. How many seconds does it take for the motion to pass through one complete cycle (2π rad) of motion? What is the natural frequency ($\omega_0/2\pi$)?

      Compare the frequency you obtain by studying the waveform with $\omega_0/2\pi$. They should be the same. Are they?

   c. Examine the values of kinetic energy K and potential energy U at various points in the oscillation. Explain in simple words why these energies assume the values that they do.

   d. Is total mechanical energy conserved in this system? Why or why not?

Figure 6.3

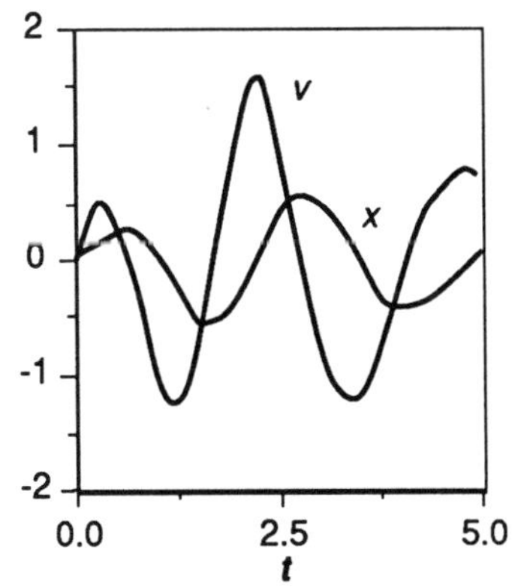

XY-PLOT

Data
Horizontal (*x*) axis: cells F17 - F67 (*t*)
Vertical (*y*) axis: cells G17 - G67 (position) and H17 - H67 (*v*)

Figure 6.4

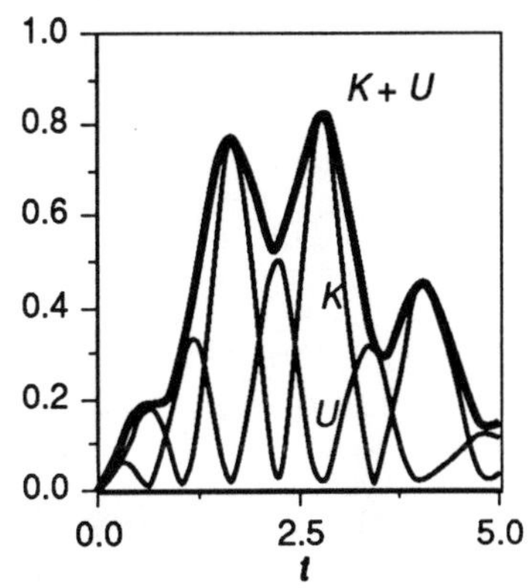

XY-PLOT

Data
Horizontal (*x*) axis: cells F17 - F67 (*t*)
Vertical (*y*) axis: cells A17 - A67 (*K*), B17 - B67 (*U*), and C17 - C67 (*K* + *U*)

It can be shown that the total mechanical energy of a simple harmonic oscillator is always equal to $kA^2/2$ where A is the amplitude of the oscillation. Verify that your model conforms to this relationship.

It is also true that total mechanical energy equals $mv_{max}^2/2$ where v_{max} is the amplitude of the velocity function. Use your model to verify this relationship.

e. How do the initial conditions x_0 and v_0 affect model behavior?

f. Explore your model by changing values for the mass and spring constant. Compare the behavior of the model with your intuition for the behavior of a mass oscillating at the end of a horizontal spring.

2. *A damped harmonic oscillator* may be represented by your model when the damping coefficient γ has a nonzero value ($b > 0$) and there is no driving force ($F_0 = 0$).

a. Reinitialize your model to the values shown in Model 6.1 and then set $F_0 = 0$ N and $b = 0.3$ N m^{-1} s.

b. Notice that the amplitude of the oscillations now decrease with time. Set A equal to x_0 and plot the function

$$A(t) = A\exp(-\gamma t/2) \quad (6.16)$$

and verify that it just skims the peaks of $x(t)$. This shows that, in this case, this exponential is a good approximation to the envelope of $x(t)$. You will soon see that it is not a good approximation when the damping is heavy.

Find a similar function for $v(t)$.

c. What happens to the total mechanical energy of the system as time progresses? Why?

Verify that the total mechanical energy oscillates around the decreasing function

$$E(t) = E_0\exp(-\gamma t/4) \quad (6.17)$$

where E_0 is the total mechanical energy at $t = 0$. This is not an approximation to the envelope of the total energy function, but nonetheless it gives a good representation of the decay of the system's mechanical energy when the damping is not too large.

d. Reduce the spring constant from 5 N m^{-1} to 1 N m^{-1} and compare the exponential functions (Equations 6.16 and 6.17) to the motion. Do they behave as nicely as your earlier observations with this smaller restoring force? Notice that reducing the restoring force without changing the damping force has made the damping relatively heavier.

e. Gradually increase the amount of damping in your oscillator from $b = 0.3$ to 1.5 N m^{-1} s. If necessary, increase the number of rows in your calculation to look at the motion for about 15 s. Although the amplitude of the motion becomes very tiny rather quickly, you should look at the position and velocity functions in the tables of the model itself and observe the time intervals between zeros (or when these functions change sign). Do the zeros of $x(t)$ and $v(t)$ always occur regularly? Are the time intervals between the zeros of $x(t)$ the same as those for $v(t)$? How do these intervals compare with π/ω_0, the half period of the undamped motion? How would you characterize the exponentials of Equations 6.16 and 6.17 when the damping is heavy?

f. An oscillator is said to be critically damped when the damping is large enough that it just fails to oscillate at all. Verify that critical damping occurs when $b = 2(km)^{0.5}$.

When $b > 2(km)^{0.5}$, the system is said to be overdamped and does not oscillate. When $b < 2(km)^{0.5}$, the oscillator is underdamped.

Notice that critical damping corresponds to $\gamma = 2\omega_0$.

g. An oscillating system is often characterized by its quality factor Q (Equation 6.18):

$$Q = \frac{2\pi}{\ln\left[\dfrac{E(t)}{E(t+T)}\right]} \tag{6.18}$$

where $E(t)$ is the mechanical energy in the oscillator at time t and T is the period of the oscillation. When the damping is very small, expand the natural logarithm function and verify that Q can be approximated by

$$Q = \frac{2\pi \ E(t)}{E(t\ ) - E(t\ + T)} \tag{6.19}$$

Although Q appears to depend on time t, it in fact does not, as you will see.

A lightly damped oscillator displays values of Q much greater than unity and it "rings" for a long period of time. Tuning forks can have Q values of about 1700. A microwave cavity might display $Q = 3 \times 10^7$.

What is the Q of your oscillator when $k = 0.1$ N m^{-1} and $b = 0.145$ N m^{-1} s?

Verify that the Q you calculate is independent of time by adding a column to your model that calculates Q as a function of time.

What is the Q when $b = 0.7677$ N m^{-1} s? When $b = 0.01$?

Verify that $Q = 2\pi/\gamma T$ where T is the period of the motion.

3. *A driven damped harmonic oscillator* can be represented by your model when both the damping coefficient γ and the driving force F_0 have nonzero values ($b > 0$, $F_0 > 0$).

   a. Reinitialize your model to the values shown in Model 6.1 and then (if necessary) extend your model by adding enough rows to include 15 s of motion (if you wish, reduce calculation time by increasing the step size, but this will result in rougher graphics). Set $x_0 = 0.04$ m, $v_0 = 0$ m s^{-1}, $k = 5$ N m^{-1}, $b = 0.3$ N m^{-1} s, and $F_0 = 0$.

   Add two columns to your model to calculate

$$P_{ext}(t) = F_0 \cos(\omega t)\ v(t) \tag{6.20}$$

   and

$$W_{ext}(t) = \int_0^t P_{ext}(t)dt \tag{6.21}$$

   $P_{ext}(t)$ is the rate at which the external force supplies energy to the system, or the *applied power*, and $W_{ext}(t)$ is the total work done on the system by the external force. Plot the applied power and the total work done by the external force (this force is often called the *drive force* or the *driver*) either separately or, if you wish, with the other energies of the problem.

By what time would you judge that the energy in the undriven oscillator has become negligible (say less than 1% of its initial value)? Call this time t_1. Now set F_0 to 1.3 N, and carefully inspect the motion, its energies, and the work done on the system by the driver. How would you characterize these for times less than t_1 and for times greater than t_1? You should notice the time intervals between successive zeros of the position and velocity graphs and the general behavior of the peak values of these functions. Is the motion periodic for $t < t_1$? For $t > t_1$? How does the frequency during $t > t_1$ compare to that of the applied force? Does energy flow only from the driver to the oscillator, or is there a sloshing of energy back and forth between the two?

The early portion of the motion is said to display *transient* behavior, and the subsequent portion is called its *steady-state* motion. During the transient, the effects of the free motion (which oscillates at one frequency) are said to *interfere* with those of the driven motion (which oscillates at another frequency).

b. To observe the various interesting motions that result during the transient before the system settles into its steady-state motion, set the initial position x_0 to 0.4 m, and vary F_0 from 0.0 N to 1.0 N in steps of 0.1 N. Observe the irregular nature of the motion and the energetics during the transient.

c. Now study the effect of the driver's frequency. Set F_0 to 0.3 N and then carefully observe the motions and energies as you increase the angular frequency of the driver from 2.3 to 4.5 rad s^{-1} in steps of 0.2 rad s^{-1}.

Complete the following table, which gives the amplitude of the steady-state motion (defined as the maximum value of $x(t)$ for $t > 10$ s) and the maximum energy in the oscillator in the steady state.

	Steady State	
ω	Amplitude	Maximum Energy
2.3	0.104	0.027
2.5	0.117	0.035
2.7	0.135	0.046
2.9	____	____
3.1	____	____
3.3	____	____
3.5	____	____
3.7	____	____
3.9	____	____
4.1	____	____
4.3	____	____
4.5	____	____

Plot the amplitude of the steady-state motion and its maximum energy as a function of the driver's angular frequency. These two curves have simple analytic representations and are very important functions in physics. They arise in many different contexts and consequently have many different names. The amplitude function is called an oscillator's *frequency response function*, or its resonance function. In the important case of light damping, the energy function is called the Lorenzentian line shape, the Cauchy distribution, the Breit-Wigner function, and probably other things. They merit your careful study as you will meet them often in science and engineering.

What is special about $\omega = 3.5$ rad s^{-1}? [HINT: Set $F_0 = 0$ and carefully find the period of the undriven motion, using it to calculate the corresponding angular frequency.] This is called the *resonant frequency* of the oscillator, and the oscillator is said to be driven *at resonance* when the driver has this frequency.

"Tune" the frequency of the driver to resonance as carefully as you can and observe the motions and energetics. How would you characterize P_{ext} and W_{ext} at resonance? Notice that energy flows in only one direction at resonance, namely, from the driver to the oscillator, much as you might add energy to a child in a swing by pushing her just in time with her natural oscillation.

d. Study the effect of the amplitude of the driving force on the amplitude of the steady-state motion at resonance by completing the following table.

F_0	Amplitude
0	0.007
0.1	0.094
1	_____
10	_____
100	_____

Make similar studies at frequencies below and above resonance. What do you conclude about the dependence of the amplitude of the steady-state motion on the amplitude of the driver? How does the transient part of the motion behave as F_0 increases?

e. Reinitialize your model to the values requested in Lesson 3a. Then study the effect of the size of the viscous force by changing the value of b from 0.0 to 0.4 N m^{-1}s in steps of 0.05 N m^{-1}s. To understand better the relatively complex motions you will see with the driven oscillator, compare what happens when $F_0 = 0$ with what you observe with $F_0 = 1.3$ N. Does the oscillator ever settle down to a nice, regular steady-state motion when $b = 0.0$? Discuss.

4. You've only touched on the behavior of driven, damped oscillators with the lessons above. Design investigations of your own and explore their consequences.

PROBLEMS

1. In Lesson 3 you examined both the Q of an undriven oscillator and its maximum steady-state energy when it is driven sinusoidally. When the oscillator was driven by a constant amplitude force, you saw that its energy function was a bell-shaped curve when plotted as a function of the drive frequency. Under these circumstances, you can characterize the bell-shaped curve with its peak height and its "full-width-at-half-max," or as it is frequently called, its FWHM. The FWHM of the bell-shaped energy function is the frequency difference between the two points on the curve that have value equal to half the curve's maximum value (Figure 6.5). By systematically changing the parameters of your oscillator, work out the relationship of these two parameters on the Q of the oscillator. [HINT: Since it is rather tedious to get the Q of the oscillator directly from its definition, use the result given above relating it to the oscillator's period and damping coefficient, γ.]

2. In Lesson 3 you explored a model of a driven, damped harmonic oscillator in which you could control the amplitude and frequency

Figure 6.5

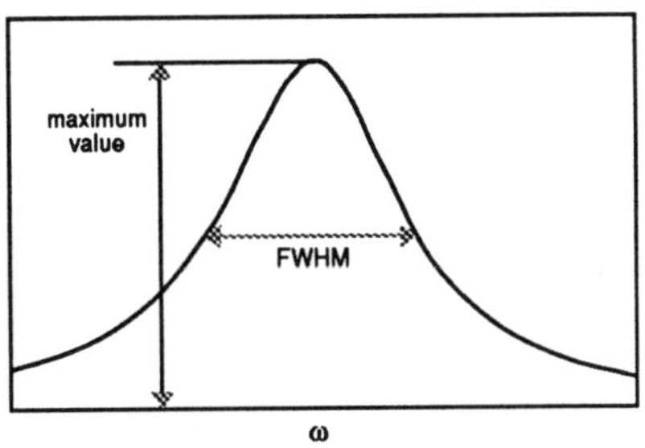

The steady-state energy function for a sinusoidally driven damped simple harmonic oscillator is bell-shaped when plotted as a function of the drive frequency. The maximum value of the function and its FWHM (full-width-at half-max) characterize the function.

It is interesting to plot the FWHM vs the maximum value and each of them vs Q of the undamped oscillator.

of the driving force but not the phase of the force function. Modify Model 6.1 so that you can vary the phase of the driving force and study how this parameter affects both the transient and steady-state parts of the motion. Describe your results qualitatively.

3. Real springs are not usually perfectly linear. Often, the spring constant gets larger as the spring is compressed ($x < 0$) and smaller as it is extended ($x > 0$). You can approximate this behavior with the force function

$$F_{\text{restore}} = -k\left(1 - \frac{x}{d}\right)x \tag{6.22}$$

where $d > 0$.

Qualitatively, how would you expect the motion of an undamped, undriven oscillator with this nonlinearity to differ from that with an ideal linear spring? Build a model to compare your intuition with the actual motion. What happens if the positive excursion of the motion exceeds $x = d$? Calculate and plot the potential energy of this force law and use it to guide your investigations.

4. Viscous friction is not the only kind of friction (as you have seen in earlier exercises). Suppose your oscillator is hindered not by viscous friction but by sliding friction. In this case, the damping force is

$$F_{\text{frict}} = -f\,\varepsilon(v) \tag{6.23}$$

where f is a constant and ε is the function given in Equation 4.58 of Exercise 4.3. Investigate both an undriven and a driven oscillator (with a perfectly linear spring) hindered by sliding friction. Is Q (Equation 6.18) independent of time in this case?

5. Another type of damping in real mechanical oscillators is the inertial effect of the material in which the oscillator moves. This effect occurs because the oscillator must push stuff out of its way and so imparts some of its motion to that stuff. This provides a damping force proportional to the square of the velocity. Suppose the damping force including this effect is

$$F_{\text{damp}} = -b\left(1 + \frac{|v|}{v_0}\right)v \tag{6.24}$$

where $v_0 > 0$ and $|v|$ is the absolute value of the oscillator's velocity. Intuitively, what would you expect the effect of this added damping term to be. Verify (or correct!) your intuition. Is Q independent of time?

6. The simple sinusoidal driver in the lesson is sometimes an accurate description of an actual physical situation, but more often the force driving an oscillator has a more complicated time structure. Suppose the oscillator has an ideal spring and has only viscous damping, but the driving force (due to an earthquake for example) is

$$F(t) = F_0 \exp(-t/\tau) \sin(\omega t) \tag{6.25}$$

Let the oscillator have a mass m of 0.4 kg, a viscous coefficient b of 0.1 N m^{-1}s, and a spring constant k of 5 N m^{-1} and assume it starts from $x = 0$ at rest. Let F_0 be 1 N and investigate the response of the system when ω = 1 rad s^{-1} (below resonance), 3 rad s^{-1} (near resonance), and 5 rad s^{-1} (above resonance), and for decay times τ of 1 s, 5 s, and 10 s.

Observe the motion for about 25 s using a step size of about 0.15 s. Write a brief report giving a qualitative description of the oscillator's response to this "shock." If necessary, investigate other values of τ and ω to help you understand more clearly what is happening.

Model 6.1

BUILD and verify Model 6.1.

ENTER data using cells B4 - B6 (Δt, x_0, v_0), and cells E4 - E8 (mass, spring constant ($b = \gamma/m$), magnitude of driving force, frequency of driving force).

READ output from the model at cells I2, I4 (damping coefficient, natural frequency of oscillator) and from columns A - H (kinetic energy, potential energy, total mechanical energy, difference between analytic and numeric solution, analytic solution for $x(t)$, time, position, velocity).

NOTICE that, just for fun, column E includes the analytic solution to $x(t)$ and that column D contains the difference between the analytic solution (column E) and the numeric solution (column G). You may wish to eliminate these columns.

MODEL

FORMULAS

	A	B	C	D	E	F	G	H	I	J	K	L	M
1	Driven damped oscillator												
2	DDX + gam Dx + w0^2 x = F0/m cos wt							gam =	=E6/E4				
3								w0^2 =	=E5/E4		=(B6+I9/L6-E8*I12)/L7		
4	Δt =	0.1	s	m =	0.4	kg		w0 =	=SQRT(I3)				
5	x0 =	0.04	m	k =	5	N/m							
6	v0 =	0	m/s	b =	0.3	N/(m/s)		dif 2 =	=I3-E8^2		tau =	=2/I2	
7				F0 =	1.3	N		g o =	=I2*E8		w' =	=SQRT(I3-(1/L6)^2)	
8				w =	2.3	rad/s		denom =	=I6^2+I7^2		T' =	=2*PI()/L7	
9								A =	=B5-I11		T =	=PI()*2/E8	
10								B =			maxerr =	=MAX(MAX(D17:D167),-MIN(D17:D167))	
11								C =	=I6/I8*E7/E4		maxamp =	=MAX(G17:G167)	
12								D =	=I7/I8*E7/E4		minamp =	=MIN(G17:G167)	
13													
14									α				
15	K	U	K+U	error	xexact	t	x	v	kx	kv	t	x	v
16	(J)	(J)	(J)	(m)	(m)	(s)	(m)	(m/s)					
17	Form Set I	Form Set II	Form Set III	Form Set IV	Form Set V	0	=B5	=B6	Form Set IX	Form Set X	Form Set XI	Form Set XII	Form Set XIII
18–20						Form Set VI	Form Set VII	Form Set VIII					

FORMULAS (Contd.)

	N	O	P	Q	R	S	T	U	V	W	X	Y
14	1					2					3	
15	kx	kv	t	x	v	kx	kv	t	x	v	kx	kv
16												
17–20	Form Set XIV	Form Set XV	Form Set XVI	Form Set XVII	Form Set XVIII	Form Set XIX	Form Set XX	Form Set XXI	Form Set XXII	Form Set XXIII	Form Set XXIV	Form Set XXV

VALUES

	A	B	C	D	E	F	G	H	I	J	K	L	M
1	Driven damped oscillator												
2	DDX + gam Dx + w0^2 x = F0/m cos wt							gam =	0.75				
3								w0^2 =	12.5				
4	Δt =	0.1	s	m =	0.4	kg		w0 =	3.5355	rad/s			
5	x0 =	0.04	m	k =	5	N/m							
6	v0 =	0	m/s	b =	0.3	N/(m/s)		dif 2 =	7.21		tau =	2.6667	
7				F0 =	1.3	N		g o =	1.725		w' =	3.5156	
8				w =	2.3	rad/s		denom =	54.96		T' =	1.7872	
9								A =	-0.386		T =	2.7318	
10								B =	-0.108		maxerr =	0.0003	
11								C =	0.4264		maxamp =	0.5761	
12								D =	0.102		minamp =	-0.555	
13													
14									0				
15	K	U	K+U	error	xexact	t	x	v	kx	kv	t	x	v
16	(J)	(J)	(J)	(m)	(m)	(s)	(m)	(m/s)					
17	0	0.004	0.004	-2E-17	0.04	0	0.04	0	0	0.275	0.05	0.04	0.1375
18	0.0132	0.0071	0.0202	5E-06	0.0532	0.1	0.0532	0.2566	0.0257	0.2307	0.15	0.066	0.372
19	0.0401	0.0199	0.06	2E-05	0.0891	0.2	0.0891	0.4479	0.0448	0.1462	0.25	0.1115	0.521
20	0.0584	0.0486	0.107	3E-05	0.1395	0.3	0.1394	0.5404	0.054	0.0358	0.35	0.1665	0.5583

VALUES (Contd.)

	N	O	P	Q	R	S	T	U	V	W	X	Y
14	1					2					3	
15	kx	kv	t	x	v	kx	kv	t	x	v	kx	kv
16												
17	0.0138	0.2625	0.05	0.0469	0.1313	0.0131	0.2544	0.1	0.0531	0.2544	0.0254	0.231
18	0.0372	0.1954	0.15	0.0718	0.3543	0.0354	0.1895	0.2	0.0886	0.4462	0.0446	0.147
19	0.0521	0.0943	0.25	0.1152	0.495	0.0495	0.0916	0.3	0.1386	0.5395	0.054	0.0369
20	0.0558	-0.025	0.35	0.1674	0.528	0.0528	-0.024	0.4	0.1923	0.5168	0.0517	-0.082

Model 6.1 (continued)

Formula Set I	Prototype Cell is A17	=0.5*E4*H17^2
Formula Set II	Prototype Cell is B17	=0.5*E5*G17^2
Formula Set III	Prototype Cell is C17	=A17+B17
Formula Set IV	Prototype Cell is D17	=E17-G17
Formula Set V	Prototype Cell is E17	=EXP(-F17/L6)*(I9*COS(L7*F17)+I10*SIN(L7*F17))+I11*COS(E8*F17)+I12*SIN(E8*F17)
Formula Set VI	Prototype Cell is F18	=F17+B4
Formula Set VII	Prototype Cell is G18	=G17+(I17+2*(N17+S17)+X17)/6
Formula Set VIII	Prototype Cell is H18	=H17+(J17+2*(O17+T17)+Y17)/6
Formula Set IX	Prototype Cell is I17	=B4*H17
Formula Set X	Prototype Cell is J17	=(E7/E4*COS(E8*F17)-I2*H17-I3*G17)*B4
Formula Set XI	Prototype Cell is K17	=F17+B4/2
Formula Set XII	Prototype Cell is L17	=G17+I17/2
Formula Set XIII	Prototype Cell is M17	=H17+J17/2
Formula Set XIV	Prototype Cell is N17	=B4*M17
Formula Set XV	Prototype Cell is O17	=(E7/E4*COS(E8*K17)-I2*M17-I3*L17)*B4
Formula Set XVI	Prototype Cell is P17	=F17+B4/2
Formula Set XVII	Prototype Cell is Q17	=G17+N17/2
Formula Set XVIII	Prototype Cell is R17	=H17+O17/2
Formula Set XIX	Prototype Cell is S17	=B4*R17
Formula Set XX	Prototype Cell is T17	=(E7/E4*COS(E8*P17)-I2*R17-I3*Q17)*B4
Formula Set XXI	Prototype Cell is U17	=F17+B4
Formula Set XXII	Prototype Cell is V17	=G17+S17
Formula Set XXIII	Prototype Cell is W17	=H17+T17
Formula Set XXIV	Prototype Cell is X17	=B4*W17
Formula Set XXV	Prototype Cell is Y17	=(E7/E4*COS(E8*U17)-I2*W17-I3*V17)*B4

EXERCISE **6.2**

COUPLED OSCILLATORS

TO GET THE MOST OUT OF THIS EXERCISE, YOU SHOULD ALREADY KNOW

The definitions of the terms:
oscillator
simple harmonic motion

How to solve equations of motion using the Runge-Kutta method.

While single oscillators are capable of fascinating varieties of behavior, the behavior of two or more coupled oscillators is almost beyond belief.

Recall from Exercise 6.1 that the equation of motion of a damped oscillator of mass m driven by an external force F is

$$m\frac{d^2x}{dt^2} + b\frac{dx}{dt} + kx = F \tag{6.26}$$

where x is the distance of the mass from its equilibrium position.

Suppose that two such oscillators are connected to each other by an ideal spring (Figure 6.6). The force that the coupling spring exerts on m_1 is

$$F = k_{12}(x_2 - x_1) \tag{6.27}$$

where k_{12} is the constant of the coupling spring and x_1, x_2 are the deviations of the two masses from their equilibrium positions. The force on m_2 due to the coupling spring is $-F$ (Newton's third law of motion).

The two damped oscillators are said to be coupled (via the spring that exchanges energy between them) and the equations of motion of the two oscillators are

$$m_1 \frac{d^2x_1}{dt^2} + b_1\frac{dx_1}{dt} + k_1x_1 = k_{12}(x_2 - x_1) \tag{6.28a}$$

$$m_2 \frac{d^2x_2}{dt^2} + b_2\frac{dx_2}{dt} + k_2x_2 = -k_{12}(x_2 - x_1) \tag{6.28b}$$

where m_1 and m_2 are the masses of the two oscillators, b_1, b_2 are their viscous coefficients, and k_1, k_2 the constants of the springs that connect them to the "walls." These equations can be rearranged into a form that is easily generalized to any number of masses constrained to one-dimensional movement and coupled together by springs (Equation 6.29a,b):

$$m_1 \frac{d^2x_1}{dt^2} + b_1\frac{dx_1}{dt} + (k_1 + k_{12})x_1 = k_{12}x_2 \tag{6.29a}$$

$$m_2 \frac{d^2x_2}{dt^2} + b_2\frac{dx_2}{dt} + (k_{12} + k_2)x_2 = k_{12}x_1 \tag{6.29b}$$

Of course, more than two oscillators can be coupled together by springs. For example, for three coupled masses as shown in Figure 6.7, the three equations of motion are

$$m_1 \frac{d^2x_1}{dt^2} + b_1\frac{dx_1}{dt} + (k_1 + k_{12})x_1 = k_{12}x_2 \tag{6.30a}$$

$$m_2 \frac{d^2x_2}{dt^2} + b_2\frac{dx_2}{dt} + (k_{12} + k_{23})x_2 = k_{12}x_1 + k_{23}x_3 \tag{6.30b}$$

$$m_3 \frac{d^2x_3}{dt^2} + b_3\frac{dx_3}{dt} + (k_{23} + k_3)x_3 = k_{23}x_2 \tag{6.30c}$$

From the symmetric form of these equations, you should be able to extend them to any number of masses.

Just as with one oscillator, other forces than those supplied by the springs can act on each of the masses, or the walls can wiggle (which gives the same effect as forces applied to the two end masses). This is the generalization of the driven oscillator problem.

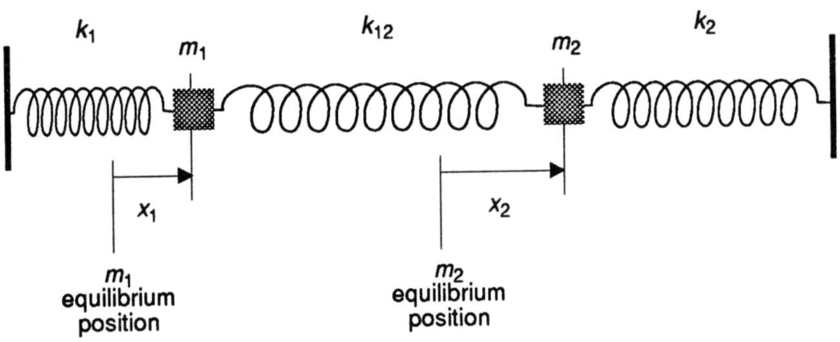

Figure 6.6

Two oscillators are said to be coupled when they can exchange energy. The oscillations of m_1 and m_2, for example, are coupled by the linear restoring force of a connecting spring. Each mass experiences a viscous retarding force with coefficients b_1 and b_2 (not shown).

Figure 6.7

Three oscillators are coupled by the linear restoring forces of connecting springs in the same manner as the two oscillators in Figure 6.6.

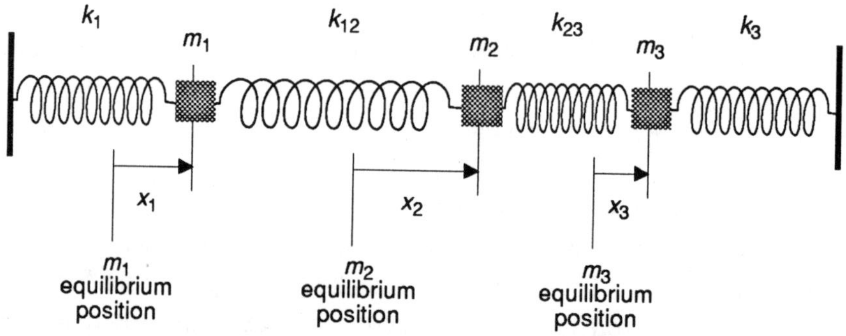

In the lessons, you will see that, when there is no damping, these systems support certain simple and interesting collective motions, called the system's *normal modes*. When the initial conditions are adjusted just right, every mass in the system oscillates steadily with the *same* frequency, but each with its own particular amplitude. Each mass has one of two phases which differ by 180°.[1]

A bit of terminology is useful:

- *Eigenvectors:* For a system of N coupled undamped oscillators, there are N different sets of initial conditions that produce N different normal modes. These N initial conditions can be written in "vector" notation and these are called the *eigenvectors* associated with the mode.

- *Eigenfrequencies:* Each normal mode of a system is associated with an oscillation frequency. These are called the system's *eigenfrequencies*, or its normal frequencies. For highly symmetric situations, two or more normal modes can have the same eigenfrequency so in these cases there are fewer than N different eigenfrequencies. Systems like this are said to be *degenerate*. In a degenerate system, any linear combination of the eigenvectors belonging to the same eigenfrequency is also an eigenvector of that frequency.[2]

You surely have noticed the similarity between this "eigen" language and the language of the inertia tensor—the general theory of the motion of these undamped coupled systems uses the mathematics of matrices and in particular that of the eigenvalue problem. Although we will use this "eigen" language in this book (because it is so common), we will not explore the underlying mathematics. However, with your personal computer and the Runge-Kutta method, you will be able to develop intuition about these interesting and important systems. It will serve you well when you study systems like this in the laboratory and when you learn the elegant mathematics associated with them.

LESSONS

1. Build and verify Model 6.2 for two nonidentical, damped oscillators coupled by a linear restoring force. Also build on-screen graphics to plot the position of the coupled oscillators vs time (Figure 6.8) and to plot the energy in each of the two oscillators, the potential energy in the coupling spring, and the total mechanical energy vs time (Figure 6.9).

   Confirm that Model 6.2 uses the Runge-Kutta method to solve Equations 6.28a,b for the position and velocity of the two oscillators.

   a. Examine your plots and notice that, using the initial values in your model, none of the energy is initially contained in oscillator 1.

   b. In the first few moments of time the energy in U_{12} is transferred to both E_2 and E_1. Which oscillator receives most of the energy that departs the coupling spring?

   c. In the next few moments the energy in oscillator 2 is put back into the coupling spring, and from there is distributed (unequally) back to E_2 and E_1. What is the evidence for this action?

   d. Follow the energy as it partitions between the two oscillators and the coupling spring. Notice that the coupling spring acts something like an idler wheel in a gear chain and mediates the "sloshing" of energy back and forth between the two oscillators.

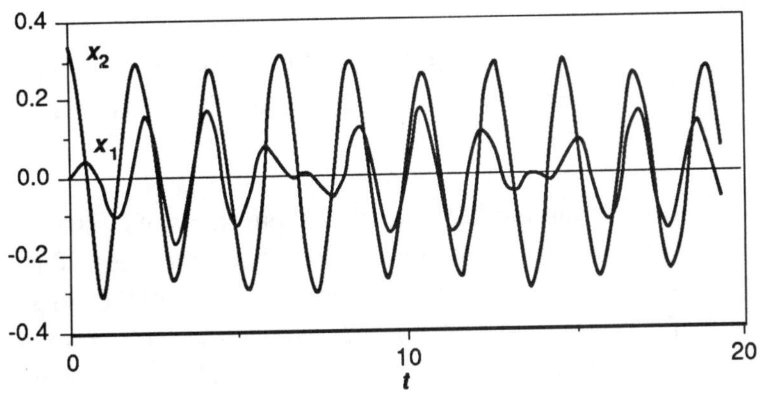

Figure 6.8

XY-PLOT

Data
Horizontal (*x*) axis: cells E13 - E110 (*t*)
Vertical (*y*) axis: cells F13 - F110 (x_1), and G13 - G110 (x_2)

Figure 6.9

XY-PLOT

Data
Horizontal (x) axis: cells E13 - E110 (t)
Vertical (y) axis: cells A13 - A110 (energy in oscillator 1), B13 - B110 (potential energy in coupling spring), C13 - C110 (energy in oscillator 2), and D13 - D110 (total mechanical energy)

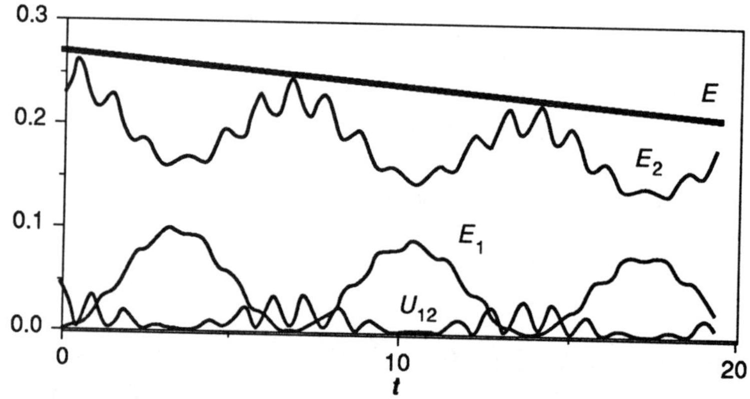

e. Cells H3 and H4 contain positive values for the damping coefficient. What happens to the total mechanical energy that is lost to viscous damping?

2. Modify your model to represent two undamped oscillators ($\gamma_1 = \gamma_2 = 0\ s^{-1}$) of identical mass ($m_1 = m_2 = 0.45$ kg) and natural frequency ($\omega_{01} = \omega_{02} = 3.6$ rad s^{-1}) coupled by means of an ideal spring of constant $k_{12} = 0.6$ N m^{-1}. Initially, both oscillators are motionless ($v_1 = v_2 = 0$), oscillator 2 is displaced 0.34 m from its equilibrium position while oscillator 1 is at its equilibrium position. Select a time interval Δt of 0.1 s.

   a. Plot x_1, x_2 vs time for the first 20 seconds or more using a time interval Δt of 0.1 s (if your computer cannot plot 200 data points or if calculation times are too slow, set $\Delta t = 0.2$ and adjust the number of rows in your model). Notice that the amplitude of oscillator 2 gradually decreases until it comes to rest temporarily at approximately 9 s. Oscillator 1 gradually increases in amplitude and then decreases—returning temporarily to rest at about 18 s. How is this behavior reflected in your energy plot? Is this system conservative (you may want to reduce the value of Δt before answering this question, especially if you're working with $\Delta t = 0.2$)? Discuss.

      Explore the effect of the coupling constant k_{12} on the length of time required for each oscillator to achieve its first moment of minimum amplitude.

   b. Return your model to the initial conditions of Lesson 2 and then displace both oscillators 0.34 m from their equilibrium positions. This is one of the collective motions of the system, one of its normal modes. Notice that both oscillators move with the same frequency and have the same phase. What is this frequency?

Explore different initial conditions where both masses have exactly the same initial conditions (you can force this with simple formulas). Observe what happens to the energy in the coupling spring. Does the motion depend on the spring constant of the coupling spring? On what does the phase depend? Discuss.

c. Explore different initial conditions where the initial conditions of m_2 are the negative of those of m_1. Again, discuss the motions and energetics of the motion. Is the motion now dependent on the strength of the coupling spring? What happens to the frequency of the oscillations as the coupling spring is strengthened?

This motion is the second, higher-frequency normal mode. Again, notice that both masses oscillate with the same frequency but now their phases differ by 180° (i.e., when one mass zigs, the other zags). What does the phase of m_1 depend upon?

Each normal mode can be concisely described by making an eigenvector out of the pair of initial positions (or the pair of initial velocities because the pattern is the same) that generate it. Thus, the normal mode (1,1) (meaning that both masses have the same value of initial position and/or velocity) has eigenfrequency $\omega_0 = (k_1/m_1)^{1/2} = (k_2/m_2)^{1/2}$ and normal mode (1,–1) has a higher eigenfrequency. Give a physical reason for the higher frequency of this mode. Notice that the dot product of these two vectors equals zero [(1,1) · (1, –1) = 1 – 1]. In other words, these two eigenvectors are orthogonal.

d. Verify that for arbitrary initial conditions (x_{10}, v_{10}, x_{20}, v_{20}), the motion of the masses can be expressed as follows:

$$x_1(t) = C\cos(\omega_0 t - \phi_0) + D\cos(\omega_1 t - \phi_1) \qquad (6.31a)$$

$$x_2(t) = C\cos(\omega_0 t - \phi_0) - D\cos(\omega_1 t - \phi_1) \qquad (6.31b)$$

where

$$C = \frac{1}{2}\sqrt{(x_{10} + x_{20})^2 + \left(\frac{v_{10} + v_{20}}{\omega_0}\right)^2}$$

$$D = \frac{1}{2}\sqrt{(x_{10} - x_{20})^2 + \left(\frac{v_{10} - v_{20}}{\omega_1}\right)^2}$$

$$\tan \phi_0 = \frac{v_{10} + v_{20}}{\omega_0(x_{10} + x_{20})}$$

$$\tan \phi_1 = \frac{v_{10} - v_{20}}{\omega_1(x_{10} - x_{20})}$$

$$\omega_0 = \sqrt{\frac{k_1}{m_1}} = \sqrt{\frac{k_2}{m_2}}$$

$$\omega_1 = \sqrt{\omega_0^2 + \frac{2k_{12}}{m_1}}$$

Make your verification by adding columns for $x_1(t)$ and $x_2(t)$ as calculated by Equations 6.31a - 6.31b to your model and comparing these results with those obtained using the Runge-Kutta method for a variety of initial conditions. In particular, what happens to the formulas when the initial conditions of the two masses are the same? When they are equal in magnitude but opposite in sign?

Notice that Equations 6.31a - 6.31b are linear combinations of the two normal modes.

e. Explore the complex motions that occur for arbitrary initial conditions. Such motions are due to the *interference* of the two normal modes that oscillate at different frequencies. Notice similarities between the character of these motions and what you saw during the transient motion of a single, driven oscillator. They are both illustrations of the interesting phenomenon of interference between sinusoidal motions. When the frequencies are different, the phenomenon is also called "beating."

3. Adjust your model to represent a system of two identical undamped oscillators in one of its normal modes (e.g., $\gamma_1 = \gamma_2 = 0$ s^{-1}, $m_1 = m_2 = 0.45$ kg, $\omega_{01} = \omega_{02} = 3.6$ rad s^{-1}, $k = 0.6$ N m^{-1}, $v_1 = v_2 = 0, x_1 = x_2 = 1$) Select a time interval Δt of 0.1 s.

Break the symmetry of your two oscillators by slightly changing the two uncoupled angular frequencies (e.g., $\omega_{01} = 3.65$ rad s^{-1} and $\omega_{02} = 3.55$ rad s^{-1}). Clearly the system is no longer in a normal mode. Now gradually increase the value of x_{20} until you once again find the characteristic behavior of a normal mode (i.e., constant amplitude motion). You should find that $(x_{10}, x_{20}) = (1, 1.306)$ is an eigenvector of the oscillator. Show that the other eigenvector of this two mass system is $(x_{10}, x_{20}) = (1, -.766)$.

What are the eigenfrequencies associated with each eigenvector? What is the vector dot product of the two eigenvectors? [HINT: It should be zero although you may have difficulty finding the eigenvectors accurately enough to show this with much precision. This is always the case when the masses of the oscillators are the same.] You may wish to explore more asymmetric situations, but increase the asymmetry gradually, finding the eigenvectors after each change, to keep from getting lost!

4. Break the symmetry by changing the values of the two masses (reset the uncoupled frequencies to $\omega_{01} = \omega_{02} = 3.6$ rad s^{-1}, but set $m_1 = .4$ kg and $m_2 = .5$ kg). Find the normal modes and give their eigenvectors and eigenfrequencies. You should be able to easily guess one eigenvector and eigenfrequency by just thinking about the situation. Are the eigenvectors orthogonal now?

   When the masses are different, the eigenvectors are not orthogonal. It can be shown, however, that if each eigenvector component is multiplied by the square root of its mass, the resulting vectors will be orthogonal. Verify this. You can always use this important fact to find a second eigenvector after you have found one.

5. Now change both the masses and the uncoupled spring constants. Let $m_1 = .4$ kg, $m_2 = .5$ kg, $\omega_{01} = 3.65$ rad s^{-1} and $\omega_{02} = 3.55$ rad s^{-1}. Find the normal modes and their frequencies. Are the eigenvectors orthogonal? Remember that you should use the result of the Lesson 4 to guide your search after you have found one eigenvector.

6. You can use your model to study the effect of damping, but let's keep things simple at first by investigating symmetric oscillators. Return your model to its condition at the beginning of Lesson 2.

   a. Set both damping coefficients (γ_1, γ_2) to 0.2 s^{-1} and set the position of m_1 at its equilibrium position while m_2 is displaced by 0.34 m. Now look at the motions produced by the initial conditions given by the eigenvectors you found in Lesson 1. Do the results agree with your intuition? Discuss.

   b. Now break the symmetry by setting $\gamma_1 = 0.3$ s^{-1} and $\gamma_2 = 0.1$ s^{-1}. On the basis of your experience so far, you might expect to be able to find initial conditions not terribly different from what you saw in Lesson 1 or Lesson 6a so that both oscillators move with the same frequency and damp out in the same way, *but this is impossible!* Convince yourself of this by looking for initial conditions with this property. You will find it interesting to plot the energy in the coupling spring on a separate graph so you can see its behavior clearly. This shows that, except in special cases, the notion of normal modes is, strictly speaking, only applicable to undamped coupled oscillators.

c. Now let oscillator 1 be undamped and oscillator 2 heavily damped. Let $\gamma_1 = 0$ and $\gamma_2 = 3\ s^{-1}$. Study the motions and discuss the energetics of the situation.

d. Let both oscillators be very lightly but differently damped, say $\gamma_1 = .01$ and $\gamma_2 = .02$. Let Δt be large enough so that you can see about 30 s of the motion (don't worry about accurate numeric calculation here). As you just saw, strictly speaking, this system has no normal modes. However, notice that for short times the normal modes you found in Lesson 1 provide a nice, intuitive, but approximate, description of the system.

To appreciate that the normal mode description is only approximate, note that the energy is transfered between the two oscillators even when you start the system in the (1,1) mode (watch U_{12}). This does not happen in this mode in the ideal, undamped case. In the ideal, undamped case, energy transfer between the oscillators is the characteristic feature of the (1,–1) mode. People say that the perturbation to the ideal, undamped system caused by the small viscous forces causes some "mode mixing," or that there is some "excitation" of the (1,–1) mode caused by the perturbation even when the system is initially purely in the (1,1) mode.

PROBLEMS

1. Study Model 6.2 with the two end springs removed, namely, set both k_1 and k_2 to zero. A glance at Figure 6.6 reveals that m_1 and m_2 are coupled but are now free to move as an isolated system.

   Is (1,1) still a normal mode? What is its "frequency"? To what kind of motion does it correspond?

   Is (1,–1) a normal mode? What is its frequency? This result should help you understand the "2" in the formulas for ω_1 in Lesson 2d that drive Equation 6.31.

   Describe qualitatively the general motion of this system.

2. In addition to a coupling spring between the two masses in Figure 6.6, there could also be a "dashpot" between them. This provides an additional coupling force $-b_{12}(v_1 - v_2)$ on m_1 and an equal and opposite force on m_2. Modify model 6.2 to include this new damping effect and study the resulting motions. Write a brief report describing the behavior of the system including discussion of such questions as the conditions under which eigenmodes exist, any mode mixing when this dashpot is the only dissipation in the system, and the effect of removing the coupling spring and leaving only the dashpot.

3. Suppose the two masses in Figure 6.6 are free to move only transversely from their equilibrium positions, i.e., only perpendicular to the line defined by the springs when the masses are motionless. Assume the distances between the left wall and the equilibrium position of m_1 is d_1, between the two equilibrium positions is d_{12}, and from m_2's equilibrium position and the right wall is d_2. Choose reasonable values for the masses, spring constants, and these distances and study the motions in the absence of damping. What can you say about the system when the transverse displacements are very small compared to the smallest of d_1, d_{12}, and d_2? Do eigenmode motions exist when the transverse displacements are large? Is the motion of either mass ever sinusoidal? Can you observe mode mixing when the motions are small?

   [NOTE: You should be able to observe the mode mixing very clearly since there is no damping. In this case, it is due to the nonlinearity of the restoring forces.]

4. Set up a model to describe the motion of the system of three masses shown in Figure 6.7; a CO_2 molecule, for example.

   a. Start by assuming that $m_1 = m_2 = m_3$, that $k_{12} = k_{23}$, and that the two end springs are removed ($k_1 = k_3 = 0$). You should be able to find two eigenvectors by carefully thinking about the system. Verify them by observing that these initial conditions in fact generate eigenmodes. What are their frequencies? Then use the orthogonality condition to find the third eigenvector; verify it and find its frequency. By carefully thinking about the forces acting on any one of the masses, find formulas for the eigenfrequencies and verify them.

   b. Now change the value of the middle mass and find the eigenmodes and eigenfrequencies.

c. Again use equal masses and replace the two end springs, assuming all four springs are equal. Find the eigenmodes and their frequencies. Can you develop formulas for the eigenfrequencies? Explore various symmetry breakings, finding the eigenmodes by following those of the fully symmetric situation.

5. In a real system of springs and masses like Figure 6.6, the springs can only be compressed so much before their coils hit each other and the spring suddenly becomes *very* stiff. For simplicity, assume that this only happens for the coupling spring. You can model this effect by letting the spring "constant" be a function of the length of the spring. One simple (but not very realistic) model for this effect is the following:

$$k_{12} = \begin{cases} k_0 & (x_1 - x_2) < 0 \\ k_0 \exp\left(\frac{x1 - x2}{d}\right) & (x_1 - x_2) > 0 \end{cases} \tag{6.32}$$

where d is some fraction of the spring's length.

Study the motion of the system, ignoring dissipation, and discuss your results. Are there normal modes for small oscillations? For large oscillations? Do you observe the phenomenon of mode mixing?

6. Suppose the left wall in Figure 6.6 were not stationary but had a sinusoidal wiggle with amplitude A and angular frequency ω. Work out the equations of motion and modify Model 6.2 to solve this problem. Assume the oscillators are damped heavily enough that the transient motion lasts only a few seconds so your calculations do not take too long. Study the steady-state energy and amplitude response of both the masses as the frequency of the wiggling wall is changed. Also observe the relative phase of the two oscillators in the steady state. Compare your results to what you observed for a single, driven damped oscillator. Can you use this method to find the modes of the undamped system?

7. Suppose an external sinusoidal force acts on the central mass of three mass system shown in Figure 6.7. Calculate the steady state amplitude, energy, and phase response of each mass as the frequency of the driving force is varied. Compare to the results you found in the single oscillator case and what you found in Problem 6.

MODEL

Model 6.2

BUILD and verify Model 6.2.

ENTER data using cells B3 - B6 (initial positions and velocities); B8 (Δt); E3 - E4 (oscillator masses), E8 (coupling spring constant), and H3 - H6 (damping coefficients and natural frequencies of oscillators). Take care to note that that the values b_1, b_2, k_1, and k_2 are calculated from the values entered into cells H3 - H6.

READ output from the model from columns A - I (oscillator 1 energy, potential energy in the coupling spring, oscillator 2 energy, total mechanical energy, time, oscillator 1 position, oscillator 2 position, oscillator 1 velocity, oscillator 2 velocity).

FORMULAS

	A	B	C	D	E	F	G	H	I	J	K	L	M
1	Coupled Oscillators												
2													
3	x1o =	0	m	m1 =	0.45	kg	g1 =	0.006	/s				
4	v1o =	0	m/s	m2 =	0.5	kg	g2 =	0.009	/s				
5	x2o =	0.34	m	b1 =	=E3*H3	N/(m/s)	w1o =	3.6	/s				
6	v2o =	0	m/s	b2 =	=E4*H4	N/(m/s)	w2o =	2.8	/s				
7				k1 =	=H5^2*E3								
8	Δt =	0.2	s	k12 =	0.8	N/m							
9				k2 =	=H6^2*E4								
10										0			
11	E1	U12	E2	E	t	x1	x2	v1	v2	kx1	kx2	kv1	kv2
12		(J)		(J)	(s)	(m)	(m)	(m/s)	(m/s)				
13					0	=B3	=B5	=B4	=B6				
14	Form Set I	Form Set II	Form Set III	Form Set IV	Form Set V	Form Set VI	Form Set VII	Form Set VIII	Form Set IX	Form Set X	Form Set XI	Form Set XII	Form Set XIII
15													
16													

FORMULAS (Contd.)

	N	O	P	Q	R	S	T	U	V	W	X	Y	Z
10					1								2
11	x1	x2	v1	v2	kx1	kx2	kv1	kv2	x1	x2	v1	v2	kx1
12	(m)	(m)	(m/s)	(m/s)					(m)	(m)	(m/s)	(m/s)	
13													
14	Form Set XIV	Form Set XV	Form Set XVI	Form Set XVII	Form Set XVIII	Form Set XIX	Form Set XX	Form Set XXI	Form Set XXII	Form Set XXIII	Form Set XXIV	Form Set XXV	Form Set XXVI
15													
16													

FORMULAS (Contd.)

	AA	AB	AC	AD	AE	AF	AG	AH	AI	AJ	AK
10								3			
11	kx2	kv1	kv2	x1	x2	v1	v2	kx1	kx2	kv1	kv2
12				(m)	(m)	(m/s)	(m/s)				
13											
14	Form Set XXVII	Form Set XXVIII	Form Set XXIX	Form Set XXX	Form Set XXXI	Form Set XXXII	Form Set XXXIII	Form Set XXXIV	Form Set XXXV	Form Set XXXVI	Form Set XXXVII
15											
16											

VALUES

	A	B	C	D	E	F	G	H	I	J	K	L	M
1	Coupled Oscillators												
2													
3	x1o =	0	m	m1 =	0.45	kg	g1 =	0.006	/s				
4	v1o =	0	m/s	m2 =	0.5	kg	g2 =	0.009	/s				
5	x2o =	0.34	m	b1 =	0.0027	N/(m/s)	w1o =	3.6	/s				
6	v2o =	0	m/s	b2 =	0.0045	N/(m/s)	w2o =	2.8	/s				
7				k1 =	5.832	N/m							
8	Δt =	0.2	s	k12 =	0.8	N/m							
9				k2 =	3.92	N/m							
10										0			
11	E1	U12	E2	E	t	x1	x2	v1	v2	kx1	kx2	kv1	kv2
12		(J)		(J)	(s)	(m)	(m)	(m/s)	(m/s)				
13	0	0.0462	0.2266	0.2728	0	0	0.34	0	0	0	0	0.1209	-0.642
14	0.0027	0.0285	0.2413	0.2725	0.2	0.0111	0.2779	0.1013	-0.6	0.0203	-0.12	0.066	-0.52
15	0.0059	0.0027	0.263	0.2716	0.4	0.0339	0.1154	0.1067	-0.973	0.0213	-0.195	-0.059	-0.205
16	0.006	0.0069	0.2575	0.2704	0.6	0.0451	-0.087	-0.011	-0.986	-0.002	-0.197	-0.164	0.1799

VALUES (Contd.)

	N	O	P	Q	R	S	T	U	V	W	X	Y	Z
10					1								2
11	x1	x2	v1	v2	kx1	kx2	kv1	kv2	x1	x2	v1	v2	kx1
12	(m)	(m)	(m/s)	(m/s)					(m)	(m)	(m/s)	(m/s)	
13	0	0.34	0.0604	-0.321	0.0121	-0.064	0.1208	-0.641	0.006	0.3079	0.0604	-0.321	0.0121
14	0.0212	0.218	0.1343	-0.86	0.0269	-0.172	0.0147	-0.403	0.0245	0.192	0.1087	-0.801	0.0217
15	0.0446	0.0181	0.0771	-1.076	0.0154	-0.215	-0.125	-0.018	0.0416	0.0078	0.0441	-0.982	0.0088
16	0.044	-0.185	-0.093	-0.896	-0.019	-0.179	-0.196	0.3654	0.0359	-0.176	-0.108	-0.803	-0.022

VALUES (Contd.)

	AA	AB	AC	AD	AE	AF	AG	AH	AI	AJ	AK
10								3			
11	kx2	kv1	kv2	x1	x2	v1	v2	kx1	kx2	kv1	kv2
12				(m)	(m)	(m/s)	(m/s)				
13	-0.064	0.0916	-0.579	0.0121	0.2759	0.0916	-0.579	0.0183	-0.116	0.0624	-0.516
14	-0.16	-0.004	-0.353	0.0329	0.1177	0.0971	-0.953	0.0194	-0.191	-0.055	-0.21
15	-0.196	-0.12	0.0003	0.0427	-0.081	-0.013	-0.973	-0.003	-0.195	-0.155	0.1684
16	-0.161	-0.168	0.3457	0.0234	-0.247	-0.179	-0.64	-0.036	-0.128	-0.157	0.4754

Model 6.2 (continued)

Formula Set I	Prototype Cell is A13	=0.5*E7*F13^2+0.5*E3*H13^2
Formula Set II	Prototype Cell is B13	=0.5*E8*(F13-G13)^2
Formula Set III	Prototype Cell is C13	=0.5*E4*I13^2+0.5*E9*G13^2
Formula Set IV	Prototype Cell is D13	=A13+C13+B13
Formula Set V	Prototype Cell is E14	=E13+B8
Formula Set VI	Prototype Cell is F14	=F13+(J13+2*(R13+Z13)+AH13)/6
Formula Set VII	Prototype Cell is G14	=G13+(K13+2*(S13+AA13)+AI13)/6
Formula Set VIII	Prototype Cell is H14	=H13+(L13+2*(T13+AB13)+AJ13)/6
Formula Set IX	Prototype Cell is I14	=I13+(M13+2*(U13+AC13)+AK13)/6
Formula Set X	Prototype Cell is J13	=B8*H13
Formula Set XI	Prototype Cell is K13	=B8*I13
Formula Set XII	Prototype Cell is L13	=B8*(E8*G13-E5*H13-(E7+E8)*F13)/E3
Formula Set XIII	Prototype Cell is M13	=B8*(E8*F13-E6*I13-(E8+E9)*G13)/E4
Formula Set XIV	Prototype Cell is N13	=F13+J13/2
Formula Set XV	Prototype Cell is O13	=G13+K13/2
Formula Set XVI	Prototype Cell is P13	=H13+L13/2
Formula Set XVII	Prototype Cell is Q13	=I13+M13/2
Formula Set XVIII	Prototype Cell is R13	=B8*P13
Formula Set XIX	Prototype Cell is S13	=B8*Q13
Formula Set XX	Prototype Cell is T13	=B8*(E8*O13-E5*P13-(E7+E8)*N13)/E3
Formula Set XXI	Prototype Cell is U13	=B8*(E8*N13-E6*Q13-(E8+E9)*O13)/E4
Formula Set XXII	Prototype Cell is V13	=F13+R13/2
Formula Set XXIII	Prototype Cell is W13	=G13+S13/2
Formula Set XXIV	Prototype Cell is X13	=H13+T13/2
Formula Set XXXVII	Prototype Cell is AK13	=B8*(E8*AD13-E6*AG13-(E8+E9)*AE13)/E4
Formula Set XXV	Prototype Cell is Y13	=I13+U13/2
Formula Set XXVI	Prototype Cell is Z13	=B8*X13
Formula Set XXVII	Prototype Cell is AA13	=B8*Y13
Formula Set XXVIII	Prototype Cell is AB13	=B8*(E8*W13-E5*X13-(E7+E8)*V13)/E3
Formula Set XXIX	Prototype Cell is AC13	=B8*(E8*V13-E6*Y13-(E8+E9)*W13)/E4
Formula Set XXX	Prototype Cell is AD13	=F13+Z13
Formula Set XXXI	Prototype Cell is AE13	=G13+AA13
Formula Set XXXII	Prototype Cell is AF13	=H13+AB13
Formula Set XXXIII	Prototype Cell is AG13	=I13+AC13
Formula Set XXXIV	Prototype Cell is AH13	=B8*AF13
Formula Set XXXV	Prototype Cell is AI13	=B8*AG13
Formula Set XXXVI	Prototype Cell is AJ13	=B8*(E8*AE13-E5*AF13-(E7+E8)*AD13)/E3

DISCRETE SYSTEMS WITH A PERIODIC RESTORING FORCE

TO GET THE MOST OUT OF THIS EXERCISE, YOU SHOULD ALREADY KNOW

The definitions of the terms:
oscillator
simple harmonic motion

How to solve equations of motion using the Runge-Kutta method.

Rotating systems are usually periodic because they return to their original angular position repeatedly. Other systems mimic such repetitive behavior because of restoring forces. In Exercises 6.1 and 6.2, you explored the periodic motions of systems subject to linear restoring forces (Equation 6.33):

$$m\frac{d^2x}{dt^2} + k_0x = 0 \tag{6.33}$$

where k_0 is a constant coefficient for the restoring force. This equation arises in many other parts of science and engineering in addition to the mechanics of point masses but with x, m, and k_0 interpreted differently. For example, if the system is rotary, then m represents a moment of inertia, x an angle, and k_0x is the system's restoring torque. In other physical contexts, the terms in the equation will represent other physical quantities, but it is customary to refer to the k_0x term as a restoring force.[1]

This final exercise explores oscillators subject to periodic restoring forces (i.e., forces in which k_0 is not a constant but is itself subject to periodic variation as a function of time t or displacement x). Examples of such oscillators include:

- A child pumping herself on a swing.

- The orbit of the Moon about the Earth as it experiences the periodic gravitational effects of the Sun.

- The motion of an electron in a crystal lattice as it periodically experiences the atoms at the lattice sites.

In some cases, the periodic restoring force is simply sinusoidal, and so the equation of motion takes the form

$$m\frac{d^2x}{dt^2} + k[1 + \alpha \cos(\omega t)]x = 0 \tag{6.34}$$

where k is a constant, α is a measure the strength of the restoring force's sinusoidal "modulation," and ω is the angular frequency of the modulation. Although the total force is not conservative, it is still useful to consider the time-varying energy of the underlying simple harmonic oscillator, namely, $E = 0.5\ mv^2 + 0.5\ kx^2$.

A useful way to write the constant k is

$$k = m(\beta\omega)^2 \tag{6.35}$$

The value of β tells you the ratio of the frequency of the unmodulated simple harmonic oscillator $(k/m)^{0.5}/2\pi$ to the angular frequency of the modulation. In an FM radio station, for example, the basic frequency (called the carrier frequency) is around 100 MHz, whereas the modulation frequency is, at most, 20 kHz. In this case, β is 5000 or greater.[2]

Next, it is very convenient to use a "normalized" independent variable instead of time. Let

$$\theta = \omega t \tag{6.36}$$

With these two substitutions you get

$$\frac{d^2x}{d\theta^2} + \beta^2(1 + \alpha \cos \theta)x = 0 \tag{6.37}$$

the equation actually used in the lessons that follow.

Notice that this equation is totally unchanged if $\theta + 2\pi$ is substituted for θ. You will discover an elegant use for this symmetry shortly.

LESSONS

Build and verify Model 6.3. Model 6.3 simulates an oscillator subject to a periodic restoring force by numerically solving Equation 6.37 using the Runge-Kutta method. Also build an on-screen graphic such as Figure 6.10 to plot oscillator motion vs time. Explore your model and verify that you fully understand its structure.

Although this model can, in principle, be used to study the behavior of an oscillator with a periodic restoring force, such oscillators are best studied over a wide range of time t (or normalized time, $\theta = \omega t$). As we write this book, most personal computers lack the speed and the large amounts of memory required to implement such solutions on electronic spreadsheets for the long time intervals you will find interesting. It turns out, however, that Equation 6.37 has two important properties that make it possible to extend a solution to long time intervals using only the solution in a short interval.

Therefore, before studying the details of the motion, you will use Model 6.3 to explore some theory that will allow you to make your investigations of the motions of these oscillators very rapid and efficient. The resulting techniques are interesting in their own right and are much used in science and engineering.

1. Initialize your model with $\alpha = 0$, $\beta = 1/2$, and count = 20. When α equals zero, the periodic restoring force is zero, and Model 6.3 represents simple harmonic motion. Use your plots to qualitatively verify this fact.

   Set your model to the initial conditions $(x_0, v_0) = (1,0)$. The initial conditions (1,0) generate what is called the "cosinelike" solution to your equation.

   a. Take the *values only* (not formulas) from column C and put them in column V by using either your spreadsheet's VALUE or PASTE SPECIAL (values only) command.[3] Label this column "xc(th)" (for "cosinelike"). In fact, for this equation, it is exactly the cosine function (check your plots to see this).

   b. Next, change the values of (x_0, v_0) to (0,1) and put the *values* of the C column in column W and label it "xs(th)," because it is the called the"sinelike" solution. Again, in this case, it is exactly the sine function.

   c. Finally, in column X, prepare formulas that correspond to Equation 6.38:

$$x(\theta) = x_0 x_c(\theta) + v_0 x_s(\theta) \tag{6.38}$$

Figure 6.10

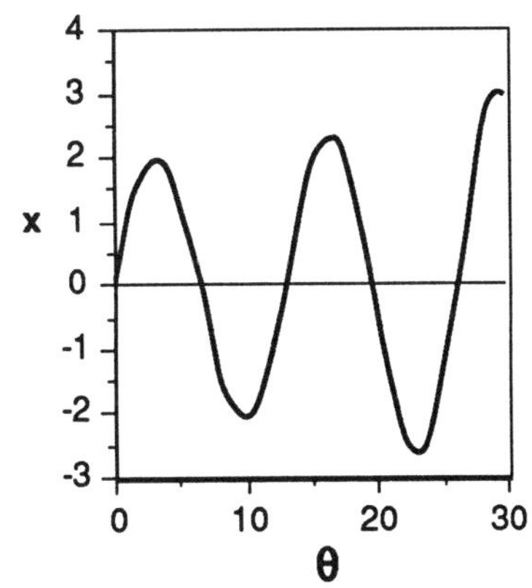

XY–PLOT

Data
Horizontal (*x*) axis: cells B10 - B110 (θ)
Vertical (*y*) axis: cells C10 - C110 (*x*)

(The formula in cell X10 is =B5*V10+B6*W10 and can be copied down the column provided that the references to B5 and B6 are absolute addresses.)

Unless you know a good deal about the theory of linear differential equations you will be astonished to discover that the values in column X are the same as in column C and will remain the same for *any* initial conditions (x_0, y_0).

d. You can verify Lesson 1c quickly by writing a formula in column Y that subtracts column C from column X. No matter what values you place in cells B5 and B6 this column will always contain values very close to zero.

The results of this lesson indicate that you need only solve equations of simple harmonic motion for *two* initial conditions $(x_0, v_0) = (1,0)$ and $(x_0, v_0) = (0,1)$ to have enough information to find the solution for *any* set of initial conditions. In words, any solution is a linear combination of the cosinelike and the sinelike solutions.

2. Now let's see if the useful property of simple harmonic oscillators discovered in Lesson 1 applies also to oscillators with a periodic restoring force. Set $\alpha = 0.3$ and repeat Lesson 1.

In all cases, the solution of the motion is a linear combination of the cosinelike and sinelike solutions. Thus, you need only solve Equation 6.37 twice to effectively solve it for all initial conditions. This is not true of most of the other mechanics problems you have examined.

This very important property of Equation 6.37 follows from the fact that if $x_1(\theta)$ and $x_2(\theta)$ are solutions, then so also is $Ax_1(\theta) + Bx_2(\theta)$, where A and B are constants. You can easily verify this. The initial conditions for the cosinelike and sinelike solutions of Lesson 1 were slyly chosen so that A and B are just the initial conditions of a particular problem. Differential equations with this property are said to be *linear*. They are very important in the physical sciences because many important problems involve them, including some of the most fundamental theories of Nature.

3. Lessons 1 and 2 have developed one of the two techniques that you will use to streamline Model 6.3. The other technique comes from the fact that Equation 6.37 is unchanged by substituting $\theta + 2\pi$ for θ. This property has the following interesting consequence:

- Find a solution of Equation 6.37 for some set of initial conditions and call it $x_1(\theta)$. [e.g., let $\alpha = 0.2$ and $\beta = 0.5$ then enter $(x_0, v_0) = (0.3, 0.5)$ into cells B5 and B6 and copy your solution (values only) from column C to column Z].

- Now re-solve the *same* equation of motion using for initial conditions $x_1(2\pi)$ and $v_1(2\pi)$. Call this solution $x_2(\theta)$ (i.e., enter the $x_1(\theta)$ and $v_1(\theta)$ values found in row 30 of your model into cells B5 and B6).

- It must be true that $x_1(\theta + 2\pi) = x_2(\theta)$ (i.e., the result in cell Z30 equals the result in cell C10, Z31 = C11, etc.).

4. The results of Lessons 2 and 3 tell you that you only need to know the cosinelike and sinelike functions $x_c(\theta)$ and $x_s(\theta)$ over the θ range $(0, 2\pi)$ to solve Equation 6.37 for *all* θ and for *any* initial conditions. You "bootstrap" your way up:

   - First use the Runge-Kutta method to calculate x_c and x_s and their corresponding velocities v_c and v_s between $\theta = 0$ and $\theta = 2\pi$.

   - Then to get x_c and x_s (you need both!) in the range 2π to 4π, use the equations

$$x_c(\theta + 2\pi) = x_c(2\pi)x_c(\theta) + v_c(2\pi)x_s(\theta) \qquad (6.39)$$

$$v_c(\theta + 2\pi) = x_c(2\pi)v_c(\theta) + v_c(2\pi)v_s(\theta) \qquad (6.40)$$

   The corresponding formulas for x_s and v_s are similar.

   To get solutions for the next 2π interval, use these same equations replacing 2π with 4π, etc.

   a. Verify Equations 6.39 and 6.40 by a suitable modification to Model 6.3.

   b. Build and verify Model 6.4. This model has a more complex structure than others you have built. Read carefully the directions given with the model.

   Study this model until you understand how it implements the strategies just discussed.

5. Now that you can solve Equation 6.37 efficiently for long time intervals, let's make a survey of the nature of the motions of an

Figure 6.11

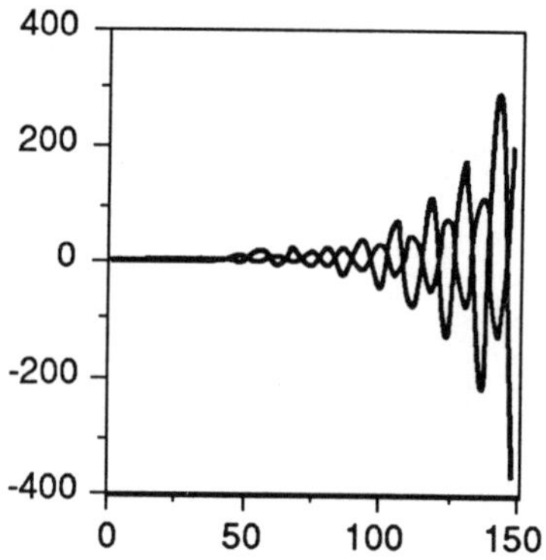

XY-PLOT

Data
Horizontal (*x*) axis: cells D11 - D491 (θ)
Vertical (*y*) axis: cells E11 - E491 (cosinelike solution), and Y11 - Y491 (sinelike solution)

Figure 6.12

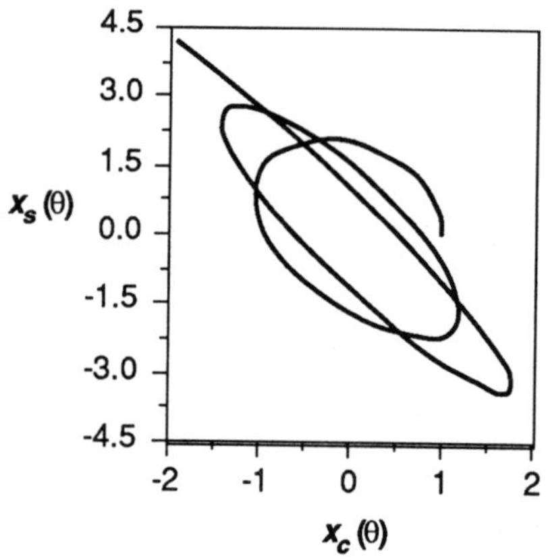

XY-PLOT

Data
Horizontal (*x*) axis: cells E11 - E111 (cosinelike solution)
Vertical (*y*) axis: cells Y11 - Y111 (sinelike solution)

Note
For other values of α and β, you should extend this plot for the entire θ range. We've reduced the range here to keep the plot from appearing boring.

oscillator with a periodic restoring force. Although your model can plot $x(\theta)$ for any initial conditions (column A), we'll confine ourselves to studies of the motions of the sinelike and cosinelike solutions (columns E and Y). Build on-screen graphics that plot $x_c(\theta)$, $x_s(\theta)$ vs θ (Figure 6.11); $x_c(\theta)$ vs $x_s(\theta)$, which is interesting to look at (Figure 6.12); and the energy of the underlying oscillator vs *t* (Figure 6.13).

a. Begin by setting $\alpha = 0$ and $\beta = 1$ and examining your plots. You are now looking at equal-amplitude sinusoids.

b. Now set $\beta = 0.5$. Describe the shapes of your two plots. What is the angular frequency of the oscillator?

c. Turn on some gentle modulation for the restoring force by setting $\alpha = 0.05$ and look at your three graphics. Can you guess at the susequent behavior of the oscillator? Recall the meaning of β and note that $\beta = 0.5$ corresponds to a condition where the modulation frequency is twice that of the underlying simple harmonic oscillator. Clearly, this frequency ratio and modulation corresponds to a resonance condition. The motion is said to be *unstable.*

   Your experience with the driven simple harmonic oscillator might lead you to expect resonance when the modulation frequency is equal to the underlying oscillator (i.e., $\beta = 1$). Try it. Are you surprised? Try to understand these results physically.

   In this case, the motion is simply a slightly modified sinusoid and is said to be *stable.*

d. Holding β at 1, vary the value of α and search for a new unstable condition. (Although you cannot see it clearly by looking at the functions over the θ range available on your personal computer, the value of α that divides stable from unstable solutions is about 0.17. In Problem 1 you will see an efficient way to calculate this result.)

   The following table indicates various stable and unstable motions. Verify the table and explore the motions (some of them are quite surprising).

	α				
β	0.0	0.5	1.0	1.5	2.0
0.25	stb	stb	stb	stb	stb
0.50	stb	unstb	unstb	unstb	unstb
0.75	stb	stb	stb	unstb	unstb
1.00	stb	unstb	unstb	unstb	unstb
1.25	stb	stb	stb	unstb	unstb

From your observations, you can see that some regions in the α, β-plane yield stable solutions and others yield unstable ones. Figure 6.14 shows you these regions at a higher degree of resolution. All systems governed by periodic restoring forces display this character—it is crucial to understanding the nature of such motions:

- A child on a swing who does not pump at just the right frequency pumps in vain. (Perhaps you've observed a small child who has not yet learned "to do it.")

- In a synchrotron, the magnetic field in which an electron moves varies slightly as a function of its position and constitutes a periodic restoring force. Operators of the machine control parameters analogous to α and β to tune it. If the tune gets misadjusted (because of a fault in a magnet, for example), the electron orbits are unstable, and the machine does not work.

- The output voltage of a parametric amplifier is amplified by circuits that are driven sinusoidally with parameters adjusted so that the solution is unstable. In real systems, of course, damping prevents the "motion" from becoming infinite.

- The motions of electrons in a crystal are governed by equations with periodic restoring forces. The energy of the electron tunes the parameters of the periodic restoring force and so only certain bands of energy are allowed (unstable solutions correspond to motions that pop out of the material and so do not occur). It is the special arrangement of these bands in silicon that makes it useful for building transistors, hundreds of thousands of which are inside the computer you are using!

Figure 6.13

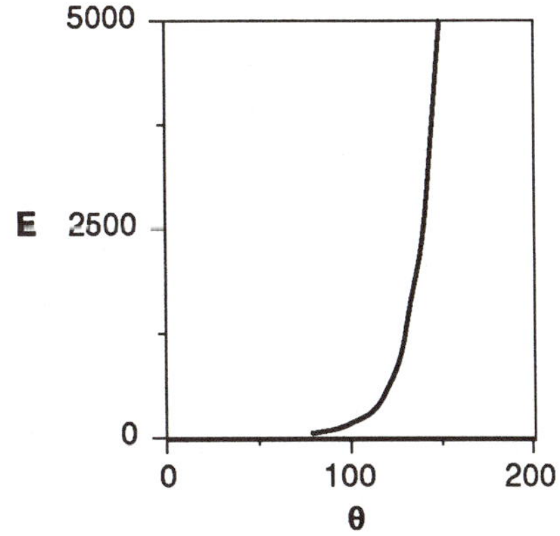

XY-PLOT

Data
Horizontal (*x*) axis: cells D11 - D491 (θ)
Vertical (*y*) axis: cells C11 - C491 (energy)

Figure 6.14

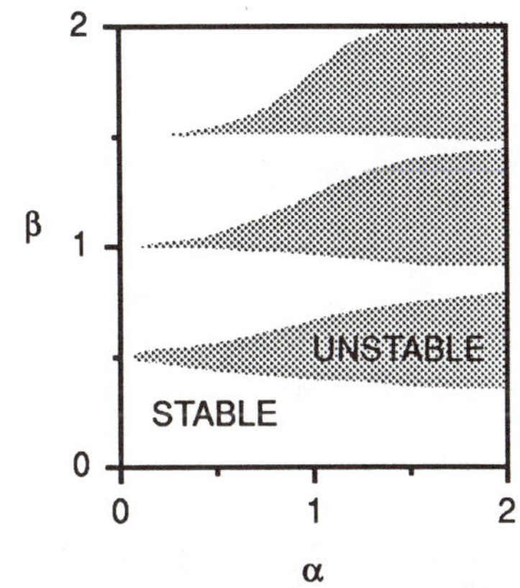

Regions in a portion of the (α, β) plane in which Equation 6.37 has stable and unstable motions.

Notice that the β = 0 line corresponds to a very simple differential equation that is unstable (i.e., one of the two independent solutions rises or falls linearly). In general, for parameter pairs in the stable region but very near an unstable region, the amplitude of the motion can get very large but ultimately gets small again. Therefore, it is difficult to pin down the boundaries accurately by examining solutions over a small range of θ values. The data for this picture were calculated by finding the eigenvalues of the matrix **T**(2π).

PROBLEMS

1. In the lessons, you found that Equation 6.37 has two types of solutions: stable and unstable. You can determine the stability of

specific examples of Equation 6.37 by finding the eigenvalues of a certain matrix using the same techniques you studied in the context of the inertia tensor.

Suppose that the position x and velocity v of an oscillator (at normalized time θ) is written as a vector $X(\theta)$ such that

$$X(\theta) = x(\theta)\boldsymbol{i} + v(\theta)\boldsymbol{j} \tag{6.41}$$

Recalling that $x(\theta)$ can be expressed as linear combinations of $x_c(\theta)$ and $x_s(\theta)$ and that $v(\theta)$ can be expressed as linear combinations of $v_c(\theta)$ and $v_s(\theta)$, you can write a matrix equation relating $X(\theta)$ to the initial conditions, $X(0) = X_0$:

$$X(\theta) = \mathbf{T}(\theta) \cdot X_0 \tag{6.42}$$

where the matrix $\mathbf{T}$ is

$$\mathbf{T} = \begin{pmatrix} x_c(\theta) & x_s(\theta) \\ v_c(\theta) & v_s(\theta) \end{pmatrix}$$

You know from the lessons that if you have the matrix $\mathbf{T}$ in the θ range from 0 to 2π you can always bootstrap a solution by using $X(2\pi)$ as new initial conditions, and then $X(4\pi)$, etc.

Suppose it turns out that $X(2\pi)$ is just a multiple times X_0. Then the motion in the range $(2\pi, 4\pi)$ is just a multiple of that in $(0, 2\pi)$. In other words, if the initial conditions are just proportional to an *eigenvector* of the matrix $\mathbf{T}(2\pi)$, the solution in the range $(2\pi, 4\pi)$ is the corresponding eigenvalue times the solution in $(0, 2\pi)$, and that in $(4\pi, 6\pi)$ is the square of the eigenvalue times the solution in $(0, 2\pi)$, etc.

If a specific periodic restoring force yields a matrix $\mathbf{T}(2\pi)$ (obtained by solving the equation of motion) with an eigenvalue of magnitude greater than 1, then the motion is unstable. As the value of the eigenvalue increases, the instability shows itself more quickly. If $\mathbf{T}(2\pi)$ does not have real eigenvalues, the motion is stable. As always with eigenvalue problems, the case of degeneracy (both eigenvalues equal) is special and will not be discussed (but you can get a hint by thinking about the two cases $\beta = 0$ and then $\beta \neq 0$, $\alpha = 0$).

a. Modify Model 6.4 to calculate the eigenvalues of the matrix $\mathbf{T}(2\pi)$ and verify the above theory. When the matrix has real eigenvalues, find the corresponding eigenvectors and plot the motions when these eigenvectors are used as the initial conditions.

b. What do you notice about the determinant of this matrix? Can you prove mathematically that this is always the case?

c. Use the 2-WAY WHAT-IF (or DATA TABLE) command of your spreadsheet to generate the table below. The word *stab* refers to stable solutions, and numeric values indicate the rate at which the solution increases without bound (the eigenvalue). You can speed your calculations by deleting all of your model below row 31.

β \ α	0	0.2	0.4	0.6	0.8	1	1.2	1.4	1.6	1.8	2
0.1	stab	stab	stab	stab	stab	stab	stab	stab	stab	stab	stab
0.2	stab	stab	stab	stab	stab	stab	stab	stab	stab	stab	stab
0.3	stab	stab	stab	stab	stab	stab	stab	stab	stab	stab	stab
0.4	stab	stab	stab	stab	stab	stab	1.316	1.653	1.936	2.222	2.520
0.5	stab	1.169	1.367	1.594	1.853	2.146	2.474	2.839	3.241	3.680	4.155
0.6	stab	stab	stab	stab	1.543	2.011	2.487	2.999	3.551	4.144	4.777
0.7	stab	stab	stab	stab	stab	stab	1.280	2.048	2.682	3.301	3.914
0.8	stab	stab	stab	stab	stab	stab	stab	stab	stab	stab	stab
0.9	stab	stab	stab	stab	stab	stab	stab	stab	stab	stab	–1.74
1	stab	–1.01	–1.09	–1.23	–1.44	–1.76	–2.24	–2.91	–3.86	–5.16	–6.89
1.1	stab	stab	stab	stab	–1.60	–2.30	–3.21	–4.44	–6.09	–8.27	–11.0
1.2	stab	stab	stab	stab	stab	–1.73	–3.08	–4.73	–6.85	–9.56	–12.8
1.3	stab	stab	stab	stab	stab	stab	–1.70	–3.54	–5.62	–8.08	–10.8
1.4	stab	stab	stab	stab	stab	stab	stab	stab	–2.09	–3.35	–4.16
1.5	stab	stab	stab	stab	stab	stab	stab	stab	1.535	3.180	6.041
1.6	stab	stab	stab	stab	1.508	2.141	3.067	4.609	7.234	11.60	18.58
1.7	stab	stab	stab	stab	stab	2.227	4.040	6.873	11.41	18.51	29.17
1.8	stab	stab	stab	stab	stab	stab	3.748	7.373	13.00	21.44	33.48
1.9	stab	stab	stab	stab	stab	stab	2.112	5.735	10.95	18.28	27.78
2	stab	stab	stab	stab	stab	stab	stab	1.845	5.029	8.276	10.49

d. Equation 6.37 is unchanged by substituting $\theta + 2\pi$ for θ. Equation 6.37 has another symmetry, namely that it is unchanged when θ is replaced by $-\theta$. Try to develop a hypothesis about how this symmetry manifests itself in oscillator behavior. Test your hypothesis by studying the actual motions.

2. In the lessons you exploited the symmetry of Equation 6.37 under the replacement of θ by $\theta + 2\pi$ to bootstrap a solution for all θ from a solution in the range $(0, 2\pi)$. The simple harmonic oscillator has an even stronger symmetry—you can replace θ by θ plus *anything* and the equation is unchanged.

   a. Set $\alpha = 0$ in Model 6.4 and $\Delta\theta$ equal to any small angle (in particular, not $2\pi/20$) and verify that you can generate the sine and cosine functions for any value of θ using Equations 6.39 and 6.40 with 2π replaced by $20\Delta\theta$. In the case of the simple harmonic oscillator, what well-known trignometric identities do these equations correspond to?

b. Try this technique on the damped harmonic oscillator. Does it work?

c. Finally, try it on the damped harmonic oscillator with a sinusoidal driving force.

3. A synchrotron is a particle accelerator in which electrons, protons or other charged particles must stay in a small pipe that curves around in a roughly circular closed loop in a horizontal plane. For example, at the proposed SSC (superconducting supercollider), the pipe will be more than 50 miles long and just a few centimeters in diameter. It is installed in a magnetic field, suitably arranged so that when a proton is launched exactly at its center and in exactly the right direction, it remains in the center of the pipe as it traverses around the machine. Almost no protons can be launched perfectly, however; most are launched both off-center and in the wrong direction. A proton's deviation from the center of the pipe when it has imperfect launch conditions is governed by differential equations with periodic restoring forces.

Let θ be an angle that measures distance around the pipe, $x(\theta)$ be the proton's horizontal distance from the center of the pipe, and $y(\theta)$ be its vertical distance from the center. Then the equations of motion for these deviations are (approximately)

$$\frac{d^2x}{d\theta^2} + s(\theta)x = 0 \tag{6.43a}$$

$$\frac{d^2y}{d\theta^2} - s(\theta)y = 0 \tag{6.43b}$$

where the "restoring force" $s(\theta)$ is piecewise constant and periodic with period 2π. In practical machines, $s(\theta)$ (called the "lattice" of the machine) is quite complex with many pieces, but for this problem assume that it consists of just four nonzero pieces as follows:

$$s(\theta) = \begin{cases} 0 & 0° < \theta < 30° \\ s_1 & 30° < \theta < 60° \\ 0 & 60° < \theta < 120° \\ -s_2 & 120° < \theta < 150° \\ 0 & 150° < \theta < 210° \\ s_1 & 210° < \theta < 240° \\ 0 & 240° < \theta < 300° \\ -s_2 & 300° < \theta < 330° \\ 0 & 330° < \theta < 360° \end{cases} \tag{6.44}$$

The regions of the machine where s is not zero are called lenses and the constants s_1 and s_2 are their strengths. Lattice designers design the lattice of the synchrotron by choosing the strengths $s(\theta)$ of its pieces. In the real world, a lens strength cannot be made arbitrarily large, and in this problem you must assume that lenses cannot be made stronger than $s = 2$.

a. Because both the horizontal and vertical motions of the protons must be stable, find the stability regions in the (s_1, s_2) plane, consistent with the maximum strength requirement above.

b. To keep the cost of the expensive magnets as small as possible, you want the beam inside the pipe to be as small as possible [i.e., for any initial condition (x_0, v_{x0}) or (y_0, v_{y0}) in a 2 cm by 2 cm/radian square centered on (0,0), you want to choose the strengths so that the largest deviation from the center of the pipe is as small as possible]. If the billions of protons in the machine actually have initial conditions that fill this 2 cm by 2 cm/radian square, what diameter must the pipe have, and at what value(s) of θ does the beam just graze the inside of the pipe?

4. You can explore the act of "pumping" a swing by noting the equation of motion of a pendulum in a uniform gravitational field g (Equation 6.45):

$$\frac{d}{dt}\left(ml^2\frac{d\theta}{dt}\right) = -\,mgl \sin\theta \tag{6.45}$$

where θ is the angle the string of the pendulum makes with the vertical, m is its mass, and l is the length of the string. Suppose that the length of the string is being periodically changed so that

$$l(t) = l_0[1 + \alpha \cos(\omega t)] \tag{6.46}$$

where $-1 < \alpha < 1$.

a. Verify that the equation of motion is approximately of the form of Equation 6.34 when q and a are very much smaller than 1, but not when either is relatively large.

b. Modify Model 6.3 to solve the exact problem, not the small-angle approximation. What pumping frequency w ensures amplitude increase? Can you explain this physically? Will the pendulum start swinging if it is initially at rest at its equilibrium position? Can you arrange things so that the pendulum ends up going in circles (becomes a "revolving pendulum")? In practice, when you pump yourself on a swing, do you keep the pumping frequency constant as in this problem?

5. A parametric amplifier can be studied by adding viscous damping and a sinusoidal drive force to Equation 6.37. Modify Model 6.3 to deal with this case, observing the steady-state motion. You can define the amplification of the system as the ratio of the steady-state amplitude with modulation ($\alpha \neq 0$) to that without it ($\alpha = 0$). Study the amplification of the system as you change its parameters and write a brief report describing what you find.

MODEL

Model 6.3

BUILD and verify Model 6.3.

ENTER data using cells B5 - B7 (initial position, initial velocity, number of rows required to span 2π); cells E4 and E5 (α, β parameters of Equation 6.37).

READ output from the model from columns A - D (total energy, normalized time $\theta = \omega t$, position, velocity).

FORMULAS

	A	B	C	D	E	F	G	H	I	J	K	L
1	Model 6.3											
2	Sinusoidal periodic restoring force											
3												
4	Δ[theta] =	=2*PI()/B7		alpha =	0.2							
5	x0 =	0		beta =	0.5							
6	v0 =	1										
7	count =	20		beta^2 =	=E5^2							
8					0					1		
9	E	theta	x	v	kx	kv	t+Δt/2	x	v	kx	kv	t+Δt/2
10		0	=B5	=B6								
11	Form Set I	Form Set II	Form Set III	Form Set IV	Form Set V	Form Set VI	Form Set VII	Form Set VIII	Form Set IX	Form Set X	Form Set XI	Form Set XII
12												
13												

FORMULAS (Contd.)

	M	N	O	P	Q	R	S	T	U
8			2					3	
9	x	v	kx	kv	t+Δt/2	x	v	kx	kv
10									
11	Form Set XIII	Form Set XIV	Form Set XV	Form Set XVI	Form Set XVII	Form Set XVIII	Form Set XIX	Form Set XX	Form Set XXI
12									
13									

VALUES

	A	B	C	D	E	F	G	H	I	J	K	L
1	Model 6.3											
2	Sinusoidal periodic restoring force											
3												
4	Δ[theta] =	0.3142		alpha =	0.2							
5	x0 =	0		beta =	0.5							
6	v0 =	1										
7	count =	20		beta^2 =	0.25							
8					0					1		
9	E	theta	x	v	kx	kv	t+Δt/2	x	v	kx	kv	t+Δt/2
10	0.5	0	0	1	0.3142	0	0.1571	0.1571	1	0.3142	-0.015	0.1571
11	0.4976162	0.3142	0.3126	0.9853	0.3095	-0.029	0.4712	0.4674	0.9707	0.3049	-0.043	0.4712
12	0.4914137	0.6283	0.6161	0.9423	0.296	-0.056	0.7854	0.7641	0.9142	0.2872	-0.069	0.7854
13	0.4838311	0.9425	0.902	0.8742	0.2746	-0.079	1.0996	1.0394	0.8346	0.2622	-0.089	1.0996

VALUES (Contd.)

	M	N	O	P	Q	R	S	T	U
8			2					3	
9	x	v	kx	kv	t+Δt/2	x	v	kx	kv
10	0.1571	0.9926	0.3118	-0.015	0.3142	0.3118	0.9852	0.3095	-0.029
11	0.4651	0.9637	0.3027	-0.043	0.6283	0.6154	0.9423	0.296	-0.056
12	0.7597	0.9081	0.2853	-0.068	0.9425	0.9014	0.8742	0.2746	-0.079
13	1.0331	0.8297	0.2607	-0.089	1.2566	1.1627	0.7857	0.2468	-0.097

Model 6.3 (continued)

Formula Set I	A10	=0.5*(D10^2+**E7** *C10^2)
Formula Set II	B11	=B10+**B4**
Formula Set III	C11	=C10+(E10+2*(J10+O10)+T10)/6
Formula Set IV	D11	=D10+(F10+2*(K10+P10)+U10)/6
Formula Set V	E10	=**B4** *D10
Formula Set VI	F10	=-**B4** ***E7*** *(1+**E4** *COS(B10))*C10
Formula Set VII	G10	=B10+**B4**/2
Formula Set VIII	H10	=C10+E10/2
Formula Set IX	I10	=D10+F10/2
Formula Set X	J10	=**B4** *I10
Formula Set XI	K10	=-**B4** ***E7*** *(1+**E4** *COS(G10))*H10
Formula Set XII	L10	=B10+**B4**/2
Formula Set XIII	M10	=C10+J10/2
Formula Set XIV	N10	=D10+K10/2
Formula Set XV	O10	=**B4** *N10
Formula Set XVI	P10	=-**B4** ***E7*** *(1+**E4** *COS(L10))*M10
Formula Set XVII	Q10	=B10+**B4**
Formula Set XVIII	R10	=C10+O10
Formula Set XIX	S10	=D10+P10
Formula Set XX	T10	=**B4** *S10
Formula Set XXI	U10	=-**B4** ***E7*** *(1+**E4** *COS(Q10))*R10

Model 6.4 (Part 1)

Model 6.4 comes in two parts. Both must be entered into your computer before the model will function correction.

BUILD and verify Part 1 of Model 6.4. This part of the model solves Equation 6.37 between $\theta = 0$ and $\theta = 2\pi$ using the Runge-Kutta method. The cosinelike solution is found in columns E and F, the sinelike solution is found in columns Y and Z. Solutions for any initial conditions are present in columns A and B.

ENTER data using cells E4 and E5 (initial position and velocity); cells H4 and H5 (α, β parameters of Equation 6.37). The model is built assuming that 20 rows corresponds exactly to 2π of normalized time θ and so the value in cell A5 should not be changed (except for Problem 2 where other issues come into play).

READ output from the model from columns A - F position, velocity, mechanical energy, normalized time, cosine-like solutions for position and velocity).

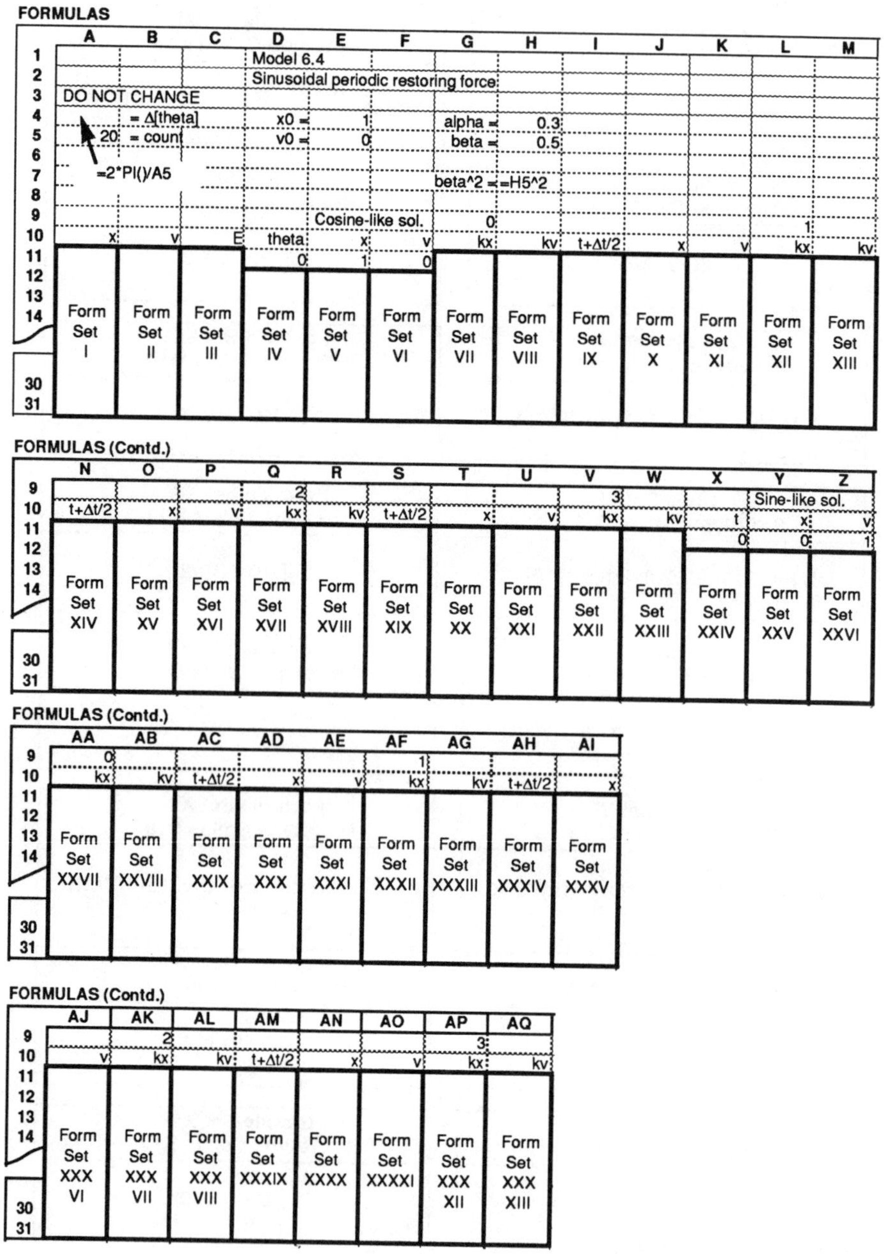

Model 6.4 (Part 1 continued)

VALUES

	A	B	C	D	E	F	G	H	I	J	K	L	M
1				Model 6.4									
2				Sinusoidal periodic restoring force									
3	DO NOT CHANGE												
4	0.3142	= Δ[theta]		x0 =	1		alpha =	0.3					
5	20	= count		v0 =	0		beta =	0.5					
6													
7							beta^2 =	0.25					
8													
9					Cosine-like sol.		0					1	
10	x	v	E	theta	x	v	kx	kv	t+Δt/2	x	v	kx	kv
11	1	0	0.125	0	1	0	0	-0.102	0.1571	1	-0.051	-0.016	-0.102
12	0.984	-0.101	0.1262	0.3142	0.984	-0.101	-0.032	-0.099	0.4712	0.9681	-0.151	-0.047	-0.096
13	0.937	-0.197	0.1291	0.6283	0.937	-0.197	-0.062	-0.091	0.7854	0.9061	-0.243	-0.076	-0.086
14	0.8613	-0.283	0.1327	0.9425	0.8613	-0.283	-0.089	-0.08	1.0996	0.8169	-0.323	-0.101	-0.073
30	-1.049	0.0158	0.1378	5.969	-1.049	0.0158	0.005	0.1059	6.1261	-1.047	0.0688	0.0216	0.1066
31	-1.028	0.1217	0.1394	6.2832	-1.028	0.1217	0.0382	0.1049	6.4403	-1.009	0.1742	0.0547	0.1027

VALUES (Contd.)

	N	O	P	Q	R	S	T	U	V	W	X	Y	Z
9				2					3			Sine-like sol.	
10	t+Δt/2	x	v	kx	kv	t+Δt/2	x	v	kx	kv	t	x	v
11	0.1571	0.992	-0.051	-0.016	-0.101	0.3142	0.984	-0.101	-0.032	-0.099	0	0	1
12	0.4712	0.9603	-0.149	-0.047	-0.096	0.6283	0.9371	-0.197	-0.062	-0.091	0.3142	0.3125	0.9841
13	0.7854	0.8989	-0.24	-0.075	-0.086	0.9425	0.8616	-0.283	-0.089	-0.08	0.6283	0.6151	0.9379
14	1.0996	0.8107	-0.319	-0.1	-0.072	1.2566	0.7611	-0.355	-0.112	-0.065	0.9425	0.8991	0.8659
30	6.1261	-1.039	0.0691	0.0217	0.1057	6.2832	-1.028	0.1215	0.0382	0.1049	5.969	0.7759	-0.965
31	6.4403	-1	0.1731	0.0544	0.1018	6.5973	-0.973	0.2236	0.0702	0.0983	6.2832	0.4621	-1.028

VALUES (Contd.)

	AA	AB	AC	AD	AE	AF	AG	AH	AI
9	0					1			
10	kx	kv	t+Δt/2	x	v	kx	kv	t+Δt/2	x
11	0.3142	0	0.1571	0.1571	1	0.3142	-0.016	0.1571	0.1571
12	0.3092	-0.032	0.4712	0.4671	0.9683	0.3042	-0.046	0.4712	0.4646
13	0.2947	-0.06	0.7854	0.7625	0.9079	0.2852	-0.073	0.7854	0.7578
14	0.272	-0.083	1.0996	1.0351	0.8243	0.259	-0.092	1.0996	1.0286
30	-0.303	-0.078	6.1261	0.6243	-1.004	-0.315	-0.064	6.1261	0.6182
31	-0.323	-0.047	6.4403	0.3007	-1.051	-0.33	-0.031	6.4403	0.297

VALUES (Contd.)

	AJ	AK	AL	AM	AN	AO	AP	AQ
9		2					3	
10	v	kx	kv	t+Δt/2	x	v	kx	kv
11	0.992	0.3116	-0.016	0.3142	0.3116	0.984	0.3091	-0.031
12	0.9609	0.3019	-0.046	0.6283	0.6143	0.9379	0.2946	-0.06
13	0.9016	0.2833	-0.072	0.9425	0.8984	0.8658	0.272	-0.083
14	0.8197	0.2575	-0.092	1.2566	1.1566	0.7741	0.2432	-0.099
30	-0.996	-0.313	-0.063	6.2832	0.4628	-1.028	-0.323	-0.047
31	-1.043	-0.328	-0.03	6.5973	0.1344	-1.058	-0.332	-0.014

Model 6.4 (Part 1 continued)

Formula Set	Prototype Cell	Formula
I	A11	=E4*E11+E5*Y11
II	B11	=E4*F11+E5*Z11
III	C11	=0.5*(F11^2+H7*E11^2)
IV	D12	=D11+A4
V	E12	=E11+(G11+2*(L11+Q11)+V11)/6
VI	F12	=F11+(H11+2*(M11+R11)+W11)/6
VII	G11	=A4*F11
VIII	H11	=-A4*H7*(1+H4*COS(D11))*E11
IX	I11	=D11+A4/2
X	J11	=E11+G11/2
XI	K11	=F11+H11/2
XII	L11	=A4*K11
XIII	M11	=-A4*H7*(1+H4*COS(I11))*J11
XIV	N11	=D11+A4/2
XV	O11	=E11+L11/2
XVI	P11	=F11+M11/2
XVII	Q11	=A4*P11
XVIII	R11	=-A4*H7*(1+H4*COS(N11))*O11
XIX	S11	=D11+A4
XX	T11	=E11+Q11
XXI	U11	=F11+R11
XXII	V11	=A4*U11
XXIII	W11	=-A4*H7*(1+H4*COS(S11))*T11
XXIV	X12	=X11+A4
XXV	Y12	=Y11+(AA11+2*(AF11+AK11)+AP11)/6
XXVI	Z12	=Z11+(AB11+2*(AG11+AL11)+AQ11)/6
XXVII	AA11	=A4*Z11
XXVIII	AB11	=-A4*H7*(1+H4*COS(X11))*Y11
XXIX	AC11	=D11+A4/2
XXX	AD11	=Y11+AA11/2
XXXI	AE11	=Z11+AB11/2
XXXII	AF11	=A4*AE11
XXXIII	AG11	=-A4*H7*(1+H4*COS(AC11))*AD11
XXXIV	AH11	=X11+A4/2
XXXV	AI11	=Y11+AF11/2
XXXVI	AJ11	=Z11+AG11/2
XXXVII	AK11	=A4*AJ11
XXXVIII	AL11	=-A4*H7*(1+H4*COS(AH11))*AI11
XXXIX	AM11	=X11+A4
XXXX	AN11	=Y11+AK11
XXXXI	AO11	=Z11+AL11
XXXXII	AP11	=A4*AO11
XXXXIII	AQ11	=-A4*H7*(1+H4*COS(AM11))*AN11

FORMULAS

$x(\theta + 2\pi) = x_c(2\pi)x_c(\theta) + v_c(2\pi)x_s(\theta)$

$v(\theta + 2\pi) = x_c(2\pi)v_c(\theta) + v_c(2\pi)v_s(\theta)$

	E	F		Y	Z
32	=E31*E12+F31*Y12	=E31*F12+F31*Z12	32	=Y31*E12+Z31*Y12	=Y31*F12+Z31*Z12
33	=E31*E13+F31*Y13	=E31*F13+F31*Z13	33	=Y31*E13+Z31*Y13	=Y31*F13+Z31*Z13
34	=E31*E14+F31*Y14	=E31*F14+F31*Z14	34	=Y31*E14+Z31*Y14	=Y31*F14+Z31*Z14
35	=E31*E15+F31*Y15	=E31*F15+F31*Z15	35	=Y31*E15+Z31*Y15	=Y31*F15+Z31*Z15
36	=E31*E16+F31*Y16	=E31*F16+F31*Z16	36	=Y31*E16+Z31*Y16	=Y31*F16+Z31*Z16
37	=E31*E17+F31*Y17	=E31*F17+F31*Z17	37	=Y31*E17+Z31*Y17	=Y31*F17+Z31*Z17
38	=E31*E18+F31*Y18	=E31*F18+F31*Z18	38	=Y31*E18+Z31*Y18	=Y31*F18+Z31*Z18
39	=E31*E19+F31*Y19	=E31*F19+F31*Z19	39	=Y31*E19+Z31*Y19	=Y31*F19+Z31*Z19
40	=E31*E20+F31*Y20	=E31*F20+F31*Z20	40	=Y31*E20+Z31*Y20	=Y31*F20+Z31*Z20
41	=E31*E21+F31*Y21	=E31*F21+F31*Z21	41	=Y31*E21+Z31*Y21	=Y31*F21+Z31*Z21
42	=E31*E22+F31*Y22	=E31*F22+F31*Z22	42	=Y31*E22+Z31*Y22	=Y31*F22+Z31*Z22
43	=E31*E23+F31*Y23	=E31*F23+F31*Z23	43	=Y31*E23+Z31*Y23	=Y31*F23+Z31*Z23
44	=E31*E24+F31*Y24	=E31*F24+F31*Z24	44	=Y31*E24+Z31*Y24	=Y31*F24+Z31*Z24
45	=E31*E25+F31*Y25	=E31*F25+F31*Z25	45	=Y31*E25+Z31*Y25	=Y31*F25+Z31*Z25
46	=E31*E26+F31*Y26	=E31*F26+F31*Z26	46	=Y31*E26+Z31*Y26	=Y31*F26+Z31*Z26
47	=E31*E27+F31*Y27	=E31*F27+F31*Z27	47	=Y31*E27+Z31*Y27	=Y31*F27+Z31*Z27
48	=E31*E28+F31*Y28	=E31*F28+F31*Z28	48	=Y31*E28+Z31*Y28	=Y31*F28+Z31*Z28
49	=E31*E29+F31*Y29	=E31*F29+F31*Z29	49	=Y31*E29+Z31*Y29	=Y31*F29+Z31*Z29
50	=E31*E30+F31*Y30	=E31*F30+F31*Z30	50	=Y31*E30+Z31*Y30	=Y31*F30+Z31*Z30
51	=E31*E31+F31*Y31	=E31*F31+F31*Z31	51	=Y31*E31+Z31*Y31	=Y31*F31+Z31*Z31
52	=E51*E12+F51*Y12	=E51*F12+F51*Z12	52	=Y51*E12+Z51*Y12	=Y51*F12+Z51*Z12
53	=E51*E13+F51*Y13	=E51*F13+F51*Z13	53	=Y51*E13+Z51*Y13	=Y51*F13+Z51*Z13
54	=E51*E14+F51*Y14	=E51*F14+F51*Z14	54	=Y51*E14+Z51*Y14	=Y51*F14+Z51*Z14
55	=E51*E15+F51*Y15	=E51*F15+F51*Z15	55	=Y51*E15+Z51*Y15	=Y51*F15+Z51*Z15
56	=E51*E16+F51*Y16	=E51*F16+F51*Z16	56	=Y51*E16+Z51*Y16	=Y51*F16+Z51*Z16
57	=E51*E17+F51*Y17	=E51*F17+F51*Z17	57	=Y51*E17+Z51*Y17	=Y51*F17+Z51*Z17
58	=E51*E18+F51*Y18	=E51*F18+F51*Z18	58	=Y51*E18+Z51*Y18	=Y51*F18+Z51*Z18
59	=E51*E19+F51*Y19	=E51*F19+F51*Z19	59	=Y51*E19+Z51*Y19	=Y51*F19+Z51*Z19
60	=E51*E20+F51*Y20	=E51*F20+F51*Z20	60	=Y51*E20+Z51*Y20	=Y51*F20+Z51*Z20
61	=E51*E21+F51*Y21	=E51*F21+F51*Z21	61	=Y51*E21+Z51*Y21	=Y51*F21+Z51*Z21
62	=E51*E22+F51*Y22	=E51*F22+F51*Z22	62	=Y51*E22+Z51*Y22	=Y51*F22+Z51*Z22
63	=E51*E23+F51*Y23	=E51*F23+F51*Z23	63	=Y51*E23+Z51*Y23	=Y51*F23+Z51*Z23
64	=E51*E24+F51*Y24	=E51*F24+F51*Z24	64	=Y51*E24+Z51*Y24	=Y51*F24+Z51*Z24
65	=E51*E25+F51*Y25	=E51*F25+F51*Z25	65	=Y51*E25+Z51*Y25	=Y51*F25+Z51*Z25
66	=E51*E26+F51*Y26	=E51*F26+F51*Z26	66	=Y51*E26+Z51*Y26	=Y51*F26+Z51*Z26
67	=E51*E27+F51*Y27	=E51*F27+F51*Z27	67	=Y51*E27+Z51*Y27	=Y51*F27+Z51*Z27
68	=E51*E28+F51*Y28	=E51*F28+F51*Z28	68	=Y51*E28+Z51*Y28	=Y51*F28+Z51*Z28
69	=E51*E29+F51*Y29	=E51*F29+F51*Z29	69	=Y51*E29+Z51*Y29	=Y51*F29+Z51*Z29
70	=E51*E30+F51*Y30	=E51*F30+F51*Z30	70	=Y51*E30+Z51*Y30	=Y51*F30+Z51*Z30
71	=E51*E31+F51*Y31	=E51*F31+F51*Z31	71	=Y51*E31+Z51*Y31	=Y51*F31+Z51*Z31
72	=E71*E12+F71*Y12	=E71*F12+F71*Z12	72	=Y71*E12+Z71*Y12	=Y71*F12+Z71*Z12

$x_c(\theta + 4\pi) = x_c(4\pi)x_c(\theta) + v_c(4\pi)x_s(\theta)$

$v_c(\theta + 4\pi) = x_c(4\pi)v_c(\theta) + v_c(4\pi)v_s(\theta)$

6π

Model 6.4 (Part 2)

Model 6.4 comes in two parts. Both must be entered into your computer before the model will function correction.

BUILD and verify Part 2 of Model 6.4 by first extending columns A - D down to row 491. The subsequent procedure is quite different from your usual experience. Begin by entering the formula =**E31***E12+**F31***Y12 into cell E32 (notice that the addresses E31 and F31 copy absolutely) and copying it down to row 51. Next edit cell E32 to read =E31***E12**+F31***Y12** (now the addresses E12 and F31 are absolute references) and similarly for every cell in the range E32 - E51. *This somewhat tedious task is made easier with a MACRO if you are so enclined!* Next, use a similar technique to build the formulas in cells F32 - F51, Y32 - Y51, and Z32 - Z51. To continue the table down to row 491 you must copy the entire 20 row block (E32 - Z51) to the block (E52 - Z71) then to block (E72 - Z71), etc. until the table down to row 491 is complete.

ENTER data using using the instructions attached to Model 6.4, Part 1.

READ output using using the instructions attached to Model 6.4, Part 1.

Model 6.4 (Part 2 continued)

VALUES

	A	B	C	D	E	F
32	-0.973	0.2238	0.1434	6.5973	-0.973	0.2238
33	-0.888	0.3166	0.1487	6.9115	-0.888	0.3166
34	-0.776	0.396	0.1536	7.2257	-0.776	0.396
35	-0.641	0.4594	0.1569	7.5398	-0.641	0.4594
36	-0.489	0.506	0.1579	7.854	-0.489	0.506
37	-0.325	0.5366	0.1572	8.1681	-0.325	0.5366
38	-0.153	0.553	0.1558	8.4823	-0.153	0.553
39	0.0214	0.5571	0.1553	8.7965	0.0214	0.5571
40	0.1957	0.5509	0.1566	9.1106	0.1957	0.5509
41	0.3666	0.5354	0.1601	9.4248	0.3666	0.5354
42	0.5311	0.5105	0.1655	9.7389	0.5311	0.5105
43	0.6863	0.4753	0.1718	10.053	0.6863	0.4753
44	0.8285	0.4282	0.1775	10.367	0.8285	0.4282
45	0.9539	0.3675	0.1813	10.681	0.9539	0.3675
46	1.0579	0.292	0.1825	10.996	1.0579	0.292
47	1.1358	0.2016	0.1816	11.31	1.1358	0.2016
48	1.1832	0.0979	0.1798	11.624	1.1832	0.0979
49	1.1963	-0.016	0.179	11.938	1.1963	-0.016
50	1.1729	-0.134	0.1809	12.252	1.1729	-0.134
51	1.1125	-0.25	0.186	12.566	1.1125	-0.25
52	1.0165	-0.359	0.1935	12.881	1.0165	-0.359
53	0.8885	-0.454	0.2016	13.195	0.8885	-0.454
54	0.7333	-0.531	0.2083	13.509	0.7333	-0.531
55	0.5568	-0.589	0.2122	13.823	0.5568	-0.589
56	0.3653	-0.627	0.2133	14.137	0.3653	-0.627
57	0.1647	-0.647	0.2127	14.451	0.1647	-0.647
58	-0.04	-0.651	0.2124	14.765	-0.04	-0.651
59	-0.243	-0.643	0.2139	15.08	-0.243	-0.643
60	-0.442	-0.623	0.2185	15.394	-0.442	-0.623
61	-0.634	-0.593	0.2261	15.708	-0.634	-0.593
62	-0.814	-0.553	0.2357	16.022	-0.814	-0.553
63	-0.98	-0.501	0.2456	16.336	-0.98	-0.501
64	-1.127	-0.436	0.2538	16.65	-1.127	-0.436
65	-1.252	-0.355	0.2589	16.965	-1.252	-0.355
66	-1.349	-0.257	0.2604	17.279	-1.349	-0.257
67	-1.412	-0.143	0.2595	17.593	-1.412	-0.143
68	-1.437	-0.016	0.2584	17.907	-1.437	-0.016
69	-1.421	0.1202	0.2597	18.221	-1.421	0.1202
70	-1.362	0.2589	0.2652	18.535	-1.362	0.2589
71	-1.259	0.3925	0.2752	18.85	-1.259	0.3925
72	-1.116	0.5137	0.2877	19.164	-1.116	0.5137

	Y	Z
32	0.1336	-1.058
33	-0.199	-1.055
34	-0.526	-1.021
35	-0.838	-0.96
36	-1.127	-0.879
37	-1.389	-0.785
38	-1.619	-0.683
39	-1.817	-0.576
40	-1.981	-0.467
41	-2.11	-0.353
42	-2.202	-0.234
43	-2.256	-0.105
44	-2.267	0.0353
45	-2.232	0.1883
46	-2.147	0.3525
47	-2.01	0.5237
48	-1.818	0.6947
49	-1.574	0.8564
50	-1.282	0.9987
51	-0.95	1.1125
52	-0.587	1.1909
53	-0.206	1.2305
54	0.1821	1.2318
55	0.5647	1.1986
56	0.9323	1.1372
57	1.277	1.0545
58	1.5933	0.9571
59	1.8773	0.8496
60	2.1263	0.7341
61	2.3377	0.6104
62	2.5087	0.476
63	2.6353	0.3275
64	2.7126	0.1616
65	2.7348	-0.024
66	2.6958	-0.227
67	2.5904	-0.445
68	2.4155	-0.669
69	2.1708	-0.887
70	1.8599	-1.088
71	1.4902	-1.259
72	1.0731	-1.39

APPENDIX, ANSWER KEY, AND ENDNOTES

APPENDIX

MAKING *XY*-PLOTS WITH SPREADSHEETS

This book often asks you to use your spreadsheet program's on-screen graphics capability to generate *xy*-plots of your data (i.e., plots of y vs x). Some programs call this kind of plot a *scatter plot*. The instructions for generating such plots are explained in the User's Manual that supports your spreadsheet program.

We've noted, however, two opportunities for confusion that we wish to draw to your attention.

LINE PLOTS VS *XY*-PLOTS

Suppose you wish to plot the function $y = x/2$ at the points shown in the table below.

x	y
0	0
2	1
4	2
8	4
16	8

There appear to be two different plotting routines that spreadsheet manufacturers can apply to this data and only one of which is valid for

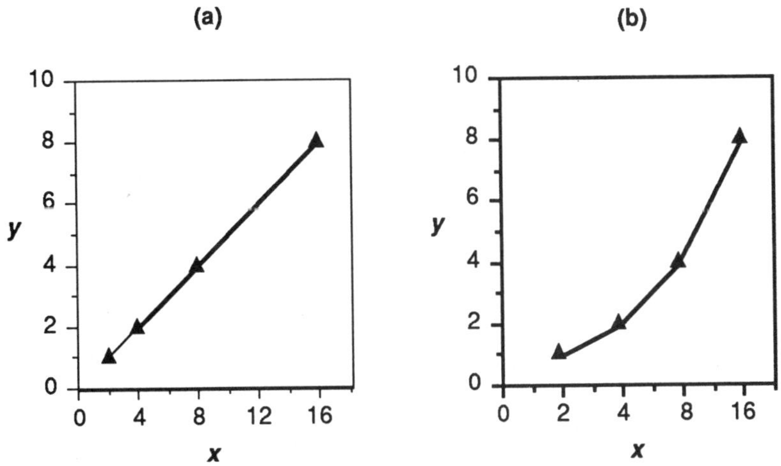

Figure A.1

The term *line chart* can be ambiguous. In (a) the *x*- and *y*-axes both represent lines in a real number system, and the linear function $y = x/2$ graphs as a straight line. In (b), however, the *x*-axis simply lists the categories 2, 4, 8, and 16. They are distributed evenly along the axis and the graph is no longer a straight line. Make sure that you invoke the graphics commands in your spreadsheet program that generate plots like those in (a).

this book. Notice that *y* is a linear function of *x* and that this fact is clearly represented in Figure A.1a. What happened in Figure A.1b? *Some programs plot spreadsheet* x*-data as if they were categories (apples, oranges, pears) and distribute the values 2, 4, 8, and 16 evenly across the* x*-axis.* Plotting commands clearly labeled "scatter plot" or "*xy*-plot" never do this, but a command named "line plot" should be examined carefully to determine which convention it follows.

XY-PLOTS IN MICROSOFT'S EXCEL (up to, at least, Version 1.5)

The graphics capabilities of Excel are very powerful, and its User's Manual clearly explains how to generate *xy*-plots (called scatter plots in Excel). Some people don't read manuals, however, and we've noticed that many first-time Excel users have trouble generating *xy*-plots. Here's how it's done on the Macintosh:

- Select the block of cells you wish to plot such that the *x*-values constitute the left-hand column of data (Figure A.2).

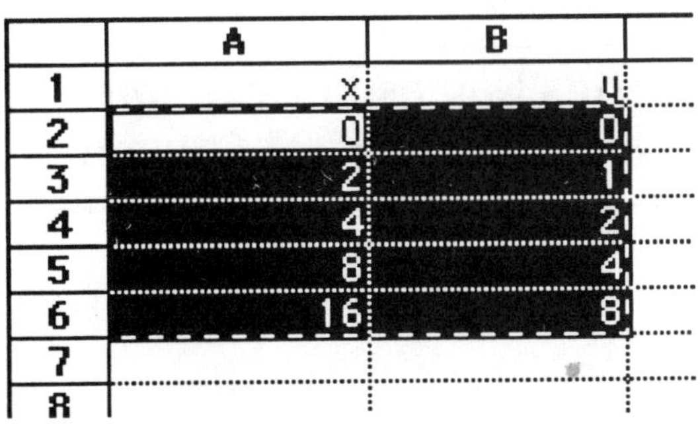

Figure A.2

In Excel, graphs are generated by selecting a block of data in your "worksheet" window and then opening a "chart" window. To generate *xy*-plots you must additionally select your data block such that the left-hand column contains data destined for the *x*-axis and place it on your clipboard with a COPY command. (Excel also supports a convention in which *x*-axis data can be in the top row, but none of the models in this book are organized in this fashion.)

- Place the contents of this block on your clipboard by using the COPY command and then open a new graphics window by selecting the NEW option in the FILE menu and requesting a "Chart" when asked. PASTE SPECIAL your data into the chart window making sure that "Categories in First Column" has been checked (Figure A.3). Don't select "Series Name in First Row" unless the first row of your selection includes identifying labels.[1]

- Use the Gallery menu to select an appropriate scatter chart.

First-time Excel users often neglect to COPY and PASTE SPECIAL. They get a chart, but the *x*-axis fails to get specified!

If your spreadsheet is organized such that the data destined for your *x*-axis lies to the right (or *very* far to the left) of your *y*-axis data, use the above instructions to plot your *x*-axis against any arbitrary data (that happens to be to the right) and then edit your formula bar. Your manual will tell you about all of the wonderful things that can be accomplished using Excel's formula bar.

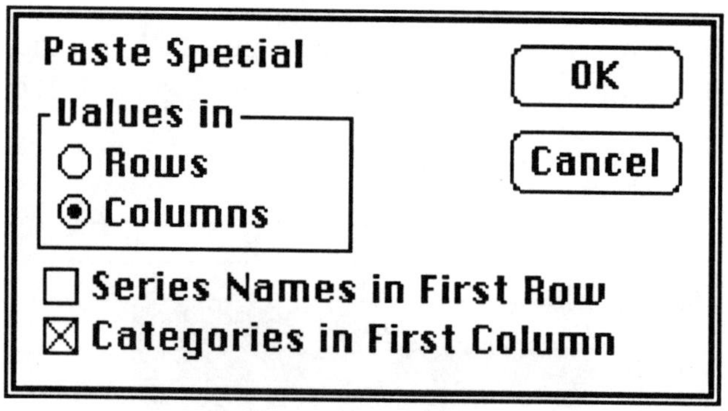

Figure A.3

The PASTE SPECIAL command permits you to specify how your data should be organized in an *xy*-plot. For a data block such as that shown in Figure A.2 you should tell Excel that your data are organized in columns and that your categories (values destined for the *x*-axis) are in the first column. If you fail to check "Categories in First Column," your first column of data will be treated as one more set of *y*-axis data.

SELECTED ANSWERS FOR LESSONS

Exercise 2.1

6c About 56 meters

6d Less, about 42 meters

7b The speed increases by a constant amount every second. The position does not increase uniformly each second because the speed is changing.

7c A uniform acceleration changes the velocity by the same amount each second. There is no constant change in the position dictated by the kinematics.

7e HINT: The maximum height occurs before the zero average velocity point!

Exercise 2.2

1a

Δt	$\Delta s/\Delta t$
2	29.4
1	24.5
0.5	22.05
0.1	20.09
0.01	19.649
0.001	19.6049
0.0001	19.60049
0.00001	19.60004

$\Delta s/\Delta t$ approaches the limiting value 19.60.

1b 39.20 m s^{-1}; 98.00 m s^{-1}

2 20 m s^{-1}; 0.40 m s^{-1}; –97.6 m s^{-1}

3a Cell B5 contains the initial value for *t* and is referred to by cell A11. Cell B6 is added to each successive row in column A to fill in a table of *t* values.

3c Choosing Δt smaller by a factor of 10 decreases the error by a factor of 10.

Exercise 2.3

1d

A	ϕ (deg)	A_x	A_y
5	0	5.00	0.00
5	30	4.33	2.50
5	45	3.54	3.54
5	90	0.00	5.00
10	30	8.66	5.00
10	90	0.00	10.00
14.14	45.00	10	10
28.28	45.00	20	20
10.00	0.00	10	0
10.00	90.00	0	10
Error	Error	0	0

2e

Vector A			Vector B			Vector C	
A_x	A_y	A_z	A_x	A_y	A_z	C	ψ
0	1	0	0	1	0	2.00	0°
0	0	1	1	0	0	1.41	90°
5	0	3	1	1	1	2.28	38°
3	5	2	–3	–5	2	4.00	142°
3	5	2	–3	–5	–2	0.00	Error
1	1	1	0	0	0	1.73	Error

3f $\boldsymbol{A} \cdot \boldsymbol{B} = 26$; $\boldsymbol{A} \times \boldsymbol{B} = -7\boldsymbol{i} - 5\boldsymbol{j} - 2\boldsymbol{k}$

5 139.08

Exercise 2.4

1c Velocity increases, stays constant, decreases.

2 Error = 0.00666

3 The error decreases as Δt decreases.

Exercise 3.1

2a $v(t)$ is decreasing when $a(t)$ is negative, $x(t)$ is not changing when $v(t)$ is zero.

2b There is a restoring force acting back toward the equilibrium position at all times, and it limits the excursion.

2c About 7.7 s. The period increases as the mass increases. The period increases as the stiffness decreases.

2d No motion ensues.

Exercise 3.2

3a They are much smaller.

3b No; no

3c The error is remarkably small, don't you think.

4a $T_{num} = 12.566373$ s and $T_{anal} = 12.566371$ s

4b About 7.7 s. The period increases as the mass increases. The period increases as the stiffness decreases.

4c No motion ensues.

Exercise 3.3

2a About 55 meters. Yes, at about 60°.

2b

θ (deg)	s_x (max) (m)
0	0
15	31
30	55.2
45	64
60	55.2
75	32
90	0

Exercise 3.4

2a The moon moves closer to the Earth at apogee. If the same initial separation between Earth and Moon is used, the orbit is eccentric.

2b No change. The acceleration does not depend on the mass of the orbiting body.

3a About 5800 m s^{-1}

3b An inward motion

Exercise 4.1

1d Both the velocity and the K decrease. The potential energy U increases. The total mechanical energy is constant.

1e When the ball is at its highest point $K = 0$ J, $U = 198$ J, $K + U = 198$ J. When the ball is about to strike the ground $K = 198$ J, $U = 0$ J, $K + U = 198$ J.

1f A 2kg ball resting in your hand 0.5 m above the ground has a potential energy U of 9.8 J; when you release the ball, this potential energy gradually converts to kinetic energy K and is fully in K just as the ball strikes the ground. $K + U$, of course, remains constant throughout the journey.

1g If the ball and the floor are perfectly elastic, the ball bounces forever; otherwise each bounce converts some of the mechanical energy into other forms such as heat, sound, and energy of deformation. In an inelastic collision (dropping sticky putty, for example), all mechanical energy is transformed into other forms.

2d Velocity and K decrease. U increases with displacement. $K + U$ remains constant.

2e When the mass is farthest from its resting position $K = 0$ J, $U =$ 6.5 J, $K + U = 6.5$ J. As the mass passes through its resting position $K = 6.5$ J, $U = 0$ J, $K + U = 6.5$ J

2g 0 J when it remains at rest at the equilibrium position.

Exercise 4.2

1c A tilted plane; a horizontal plane

1d At maximum height, $K = 0$ J, $U = 502$ J, $K + U = 502$ J; as the object strikes the ground, $K = 502$ J, $U = 0$ J, $K + U = 502$ J; yes, total mechanical energy is conserved.

Exercise 4.3

1a $K + U = 25$ J as the block starts up the ramp and as it returns (and, indeed, throughout its journey). Total mechanical energy is conserved.

1b The plot of velocity vs time is linear, and thus the time rate of change of velocity is a constant.

2a When μ_k has a value of 0.12, $K + U = 25$ J as the block starts up the ramp. $K + U$ is approximately 14 J after it returns. The system is clearly not conservative.

The work done by the friction force exactly equals the change in total mechanical energy.

3 Friction force is proportional to that component of mg that is normal to the surface on which the block moves. As the incline becomes steeper (θ increases towards 90°), the normal component of mg decreases and reaches zero at 90°. When the incline is totally vertical, the system conserves total mechanical energy.

Exercise 4.4

2b The model cannot evaluate t when y exceeds the maximum height permitted for the assigned value of U. There is no time t when this value of y is achieved.

2c When total energy E increases fourfold, maximum attainable height increases fourfold.

Exercise 4.5

1b The initial conditions of the model provide insufficient kinetic energy of relative motion to exceed the repulsive potential barrier of the spring. During the collision, the kinetic energy of relative motion is transformed into potential energy in the spring. Both kinetic energy and momentum are conserved during the collision (i.e., they have the same values long after the collision that they had long before it).

1c As m_a is increased, the kinetic energy of relative motion does not increase indefinitely but rather approaches $m_b(v_a - v_a)^2/2$. If this is not enough to surmount the repulsive potential, the pucks always "bounce off each other."

Nevertheless, when m_a is 20 kg, puck a is so massive that it just "bulls" its way through the collision. Puck b reverses direction while puck a is affected only slightly because it carries almost all of the system's momentum.

1d When the kinetic energy of relative motion exceeds 0.0478 J ($v_a > 0.198$), the pucks pass each other, otherwise not.

1e As m_b increases, the ability of the spring to "mediate" the collision decreases. By the time m_b reaches 20 kg, no apparent collision takes place because the repulsive potential barrier provided by the spring is very small relative to the kinetic energy of relative motion.

2c All the energy of relative motion has been absorbed by the shock, but the total momentum of the center of mass has not been affected. Thus, the two pucks must end up moving, but stuck together.

2d Fin 2 is now selected because the energy absorbed by the shock prevents the pucks from surmounting the potential barrier of the spring's potential energy.

2e About .665 m s^{-1}

2f As m_a is increased, the kinetic energy of relative motion does not increase indefinitely but rather approaches $m_b(v_a - v_b)^2/2$. If this is not enough to surmount the repulsive potential, the pucks always "bounce off each other." However, the heavier one eventually just "bulls" its way through with very little effect from the interaction since it carries almost all of the momentum of the system.

2g Modify the velocity of one or both of the masses to find the conditions that separate the two possible sets of final velocity demanded by energy-momentum conservation. The kinetic energy of relative motion = 0.0748 J is the same in both cases because the potential is the same in both cases.

Exercise 4.6

a Linear, and their sum is 90°, 0°, and 90°. The 0° case corresponds to a head-on collision — particle a stops and particle b moves off with its initial velocity. The 90° case corresponds to no collision.

b θ_a now can range from 0° to 180°. The corresponding values of θ_b are 90° (no collision) and 0° (head-on collision). When θ_a is near 0°, the two objects just graze one another, and m_b gets flicked to the side with a small velocity. When θ_a is near 180°, there has been a nearly head-on collision, the lighter m_a bounces back rather slowly, and the heavier m_b achieves a reasonable forward velocity.

c The angles are linearly related, with $2\theta_b + \theta_a = 180°$. As θ_a gets larger, v_{a2} decreases nonlinearly, and v_{b2} increases similarly.

d 65.4° for θ_a. θ_b is about 12.7° at this Jacobian point, as it is called. The upper part of the curve comes from the solution with a plus sign and the lower part, from the minus sign.

e About 5.73°. It is so small because the large momentum of particle a causes it to "bull" its way through the puny particle b, which gets knocked for a loop! θ_b can have any angle from 0 to 90°.

Exercise 5.1

1d The linear momentum is perpendicular to the vector ***r***.

1e Parallel to but opposite in direction to the radius vector

1f Yes, notice that $\Gamma_z = 0$.

2b As the *z* component of ***r*** becomes larger the *x* and *y* components of ***L*** become enormous. Most of the angular momentum is in the *x*-, *y*-plane. At $z_o = -0.1$ units.

2c

z_o	$L_x(t=t_o)$	$Ly(t=t_o)$	$Lz(t=t_o)$	$Lx(t=t_1)$	$Ly(t=t_1)$	$L_z(t=t_1)$
100	-6535.46	-85.5538	11.30844	85.55386	-6535.46	11.30844
10	-659.422	-8.63230	11.30844	8.632308	-659.422	11.30844
5	-332.975	-4.35888	11.30844	4.358888	-332.975	11.30844
0.2	-19.5867	-0.25640	11.30844	0.256405	-19.5867	11.30844
0.1	-13.0578	-0.17093	11.30844	0.170936	-13.0578	11.30844
0	-6.52893	-0.08546	11.30844	0.085468	-6.52893	11.30844
-0.1	0	0	11.30844	0	0	11.30844
-0.2	6.528931	0.085468	11.30844	-0.08546	6.528931	11.30844
-0.3	13.05786	0.170936	11.30844	-0.17093	13.05786	11.30844
-5	319.9176	4.187951	11.30844	-4.18795	319.9176	11.30844
-10	646.3642	8.461371	11.30844	-8.46137	646.3642	11.30844
-100	6522.402	85.38293	11.30844	-85.3829	6522.402	11.30844

2d A constant is added.

3

z_o	$L_x(t=t_o)$	$Ly(t=t_o)$	$Lz(t=t_o)$	$Lx(t=t_1)$	$Ly(t=t_1)$	$L_z(t=t_1)$
100	-24593.8	-966.293	0	-322.171	-24610.6	0
10	-24593.8	-966.293	0	-322.171	-24610.6	0
5	-24593.8	-966.293	0	-322.171	-24610.6	0
0.2	-24593.8	-966.293	0	-322.171	-24610.6	0
0.1	-24593.8	-966.293	0	-322.171	-24610.6	0
0	-24593.8	-966.293	0	-322.171	-24610.6	0
-0.1	-24593.8	-966.293	0	-322.171	-24610.6	0
-0.2	-24593.8	-966.293	0	-322.171	-24610.6	0
-0.3	-24593.8	-966.293	0	-322.171	-24610.6	0
-5	-24593.8	-966.293	0	-322.171	-24610.6	0
-10	-24593.8	-966.293	0	-322.171	-24610.6	0
-100	-24593.8	-966.293	0	-322.171	-24610.6	0

4 The components of ***r*** depend on the location of the origin. The torques depend on z and F_z, which are now zero.

5 The bearing forces behave sinusoidallyin time. Bearing #1 supplies the greater force. These forces are very large.

6a $d_1 = 0.6$ m, $d_2 = -0.4$ m to minimize problems. Bearings are now the same distance from mass m_2 position.

6b Make $d_2 = +0.1$ m.

7a L_z component only. All components of γ are equal to zero.

7b No lever arm at $\alpha = 0, 180°$. The center of force moves along the z-axis. The centripetal force increases with increasing α.

8a The axle line and both masses are in the same plane. When $\theta \neq 180°$, $F_x \neq 0$ and $X_{cm} \neq 0$.

8b 0.1366 m; on opposite sides of the axle

8e The center of mass of the system is on the axle.

Exercise 5.2

2a 5.465 in.

2b Add 1.4 g to give a total of 9.4 g

2c Doesn't move. One way to see this is to notice that in frame with c.m. at origin, the coordinates of this mass are (0,0). The c.m. moves toward the heavy mass

4a (1,3) and (1,-3); additional mass off-axis increases everything.

4b $X = 1$, $Y = -3$. Off diagonal elements I_{xy} and I_{yx} increase from 0 to 224 gm in.^{2}, and the diagonal elements also increase from 2600, 2192, and 4792 gm in.^{2}, to 3274, 2302, and 5577 gm in.^{2}. Magnitude of angular momentum and kinetic energy would increase because more mass is being moved.

4c Cell J19 with $x = 4$, $y = 11$, and $X = -1$, $Y = +3$; off-diagonal components and the diagonal elements increase again because there is more mass. Notice that the off-diagonal components do *not* vanish despite the symmetric placement of the two masses.

Exercise 5.3

3 See Figure 5.18.

Exercise 5.4

1b 2 and 6

1d 90°

1e One eigenvalue, 2, and any vector in the plane is an eigenvector

1g At 0°, value 2; at 90°, value 2.5

1h At 58.282°, value 2.809; at 148.1282°, value 1.691

1i At 31.73°, value 1.691; at -58.282°, value 2.809

1k

θ (**A**)	θ (T·**A**)
134	53.17090
134.1	54.06461
134.2	55.17546
134.3	56.59229
134.4	58.45899
134.5	61.02351
134.6	64.74764
134.7	70.57843
134.8	80.67794
134.9	100.1471
135	135
135.1	169.8528
135.2	-170.677
135.3	-160.578
135.4	-154.747
135.5	-151.023
135.6	-148.458
135.7	-146.592
135.8	-145.175
135.9	-144.064
136	-143.170

Exercise 5.5

1b The angular velocity and the linear velocity behave in a similar fashion.

2b The system changes direction of rotation.

3c The same motion results only when the net force is the same.

Exercise 5.6

2b The e_1-axis remains parallel to the z-axis as the body rotates. the e_2- and e_3-axes rotate sinusoidally in the x-, y-plane. When the initial angular velocity is in the z-direction, then the e_2-axis remains parallel to the z-axis and the e_1- and e_3-axes vary sinusoidally in the x-, y-plane. The angular momentum vector and the angular velocity vector remain parallel along the z-direction.

3b In the second case with $\omega_{20} = \pi$, e_2 is constant along z wheres e_1 and e_3 vary sinusoidally in the x-, y-plane.

3d The precession frequency is about 5 Hz.

Exercise 6.1

1b 1.8 s; 0.562 s^{-1}

1d Yes. The force is conservative.

2d The exponential envelope still appears reasonable.

2g $Q = 1.38$; 0.260; 20

3a About 5.9 s. For $t < t_1$, there is much variation in the motion.

Exercise 6.2

1e Becomes thermal energy, etc.

2b $\omega = 3.6\ s^{-1}$

2c Halving k_{12} decreases the frequency of oscillations by nearly 6%.

3 (1,1.306), (1,–.766); dot product of these vectors is about –.00004, which is pretty small.

4 (1,1) and (1,–0.8) approximately. They are not orthogonal.

5 (1,1.4) and (1,–0.6), approximately

Exercise 6.3

5b Sine; cosine

ENDNOTES

CHAPTER 1

Getting Started

1 Another elegant feature is contained in the very nature of the spreadsheet language. Complex formulas can be built with ease (as you will see) and complex relationships between formulas can be established without resort to such procedural structures as "for *i* equals 1 to 10 do ..." or "if *i* is less than or equal to 0 do ... else ... ."

CHAPTER 2

Introduction

1 Pendulum clocks had not yet been invented by Huygens.

2 It was René Descartes, a contemporary of Galileo, who first wedded algebraic expressions to the grid of a plane sheet of paper, the key notion supporting the *xy*-plots of this book.

Exercise 2.1

1 One notable exception to the rule is Multiplan, an electronic spreadsheet program that labels both columns and rows with numbers. In Multiplan, cell B3 would be written R3C2 (for row 3, column 2). Sorry about that!

2 The word *prepare* is used advisedly. We strongly recommend that you get into the habit of constructing formulas by pointing to cell addresses with your cursor rather than by typing. You can find this technique in your spreadsheet program's reference manual. You will get a better sense of the layout of your model, and you will introduce fewer errors.

You may also have noticed that you're copying FROM B12 TO B12 - B52. You can copy to B13 - B52 if you wish, but you get the same result either way.

3 When you prepare a formula by pointing, most spreadsheets assign a special key or key sequence that renders a cell address *absolute*.

4 Some spreadsheets arbitrarily assume that any character string that begins with a letter must be a label. If your spreadsheet behaves this way you must "fool" it into accepting the character string as a formula by first entering +, or (. Any algebraic expression would treat B8 the same way that it would treat +B8 or (B8). Many newer programs avoid this confusion by expecting all formulas to begin with =.

5 Some spreadsheet programs treat the first values in a series of ordered pairs such as (x,y) as labels and spread them evenly across the horizontal axis. Other programs use a true Cartesian coordinate system for both their xy/*scatter* and their *line* options. It is this latter feature that we need.

6 Computer simulations are often used to find (by trial and error) solutions that cannot be expressed in terms of simple, well-known functions. In the present case, of course, the solution is such a function. Nevertheless, get a feel for this kind of approach by inserting different values for v_0 and gradually "homing in" on a value that yields a maximum height of 200 m.

7 You'll use this technique quite a bit in this book so take a moment to explore it carefully. The value that is entered into cell B5 [t(incr)] permits you to quickly change the range of values displayed in cells A12 - A52. Cells A12 - A52 will serve as the x-axis of your on-screen graphic displays of this model, and you will find it convenient to easily change the range of values you wish to plot.

Exercise 2.2

1 The only requirement is that you can write a formula to evaluate $f(t)$ for any value of t.

2 You could, of course, also change a single "working" copy (cell B11, say) and copy this formula. The real advantage of keeping a copy of your formula in a single, easily identifiable location comes from the existence in some spreadsheets of keyboard macros (see spreadsheet technique "Keyboard Macros" earlier in this exercise). If your spreadsheet supports keyboard macros, your spreadsheet can remember all the appropriate COPY commands and execute them on your instruction. Thus, your spreadsheet truly becomes a machine for computing rates of change — change the equation of motion, execute the COPY macro, explore the behavior of the new function.

Exercise 2.3

1 A three-dimensional vector *A* also can be represented by two angles θ and ϕ and a magnitude *A*. You will quickly discover, however, that the three components A_x, A_y, and A_z often offer the more convenient treatment, mathematically.

2 Excel requires all built-in functions to have an argument list — even if the argument list is empty. Thus, π is represented as PI(). Other spreadsheets (e.g., Lotus 1-2-3) do not adhere to this convention, and π is represented as @PI.

CHAPTER 3

Introduction

1 You can assume for the moment that these pairs act simultaneously. However, when you study electromagnetic phenomena, you will need to modify Newton's third flaw.

Exercise 3.1

1 This trick gives rise to the important physical concept of energy that you will explore later.

2 Physicists and engineers often use the phrase *analytic expression* to mean one that can be written in terms of simple, well-known functions such as polynomials, ratios of polynomials, exponentials, logarithms, the trigonometric functions and their inverse, and other special functions that often appear in physical applications and have been extensively studied (Bessel functions are an example). On the other hand, mathematicians say that an expression is analytic at a point if it has a derivative at that point. We will always use the word *analytic* in the first sense.

Exercise 3.2

1 You should recall a similar replacement when you used Simpson's rule for numeric integration.

2 This model contains numerous forward references, and you must be certain that the recalculation order of your spreadsheet program is set to *natural*.

3 This is a nice place for you to build a macro, don't you think?

4 Notice that this column uses the formula for k_{v0}. You will find that you can often calculate various other interesting quantities using ingredients already present in your Runge-Kutta table.

5 Often, of course, there will be no analytic solution.

Exercise 3.4

1 Indeed, for other force laws or other initial conditions with the gravitational force, the orbit need not close.

CHAPTER 4

Introduction

1 When velocities are so large that the effects of Einstein's theory of relativity become important, Newton's expression for linear momentum must be modified and his third law is no longer satisfied. The deviations due to these effects can be ignored for most velocities in macroscopic physics, but in microscopic physics (where velocities are often very large) they are important.

2 Total energy is always conserved. Often, however, you must deal with a subset of energies (e.g., mechanical energy), and then it may or may not be conserved.

3 $v\,dt$, of course, equals dx (an element of distance along the particles path).

4 These special cases include several of Nature's fundamental forces.

5 The law of conservation of mechanical energy stands at the beginning of a path that leads to one of the grandest laws of classical physics — the law of conservation of energy. Throughout history apparent violations of this law have served as starting points for the discovery of new forms of energy.

Exercise 4.1

1 You will explore the physics of collisions later in this chapter.

Exercise 4.2

1 The fact that such forces *can* exist does not imply that they *do* exist. In Nature no fundamental forces are known for which potential energy functions cannot be written. Can you think of a way to "build" such a force (e.g., with actuators, relays, or electronically)?

2 In three dimensions, two further relations similar to Equation 4.32 are also satisfied if a force law is conservative.

3 Notice that we say "a potential energy function" rather than "the potential energy function." The fact that potential energy functions are obtained from an indefinite integral of the force should tell you that there are an infinite number of of functions U that can satisfy Equation 4.39. Of course, one such function differs from others only by an additive constant.

Exercise 4.3

1 Your automobile, for example, as you approach a stop sign.

2 Heat is actually the mechanical energy of the microscopic parts of a large-scale body, and light carries the energy of the electromagnetic field. It is usually convenient to ignore the exact details of these phenomena when studying the large-scale behavior of things much bigger than a molecule.

3 For friction, of course, this energy is virtually all in the form of heat (i.e., an increase in the kinetic energy of the molecules that make up the block).

4 You might be inclined to make some tiny velocity range around zero as your test of the $v = 0$ condition. Depending on the motion of the block (and the choice of Δt), you might still "step over" your range.

5 Strictly speaking, the actual friction force does no work (unless the tires are slipping) because it acts on the tire at just the place where the tire is stationary. Thus, the friction force does not act through a distance and so it does no work. However, the wheel transfers this force to the axle, and the reaction of the axle is essentially the same as the friction force on the road. This force *does* act through a distance and you can calculate the work done by the engine to propel the car forward as this axle force times the distance the car moves. In practice, of course, the engine must supply energy to make up for all of the friction losses in itself and the drive chain in addition to that required to move the car.

6 The assumption of a constant driving force is not very realistic and is used for simplicity only.

Exercise 4.4

1 *R* is easy to remember because the meter was defined after the French Revolution such that 10^7 of them equaled the distance from the North Pole to the Equator (through Paris!).

2 Note that 1 nm = 1 nanometer = 10^{-9} m is a convenient unit of distance in atomic problems. Similarly, 1 fs = femtosecond = 1×10^{-15} s is a convenient time scale. c is the velocity of light, and because this is the speed limit of the Universe, it is often convenient to give physical quantities in terms of it. Finally, eV is a unit of energy much used in atomic, nuclear, and subnuclear physics; it is called an electron-volt and 1 eV = 1.6×10^{-19} J. Iron atoms have approximately the mass given in the problem.

3 Recall that the integrand is just the reciprocal of the velocity.

4 To see how the formula is derived, consider

$$\int_{x_0-\Delta}^{x_0} \frac{dx}{\sqrt{E - U(x)}}$$

where $E = U(x_0)$. If Δ is small, the definition of a derivative (notated here with a prime) may be used to write

$$U(x) \sim U(x_0) + (x - x_0)\, U'(x_0)$$

for all x-values in the range of the integration. The integral then becomes (approximately)

$$\int_{x_0-\Delta}^{x_0} \frac{dx}{\sqrt{(x_0 - x)\; U'(x_0)}}$$

which can be integrated to yield

$$2\sqrt{\frac{\Delta}{U'(x_0)}}$$

Finally, in the spirit of numeric calculation, you can use your usual formula for numerically performing a differentiation to get the final formula used in the model

$$\frac{2\Delta}{\sqrt{U(x_0) - U(x_0 - \Delta)}} = \frac{2\Delta}{\sqrt{[E - U(x_0 - \Delta)] - [E - U(x_0)]}}$$

In the actual model, an absolute value function is also used to accommodate the case that the turning point is at the lower end of the integration range rather than at the upper (as is the case in the above calculation).

Exercise 4.5

1 Linear momentum $\boldsymbol{p}$ is a very important variable in dynamics and, when there is no possibility of confusion with other types of momentum (e.g., angular momentum), the linear momentum is usually called simply "the momentum."

2 There are also collisions in which kinetic energy increases because other forms of energy are converted into kinetic energy. These *superelastic* collisions are important in atomic and nuclear physics.

Exercise 4.6

1 You saw a hint of this in the one-dimensional case in Exercise 4.5 where information about the interaction is needed to select between two possible solutions of the energy-momentum conservation equations.

2 Because any inertial frame of reference is as good as any other, you really are not losing any generality by chosing a frame of reference centered on the "motionless" puck.

CHAPTER 5

Introduction

1 The angular velocity is called an *axial vector* or a *pseudovector*. All vectors you have seen until now (position, velocity, momentum, force, etc.) are called *polar vectors*. The vectorial quantities in rotational motion such as angular velocity, angular momentum, and torque are axial vectors. For our purposes, the distinction between axial and polar vectors is not important, but in some other branches of physics — particle physics, for example — the distinction is very important indeed. You will learn more about the differences between these two kinds of vectors when you study these more advanced subjects.

2 Although, in many important special cases, it is not.

Exercise 5.1

1 d_1 and d_2 are signed quantities with displacement upward along the z-axis taken as positive.

2 It is not hard to do analytic derivatives of the quantities in this mechanical system, but it is useful to do them numerically to fully appreciate exactly what the derivative of a vector is.

3 Model 5.2 actually does a better job of numeric differentiation than Model 5.1 because you can now make the time increment as small as you wish.

4 Your plot would be even more informative if it were to show an area proportional to the mass instead of a simple symbol, but we will let you design such refinements yourself, if you wish.

Exercise 5.2

1 There is one extremely important exception to the general rule that the angular momentum and angular velocity have different directions: When the axis of rotation is fixed, passes through the center of mass of the body, and is along a principal axis of the body, these two vectors are in the same direction. You will explore the question of finding principal axes in Exercise 5.3.

2 You performed the inverse of this process in the previous exercise (for steady rotary motion about a fixed axis).

3 When you are studying the motions of small objects near the surface of the Earth, a reference frame fixed to the Earth is usually a good approximation to a true inertial system. When you are studying the rotation of the Earth itself, then you might use a frame fixed to the Sun. A study of the motion of the Sun might call for a reference frame fixed to the stars. As for the motion of the stars ...

4 Actually, though, most objects in your daily life have such complicated geometries that analytic integration is impractical. When a body has a simple geometry and mass distribution (a sphere or cylinder, for example) an analytic integration is appropriate. However, the technique you will study here always works, and you can always increase the accuracy of your calculation by increasing the number of cells used to represent the object. The integration technique is just the two-dimensional generalization of the simplest numeric method for one-dimensional integration you studied in Chapter 2.

5 For a three-dimensional object, of course, you would use volume elements.

6 Equations 5.48a - 5.48g do not include the contributions of the z-component of a three-dimensional object because, for a two-dimensional object, its value is always zero. Can you work out (or just guess!) what they would be?

7 Be sure to introduce new rows in columns well within the SUM() ranges (so that they are included in the sums) and be sure to COPY the formulas outside the grid to the appropriate columns and rows. The model is designed to facilitate this by including a row and two columns of lines to mark the location of the insertions. See the model description for more detail.

Exercise 5.3

1 The center of mass of a body is that single point in the body from which (in a uniform gravitational field) it can be supported and perfectly balanced.

2 In highly symmetric bodies, every direction in the plane containing the masses has the property of being a principal axis, but usually there are only two and they are perpendicular to each other.

Exercise 5.4

1 People say that the inertia tensor "operates" on the angular velocity to produce the angular momentum and sometimes refer to it as an "operator." However, tensors are also geometric objects in their own right and need not be viewed only as operators. When people want to especially emphasize the operator nature of a tensor, they often call it a matrix or say that the "tensor is represented as a matrix." We will sometimes call a tensor operator a matrix.

2 In cases where any vector in a plane acts as a principal axis, you can *choose* a pair that are orthogonal and use them as your principal axes; any other vector in the plane can be expressed as a vector sum of multiples of these two. The most extreme situation of this kind is when any direction in a body acts as a principal axis. This happens for a uniform sphere or a uniform cube. For these, choose any three convenient orthogonal directions as your principal axes. The inertia tensors of highly symmetric bodies such as these are said to be degenerate or to have a degeneracy.

3 In algebra, one procedure for solving simultaneous linear equations is called Cramer's rule, and the above statements (in the main text) follow directly from it. For each eigenvalue λ, you can set one of the components of $\boldsymbol{\omega}$ to a nonzero value and use Cramer's rule on some pair of Equations 5.65 - 5.67 to find the other two components of the associated eigenvector $\boldsymbol{\omega}$.

4 In your plot of $\theta(\mathbf{T}\cdot A)$ versus $\theta(A)$, you will probably observe an almost vertical line from +180 degrees to –180° at $\theta(A)= \sim 166°$. This is an artifact of the inverse tangent function, which is designed to return values between $-\pi$ and π only. Therefore, the intersection of the line $y = \theta(A) - 180°$ with this almost vertical line does not signal an eigenvector as you can readily see by using the on-screen graphic associated with Figure 5.20.

Exercise 5.5

1 People say that a function is "piecewise constant" when it consists of two or more locally constant pieces joined by steps. Thus, $\varepsilon(\omega)$ is piecewise constant (Equation 5.111). Another example of a piecewise function is $f_{p.c.}(x) = 2 + 3\,\varepsilon(x-1) - \varepsilon(x+1)$. Try sketching this function.

Exercise 5.6

1 Compare these with columns K - M that give the angular momentum along the axes fixed in the body; because of the tumbling, these components vary strongly in time.

CHAPTER 6

Exercise 6.1

1 The analytic solution to Equation 6.a, for example, may be found in Lesson 1 of Exercise 3.1.

Exercise 6.2

1 Of course, the case of no damping never exactly occurs in macroscopic systems, but for high Q systems, it is often a reasonable approximation for short times. When a system is heavily damped, the character of the motion tends to be somewhat more complicated.

2 In the case of a degeneracy, you could say that the system has an infinite number of normal modes, but only N of them are really different. The situation is analogous to the principal axes of a highly symmetric body such as a sphere. Any line through its center is a principal axis, and so you could say that it has an infinite number of principal axes, but there are only three really different ones in the sense that none of the three can be written as a linear combination of the other two. It is convenient to choose them as three orthogonal directions, although any three different directions could be used. Any of the degenerate principal axes can be expressed as a linear combination of the chosen three. The case of a degenerate system of coupled oscillators is exactly analogous.

Exercise 6.3

1 Damping "forces" and external "forces," of course, can also be added as terms to this equation.

2 This case will not, however, be treated in this exercise because the huge frequency ratio makes the numeric approach difficult.

3 It is often useful to copy tables of numbers from their site of calculation to another site. If you were to simply COPY (or COPY AND PASTE) such tables, the formulas would recalculate in their new location. The VALUE command (of Lotus 1-2-3-like spreadsheets) and the PASTE SPECIAL command (of Excel-like spreadsheets) allow you to move these tables to new locations as *numbers only*. The copy, of course, will not change when model parameters are changed.

Appendix

1 Perhaps you've noticed that we've chosen to use that location for units rather than the variable name.